GENOME ANALYSIS

A LABORATORY MANUAL

VOLUME 2

DETECTING GENES

GENOME ANALYSIS
A LABORATORY MANUAL

VOLUME 1
ANALYZING DNA

VOLUME 2
DETECTING GENES

VOLUME 3
CLONING SYSTEMS

VOLUME 4
MAPPING GENOMES

Series Editors

Eric D. Green
National Human Genome Research Institute

Bruce Birren
Whitehead Institute/MIT Center for Genome Research

Sue Klapholz
Stanford, California

Richard M. Myers
Stanford University School of Medicine

Philip Hieter
University of British Columbia

GENOME ANALYSIS

A LABORATORY MANUAL

VOLUME 2

DETECTING GENES

Volume Editors

Bruce Birren
Whitehead Institute/MIT Center for Genome Research

Eric D. Green
National Human Genome Research Institute

Sue Klapholz
Stanford, California

Richard M. Myers
Stanford University School of Medicine

Jane Roskams
University of British Columbia

COLD SPRING HARBOR LABORATORY PRESS

GENOME ANALYSIS

A LABORATORY MANUAL
VOLUME 2 / DETECTING GENES

Library of Congress Cataloging-in-Publication Data

Genome analysis : a laboratory manual / editors Eric D. Green ... [et al.].
    v. <1-2 >. cm.
    Includes bibliographical references and index
    Contents: v. 1. Analyzing DNA   v. 2. Detecting genes.
        ISBN 0-87969-496-3 (comb). ISBN 0-87969-511-0 (comb).
        ISBN 0-87969-495-5 (cloth). ISBN 0-87969-510-2 (cloth).
    1. Gene mapping -- Laboratory manuals. 2. Nucleotide sequence --
Laboratory manuals. I. Green, Eric D.
QH445.2.G4645   1997
576.5'3—dc21

                             97-17117
                                CIP

Cover caption for paperback
Top: Photographs of human infant and mouse were reproduced with permission from Comstock, Inc.
Middle: (*Left*) *S. cerevisiae*; (*center*) *C. elegans*; (*right*) *Drosophila*
Bottom: E. coli

All Cold Spring Harbor Laboratory Press publications may be ordered directly from Cold Spring Harbor Laboratory Press, 10 Skyline Drive, Plainview, New York 11803-2500. Phone: 1-800-843-4388 in Continental U.S. and Canada. All other locations: (516) 349-1946. E-mail: cshpress@cshl.org. For complete catalog of all Cold Spring Harbor Laboratory Press publications, visit our World Wide Web site: http://www.cshl.org.

Contents

VOLUME 2

2 Constructing and Screening Normalized cDNA Libraries, 49
M.B. Soares, M. de Fatima Bonaldo

6 Detection of DNA Variation, 287
R.M. Myers, L. Hedrick Ellenson, K. Hayashi

APPENDICES

Companion Volumes

*The titles and authors of Volumes 3 and 4 may change.

VOLUME 4 Mapping Genomes*

*The titles and authors of Volumes 3 and 4 may change.

Preface

Although biologists have studied the genomes of numerous organisms for decades, the last ten years have brought an enormous increase in the pace of genome research. Problems of a previously unthinkable scope are now routinely solved, and the results of these studies are having a profound impact on biomedical research. Many of these advances are the direct result of the coordinated effort of the Human Genome Project, which was established to produce genomic maps and sequences for a set of well-studied organisms. Recent strategic and technological advances have made it feasible to clone, genetically manipulate, and analyze very large segments of DNA; to identify expressed sequences within large genomic regions; to identify DNA sequence variation associated with phenotypic variation; and to determine the nucleotide sequences of DNA cheaper and more efficiently than before. Indeed, to date, several bacterial genomes and that of baker's yeast have been sequenced in their entirety, and the sequencing of other genomes is well under way.

The purpose of this four-volume manual is to provide newcomers and experienced practioners alike with theoretical background, laboratory protocols, and resource materials for applying these powerful new techniques of genome analysis to the study of the very large number of genes and genomes yet to be characterized. We feel that there are several compelling reasons for producing such a manual. First, there are few sources that provide detailed information on the application of methods for genome analysis, due to the recent development of these techniques and the rapid rate at which they have evolved. Second, genome research has led to a new way of thinking that allows a completely different scale of question to be addressed. Many of these techniques, while originally developed for the systematic analysis of mammalian genomes, are now applied to specific biological questions in a variety of organisms. We hope to hasten further the increased understanding of basic biological phenomena by making this technology more widely applied. We have therefore provided step-by-step protocols with detailed explanations regarding why key manipulations are performed as described and which of these steps are most critical to success or prone to failure. We also describe how to recognize and avoid common problems and provide guidelines for troubleshooting various aspects of the protocols. Finally, we believe that a manual that covers the vast array of approaches needed at the different stages of a typical genome analysis project will find application in many laboratories.

This manual consists of chapters authored and edited by genome scientists who are experts in, and in many cases, the developers of, the described experimental techniques. The editors wrote some of the chapters themselves and worked closely with the other authors and the staff at Cold Spring Harbor Laboratory Press to develop cohesiveness, a consistent style, and substantial cross-referencing. The methods are presented with a high level of detail and completeness to allow readers without experience with these methods to evaluate the strategies and successfully implement the protocols. All of the editors have led laboratory courses at Cold Spring Harbor Laboratory, and thus have an appreciation for the importance of technology transfer and the amount of background explanation and detail that must accompany a protocol to make its use a success, especially in the hands of an inexperienced user. While we have not assumed that the user is experienced in the techniques of genome research, a basic knowledge of molecular biology techniques (i.e., recombinant DNA cloning) is a prerequisite for the successful use of most of the described methods.

Volume 1 of this manual contains seven chapters describing basic techniques in genome analysis that are applicable to most of the experimental methods appearing in subsequent volumes. Chapter 1 provides basic protocols for isolating genomic DNA and performing standard manipulations, such as gel transfer for hybridization analysis. Chapter 2 describes protocols for isolating, manipulating, and analyzing high-molecular-weight DNA, many of which are based on technologies developed specifically in response to the need to clone, map, and sequence large genomes. The polymerase chain reaction is now a standard tool in almost all biomedical research laboratories, and the method has many key roles in genome research. Chapter 3 provides an overview of these PCR-based applications in genome analysis. Chapters 4, 5, and 6 describe large-scale DNA sequencing and include general protocols for dideoxy-mediated sequencing, as well as more specific methods for shotgun and directed sequencing strategies. Chapter 7 reviews perhaps the most critical set of all genome methods—those concerned with analyzing and accessing genomic information, particularly sequence data. This chapter is associated with an electronic version available at http://www.cshl.org/books/g_a that contains supplemental information that should be useful to all researchers as well as to those interested in using the vast quantities of mapping and sequencing data generated by the genome project. Appendices containing instructions for preparing reagents, protocols for basic methods used throughout the manual, safety information, and useful reference information are included at the end of the volume.

In Volume 2, comprehensive approaches are presented for identifying, isolating, and analyzing genes. Chapter 1 reviews strategies for gene discovery in mammalian systems, including approaches for gene identification, mapping, isolation of transcribed sequences, and assessment of candidate genes. Chapter 2 provides detailed protocols for the construction and screening of normalized cDNA libraries. Methods for gene isolation are presented in Chapters 3 (Direct cDNA Selection), 4 (Exon Trapping), and 5 (Gene Detection by the Identification of CpG Islands). Chapter 6 describes a variety of methods for the detection of DNA sequence variation, including protocols for identifying alterations in electrophoretic mobility and recognizing mismatches.

Subsequent volumes of this manual detail an array of other methods for performing genome analysis. Volume 3 contains methods for using a variety of genomic cloning systems, including cosmids, bacteriophage P1, bacterial artificial chromosomes, and yeast artificial chromosomes. Finally, Volume 4 includes methods central to the generation and use of genomic maps, including the analysis of DNA polymorphisms, meiotic mapping in humans, meiotic and comparative mapping in the mouse, mapping by fluorescence in situ hybridization, generation of PCR-based markers for genome mapping, and radiation hybrid mapping.

We strongly encourage users of this manual to heed all safety cautions noted in the protocols, in the Appendices, and in the instructions provided by manufacturers. We urge all investigators to be familiar with the safe use of reagents and laboratory equipment, as well as with national, state, local, and institutional regulations regarding the use and disposal of materials described in this manual.

The 1970s and 1980s brought the revolution of molecular biology—initially a field of research but more recently a fundamental set of techniques that have come to have key roles in virtually all studies of biological systems. The "genome revolution" of the 1990s will likely evolve in a similar fashion. Our hope is that the experimental techniques described in this four-volume manual will ultimately be useful for investigators focusing on the study of genomes, as well as for those wishing to manipulate and analyze genomes as a means of gaining insight into basic biological processes.

Eric D. Green
Bruce Birren
Philip Hieter
Sue Klapholz
Richard M. Myers

Acknowledgments

The editors and authors would like to thank the following:

Doug Bassett

Art Beaudet

W.M. Keck Foundation

Tony Monaco

Jeremy Nathans

National Institutes of Health

Paul Nisson

Jeff Touchman

U.S. Department of Energy

We would also like to acknowledge the staff at Cold Spring Harbor Laboratory Press:

Executive Director: John Inglis

Assistant to the Director: Elizabeth Powers

Developmental Editor: Michele Ferguson

Production Editor: Dorothy Brown

Project Coordinator: Mary Cozza

Desktop Editor: Susan Schaefer

Book design: Nancy Ford and
  Emily Harste

Contributors

Andrea Ballabio *Telethon Institute of Genetics and Medicine, Milano, Italy*
Stephen Brown *Mouse Genome Centre, MRC, Oxfordshire, United Kingdom*
Maria de Bonaldo Fatima *College of Physicians & Surgeons of Columbia University, New York*
Sally H. Cross *Edinburgh University, Scotland*
Lora Hedrick Ellenson *Cornell University Medical College, New York*
Elizabeth Fisher *Imperial College School of Medicine at St. Mary's, London, United Kingdom*
Kenshi Hayashi *Kyushu University, Fukuoka, Japan*
Rosalind M. John *Wellcome/CRC Institute, Cambridge, United Kingdom*
David B. Krizman *National Cancer Institute, NIH, Bethesda, Maryland*
Richard M. Myers *Stanford University School of Medicine, California*
Andrew S. Peterson *Duke University Medical School, Durham, North Carolina*
Marcelo Bento Soares *University of Iowa, Iowa City*

Abbreviations and Acronyms

In addition to standard abbreviations for metric measurements (e.g., ml) and chemical symbols (e.g., HCl), the abbreviations and acronyms below are used throughout this manual.

A	adenosine (RNA) or deoxyadenosine (DNA) residue
AHC medium	acid-hydrolyzed casein medium
AMCA	7-amino-4-methylcoumarin-3-acetic acid
amp^r	β-lactamase gene conferring resistance to ampicillin
Amp^r	ampicillin-resistance phenotype
AMV	avian myeloblastosis virus
AP PCR	arbitrarily primed PCR
AT-2	Artificial Transposon-2
ATCC	American Type Culture Collection
ATP	adenosine triphosphate
ATPγS	adenosine-5′-O-(3-thiotriphosphate)
AV-FIGE	asymmetric voltage field-inversion gel electrophoresis
BAC	bacterial artificial chromosome
BAP	bacterial alkaline phosphatase
BLAST	basic local alignment search tool
bp	base pair
Bq	Becquerel
BrdU	bromodeoxyuridine
BSA	bovine serum albumin
C	cytidine (RNA) or deoxycytidine (DNA) residue
C. albicans	Candida albicans
cam^r	gene conferring resistance to chloramphenicol
C-banding	centromere banding
CCM	chemical cleavage at mismatches
CDGE	constant denaturant gel electrophoresis

cDNA	complementary DNA
C. elegans	*Caenorhabditis elegans*
CERN	European Nuclear Research Council
cfu	colony-forming units
CGH	comparative genome hybridization
CHAPS	3-([3-cholamidopropyl]-dimethylammonio)-1-propanesulfonate
CHEF	contour-clamped homogeneous electric field
Ci	Curie
CIP (also known as CIAP)	calf intestinal alkaline phosphatase
cM	centiMorgans
cpm	counts per minute
CTAB	cetyltrimethylammonium bromide
CTP	cytosine triphosphate
DAPI	4′,6-diamidino-2-phenylindole
dATP	deoxyadenosine triphosphate
dCTP	deoxycytidine triphosphate
DDBJ	DNA Database of Japan
ddNTP	dideoxynucleoside triphosphate
DD PCR	differential display PCR
DEPC	diethyl pyrocarbonate
DGGE	denaturing gradient gel electrophoresis
dGTP	deoxyguanosine triphosphate
dITP	deoxyinosine triphosphate
DMD	Duchenne muscular dystrophy
D. melanogaster	*Drosophila melanogaster*
DMEM	Dulbecco's modified Eagle's medium
DMF	*N,N*-dimethylformamide
DMSO	dimethyl sulfoxide
DNA	deoxyribonucleic acid
DNase	deoxyribonuclease
dNMP	deoxynucleoside monophosphate
dNTP	deoxynucleoside triphosphate
DOP PCR	degenerate-oligomer-primed PCR
D-PBS	Dulbecco's phosphate-buffered saline
dpm	disintegrations per minute
DTT	dithiothreitol
dTTP	deoxythymidine triphosphate
dUTP	deoxyuridine triphosphate
EBI	European Bioinformatics Institute
EBV	Epstein-Barr virus
EC number	Enzyme Commission number
E. coli	*Escherichia coli*
EDTA	ethylenediaminetetraacetic acid
EEO	electroendosmosis

EGTA	ethylene glycol-bis(β-amino-ethyl ether) N,N,N',N'-tetraacetic acid
ELISA	enzyme-linked immunosorbent assay
E-mail	electronic mail
EMBL	European Molecular Biology Laboratory
EPPS	N-(2-hydroxyethyl)piperazine-N'-(3-propanesulfonic acid)
EST	expressed sequence tag
EUCIB	European Backcross Collaborative Group
FBS	fetal bovine serum
FCS	fetal calf serum
FIGE	field-inversion gel electrophoresis
FISH	fluorescence in situ hybridization
FITC	fluorescein isothiocyanate
FPLC	fast-performance liquid chromatography
ftp	file transfer protocol
F.W.	formula weight
G	guanosine (RNA) or deoxyguanosine (DNA) residue
G-banding	Giemsa banding
gDGGE	genomic denaturing gradient gel electrophoresis
GSS division	GenBank division for genome survey sequences
GTP	guanosine triphosphate
HAP	hydroxyapatite
HBSS	Hanks' balanced salt solution
HEPES	N-(2-hydroxyethyl)piperazine-N'-(2-ethanesulfonic acid)
HGMP	Human Genome Mapping Program
HIV	human immunodeficiency virus
HMBD	histidine-tagged methyl-CpG-binding domain
HMW DNA	high-molecular-weight DNA
HPLC	high-performance liquid chromatography
HTG division	GenBank division for data from high-throughput genome sequencing centers
HTML	hypertext markup language
http	hypertext transfer protocol
H. wingei	*Hansenula wingei*
IgG	immunoglobulin G
IPTG	isopropylthio-β-D-galactoside
IRS PCR	interspersed-repetitive-sequence-based PCR
kan[r]	gene conferring resistance to kanamycin
Kan[r]	kanamycin-resistance phenotype
kb	kilobase pair
kD	kilodalton
KGB	potassium glutamate buffer
lb	pound(s)
LB medium/plate	Luria-Bertani medium/plate

LIDS	lithium dodecyl sulfate
LTR	long terminal repeat
M	molar
M13 RF	M13 replicative form
MACAW	multiple alignment construct and analysis workbench
Mb	megabase pair
MBD	methyl-CpG-binding domain
Mbytes	megabytes
α-MEM	α-minimum essential medium
MES	2-(N-morpholino)ethanesulfonic acid
MMLV	Moloney murine leukemia virus
m.o.i.	multiplicity of infection
MOPAC	mixed oligonucleotide-primed amplification of cDNA
MOPS	3-(N-morpholino)propanesulfonic acid
mRNA	messenger RNA
m.w.	molecular weight
N	normal
β-NAD	β-nicotinamide adenine dinucleotide
NCBI	National Center for Biotechnology Information
NGM	nematode growth medium
NIGMS	National Institute for General Medical Sciences
NIH	National Institutes of Health
NMR	nuclear magnetic resonance
NOR staining	nuclear organizing region staining
NP-40	Nonidet P-40
OFAGE	orthogonal field alternation gel electrophoresis
ORF	open reading frame
PAC	P1-derived artificial chromosome
PACE	programmable autonomously controlled electrodes
PAGE	polyacrylamide gel electrophoresis
PBS	phosphate-buffered saline
PCR	polymerase chain reaction
PEG	polyethylene glycol
PFG	pulsed-field gel
PFGE	pulsed-field gel electrophoresis
pfu	plaque-forming unit
PIPES	piperazine-N,N'-bis(2-ethanesulfonic acid)
PMSF	phenylmethylsulfonyl fluoride
poly(A)$^+$	polyadenosine residues
PRINS labeling	primed in situ labeling
PTT	protein truncation test
Q-banding	quinacrine banding
QFD-banding	Q-banding by fluorescence using DAPI
r_{avg}	average radius
RACE	rapid amplification of cDNA ends
RAPD	random amplified DNA polymorphism

RARE cleavage	RecA-assisted restriction enzyme cleavage
R-banding	replication or reverse banding
RC	recombinant congenic strains
RCRE	rare-cutting restriction enzyme
RDA	representational difference analysis
RF	replicative form
RFLP (also known as RFLV)	restriction fragment length polymorphism (variant)
RGE	rotating gel electrophoresis
RI strain	recombinant inbred strain
RNA	ribonucleic acid
RNase	ribonuclease
rpm	revolutions per minute
rRNA	ribosomal RNA
RT-PCR	reverse transcription followed by PCR
S. cerevisiae	*Saccharomyces cerevisiae*
SC medium	synthetic complete medium
SDS	sodium dodecyl sulfate
SINEs	short interspersed repeated DNA sequences
S. pombe	*Schizosaccharomyces pombe*
sq. in.	square inch(es)
SSCP	single-strand conformational polymorphism
SSLP	simple-sequence length polymorphism
ssp.	subspecies
STS	sequence-tagged site
S. typhimurium	*Salmonella typhimurium*
SV40	simian virus 40
T	thymidine (DNA) residue
TAE	Tris-acetate/EDTA
TAFE	transverse alternating field electrophoresis
TAK buffer	Tris-acetate/potassium acetate buffer
Taq DNA polymerase	DNA polymerase from *Thermus aquaticus*
T. aquaticus	*Thermus aquaticus*
TBE	Tris-borate/EDTA
TB medium	Terrific Broth medium
TCA	trichloroacetic acid
TE	Tris/EDTA
TEMED	N,N,N',N'-tetramethylethylenediamine
tet^r	gene conferring resistance to tetracycline
Tet^r	tetracycline-resistance phenotype
TFA	trifluoroacetic acid
T_m	melting temperature
tRNA	transfer RNA
TTP	deoxythymidine triphosphate
U	uridine (RNA) residue
UDG	uracil DNA glycosylase
UNG	uracil-N-glycosylase
URL	uniform resource locator

UTP	uridine triphosphate
UV	ultraviolet
V	volt
VLP	virus-like particle
VNTR	variable number of tandem repeats
v/v	volume/volume
w/v	weight/volume
X-gal	5-bromo-4-chloro-3-indolyl-β-D-galactoside
YAC	yeast artificial chromosome
YNB	yeast nitrogen base
ZIFE	zero integrated field electrophoresis

GENOME ANALYSIS

A LABORATORY MANUAL

VOLUME 2

DETECTING GENES

1

Strategies for Gene Discovery in Mammalian Systems

ANDREA BALLABIO, STEPHEN BROWN, AND ELIZABETH FISHER

During the last decade, exponential growth in genomic research, coupled with advances in molecular biology techniques, has resulted in the development of a variety of exquisite tools and technologies to aid both in the identification of genes responsible for a given phenotype (or human disease state) and in the subsequent characterization of the way in which a given mutation can affect the function of the predicted protein product of those genes (Figure 1). The establishment of this relationship between a gene and its product is instrumental in understanding the connection between genotype and phenotype or between mutation and pathology. The relatively new discipline of bioinformatics has made possible the rapid computer-based screening of DNA and protein databases, an approach that has become an increasingly effective strategy to aid in the prediction of function. Cross-species comparisons, between human beings and lower invertebrates and eukaryotes, can help identify functional domains of a protein that may be particularly vulnerable to mutations in inherited diseases. In vitro expression of a candidate gene in prokaryotic and eukaryotic cells allows for initial immunological and biochemical characterization of its protein product. This approach also provides an opportunity to examine the gene and its mutations in a model environment (cellular or organismal), providing an immediate means to assess how a mutation in a given gene can ultimately change its function. Functional genomics, the use of genomic structure in the prediction of function, is now an explosive and vital area of scientific endeavor.

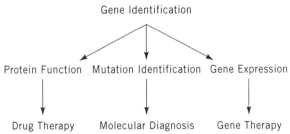

Figure 1 The three major areas in which disease gene identification has an impact in biology and medicine.

1

The identification of a disease gene is a crucial first step in the development of molecular diagnostic strategies for inherited diseases. Knowledge of the sequence and structure of a disease gene allows identification of the molecular defects in affected (and, as yet, unaffected) individuals. This information can be used for the development of molecular diagnostic tools in both post- and presymptomatic testing, prenatal diagnosis, and carrier identification. In most cases, population-based screening for genetic predisposition to common polygenic disorders, such as diabetes or hypertension, also relies on the identification of the disease gene(s) involved.

Ultimately, a variety of in vitro and in vivo approaches are needed to study the normal function of a gene and the pathogenic mechanisms underlying a disease phenotype, including both the study of gene expression and the creation and phenotypic dissection of novel mutations. Knowledge of such mechanisms and of the genetic, biochemical, and physiological pathways involved may eventually lead to the development of new strategies for therapy. Finally, the development of gene therapy approaches, which may represent the only effective cure for some severe diseases, depends entirely on that important first step—the cloning and characterization of the disease gene.

GENE IDENTIFICATION

Initial Approaches in Gene Identification

Depending on the type and amount of information available for the target gene and the genomic resources available, four major routes can be used for gene identification: functional cloning, the candidate gene approach, positional cloning, and the positional candidate approach (Figure 2). As indicated below, these different approaches do not necessarily proceed in isolation but can often be interlinked and complementary.

FUNCTIONAL CLONING

Functional cloning is an historically successful strategy for disease gene identification (Beaudet et al. 1995) that will become less important with the increase in genomic resources. This straightforward approach is based on the availability of information regarding the function of the target gene and its corresponding protein product. Knowledge of function assists investigators to develop strategies to purify the protein and link it with a phenotype (e.g., by using an enzyme assay), raise antibodies against it in laboratory animals, and derive partial amino

Functional cloning	Candidate gene	Positional cloning	Positional candidate
Identification of the target gene based on knowledge of its function.	Survey of prevoiusly identified genes that seem to perform the function altered in the disease.	Identification of the gene based on its map position in the genome.	Identification of a gene based on its map position and on the availability of candidate genes mapped to the same region.
Human: Knowledge of the phenylalanine hydroxylase amino acid sequence and screening of cDNA libraries using degenerate oligonucleotides.	*Human*: Identification of mutations in the rhodopsin gene from patients with retinitis pigmentosa.	*Human*: Mapping of the Duchenne muscular dystrophy locus to Xp21 and cloning of the dystrophin gene from this region.	*Human*: Mapping of both the Marfan syndrome locus and the fibrillin gene to 15q and identification of mutations in the fibrillin gene from patients with Marfan syndrome.
Mouse: Screening of an expression library with antibodies to identify the tyrosinase gene.	*Mouse*: Identification of mutations in the β subunit of rod cGMP phospho-diesterase in the mouse retinal degeneration mutation.	*Mouse*: Mapping of the shaker-1 locus and cloning of the mutated myosin VII gene from the critical region.	*Mouse*: Mapping of the leptin receptor to the region of the diabetes mutation and identification of abnormal splicing of this gene in diabetic mice.

Figure 2 Four major routes for disease gene identification. Appropriate examples of both human and mouse genes are provided. See text for details and references.

acid sequence information. Typically, screening of expression cDNA libraries using antibodies, tagged fusion proteins, or degenerate oligonucleotides derived from the protein sequence leads to the identification of the target gene. For example, by using a functional cloning approach, Yamamoto et al. (1987) isolated a tyrosinase cDNA that is defective in the albino mouse mutation.

CANDIDATE GENE APPROACH

The candidate gene approach identifies potential candidates for a disease or phenotypic change on the basis of assumptions that are made regarding the mechanism of the primary pathological lesion. The availability of a known gene with features (e.g., sequence domains and expression pattern) suggesting that it may be implicated in the disease allows its appraisal as a candidate gene for the disease. For promising candidates, an initial method of assessment is to ask whether or not a candidate gene maps in the vicinity of the disease locus. In this way, the candidate gene approach is clearly the forerunner of disease gene identification by the positional candidate approach (see below) in which candidate genes are derived primarily, but not solely, on the basis of their map position. Mutation analysis, within the context of the candidate gene approach (based on understanding the normal function of the protein), can be used to confirm the candidacy of the gene.

POSITIONAL CLONING

In positional cloning (Collins 1992), originally referred to as reverse genetics (Ruddle 1984; Orkin 1986), the identification of the gene underlying a mutant phenotype is made without any prior knowledge of the function of its protein product and relies exclusively on determining the position of the mutant locus in the genome. Subsequently, genes within the appropriate genetic interval are identified and tested for the presence of causal mutations within disease patients. For most genetic diseases, no information is available on the function of the defective gene and its product, and positional cloning has therefore become an important method for disease gene identification. This has been greatly facilitated by the extraordinary availability of information, technology, and reagents provided by the human, mouse, and other genome projects, which have considerably improved physical and genetic mapping, chromosome walking, and transcript identification efforts. However, positional cloning is still a difficult and time-consuming approach.

POSITIONAL CANDIDATE APPROACH

Recent successes in disease gene identification have relied on the development within the human and mouse genome projects of research programs for building

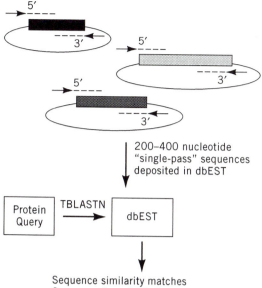

200–400 nucleotide
"single-pass" sequences
deposited in dbEST

| Protein Query | TBLASTN → | dbEST |

Sequence similarity matches
Corresponding cDNA clone availability

Figure 3   What is an EST? Expressed sequence tags are partial, single-pass sequences (*hatched lines*) from either end of a cDNA clone (*shaded box*). The EST strategy was developed to allow rapid identification of expressed genes by sequence analysis by generating unique sequence tags for each clone in a normalized library. EST sequences are immediately deposited into dbEST and are available for sequence comparison queries.

genome-wide gene or EST maps (see Figures 3 and 4) (Adams et al. 1995). Extensive databases of expressed sequences presently exist for human beings, and a large-scale EST effort is under way for the mouse that will provide a combined database of human and mouse ESTs. EST databases help identify disease genes by efficiently combining knowledge of the map position of a disease locus with the assessment of cloned candidate gene sequences, either cDNAs or ESTs, that map to the same chromosome region.

This new way of identifying genes, known as the positional candidate approach (Ballabio 1993), is rapidly becoming the predominant method because of the increasing number of cloned and mapped genes and ESTs available (Collins 1995). The identification and isolation of the mouse diabetes mutation in the leptin receptor gene provide a recent example of the positional candidate approach (Lee et al. 1996).

Genomics Programs

Current paradigms for disease gene identification and study are multiorganismal. The isolation of all genes is heavily dependent on the availability of genomic resources. In addition to the Human Genome Project (in which the primary goal is to identify all known coding sequences), a number of other so-called model organisms have extensive genomic mapping/sequencing programs.

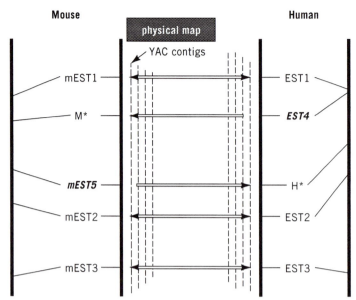

Figure 4 Schematic representation of a conserved ordered segment in the mouse and human genomes. Genomic maps of various mammals, demonstrate ordered segments in which gene content and gene order are conserved. In the example shown here, three human genes (EST1, EST2, and EST3) and their mouse homologs (mEST1, mEST2, and mEST3) reciprocally define a syntenic region of the mouse and human genomes, which demonstrates both conserved gene content and order and delineates a conserved ordered segment in the two species. A conserved ordered segment can be defined at the level of the genetic map; alternatively, it can be defined at higher resolution at the physical mapping level by assigning the ESTs on YAC contigs across the region as shown here. Identification of a new human gene mapping to this conserved ordered segment (*EST4*) defines a homologous mouse locus mapping in the equivalent region and identifies this EST as a potential candidate for a mouse mutant (M*). Conversely, the mapping of a new mouse gene (*mEST5*) adds a further locus to the human map (H*), which may ultimately provide a potential candidate for a newly mapped disease locus in the human genome.

E. COLI

The genomic sequencing of *E. coli* is now complete, as is the sequencing of a number of other bacteria (Koonin 1997).

S. CEREVISIAE

The sequence of the yeast *S. cerevisiae* is now complete (Dujon 1996; Clayton et al. 1997; Mewes et al. 1997), representing the first eukaryotic genome to be completely sequenced. Comparison of genes conserved between the yeast and human genomes has already provided remarkable insights into the genes' function (Tugendreich et al. 1993; Bassett et al. 1996). The availability of the complete sequence of *S. cerevisiae* provides a rich source of additional material for gene function studies using sophisticated mutagenesis approaches available in the yeast. More than 30% of genes identified from the yeast sequencing program are so-called "orphan" genes, which have no identifiable homology with known genes in other species and are of unknown function. However, genes showing homology with the yeast orphan set are increasingly being discovered in other species including mammals. The determination of the function of these orphan genes in yeast will provide valuable information for gene identification and gene function studies in other organisms.

C. ELEGANS

The complete sequence of the nematode *C. elegans* will be available by the end of 1998. To date, more than 60 Mb of the 100-Mb genome is complete, including the bulk of the gene-rich regions. Determination of the complete sequence of the genome and the identification of the full nematode gene set will speed functional studies relating known and newly derived mutations to their underlying genes. Future studies will thus be able to take advantage of the exquisite detail with which development in the nematode has been described (Waterston and Sulston 1995). As with the yeast, gene identification and gene function studies in the nematode will also enhance the ability to investigate the function of candidate genes in higher organisms.

D. MELANOGASTER

Among model organisms, the fruit fly *Drosophila* represents a powerful source of information since there are a large number of well-characterized mutants displaying interesting and diverse phenotypes (Rubin 1988; Lindsley and Zimm 1992; Bellen and Smith 1995). The systemic identification of human genes homologous to *Drosophila* genes responsible for these phenotypes represents an efficient strategy to identify novel genes of high biological interest and potentially involved in human disease (Banfi et al. 1996). The YAC and bacteriophage P1 maps of the *Drosophila* genome that are now available will be the basis for extensive sequence analysis. Mapping data are available via FlyBase at:

```
http://morgan.harvard.edu
```

In addition, considerable progress has been made toward the development of a cosmid physical map. For example, the X chromosome has been extensively covered with cosmid contigs (Madueno et al. 1995), and the mapping of other chromosomes is approaching fruition.

MUS MUSCULUS

The genetic and physical mapping of the mouse genome is well advanced (Deitrich et al. 1995). A genetic map at intermediate resolution (~1 cM) of more than 6000 microsatellite markers has been constructed and integrated with the gene map (Deitrich et al. 1996). More recently, a high-resolution genetic map incorporating more than 3300 microsatellite markers resolved to 0.3 cM has been completed (S. Brown, unpubl.), using the European backcross high-resolution mapping resource (EUCIB) (Breen et al. 1994), and can be found at:

```
http:/www.hgmp.mrc.ac.uk/MBx/MBxHomepage.html
```

Dense genetic maps at high resolution consisting of large numbers of ordered markers provide a framework for building physical maps using an STS content approach (Brown 1996). The physical mapping of the mouse genome is under way, and the latest physical maps are available from the MIT Genome Center for Genome Research at:

```
http://www-genome.wi.mit.edu
```

and the UK Mouse Genome Centre at:

```
http://www.mgc.har.mrc.ac.uk
```

Comparison of the detailed genome-wide genetic maps of mice and human beings has already demonstrated a large number of regions where both gene content and gene order are conserved—the so-called conserved ordered segments (Figure 4). To date, 113 conserved segments have been identified between the mouse and human genomes (Andersson et al. 1996). The value of these comparative maps between mice and human beings as well as between human beings and other species lies in the ability not only to use model organisms to identify candidate genes, but also to manipulate their function in an easily interpretable environment (Brown 1994).

OTHER ORGANISMS

In addition to mapping the genomes of the organisms above, considerable progress has been made in mapping the genomes of a number of farm animals, including pigs, sheep, cows, and chickens (see Anderson et al. 1996). The rat is also an increasingly powerful model organism for the dissection of genetic disease models (James and Lindpainter 1997). A genome-wide genetic map of the rat incorporating 432 markers with an average spacing of 3.7 cM has recently been completed (Jacob et al. 1995), and YAC libraries are now available (Haldi et al. 1997). With the mapping of disease phenotypes in these organisms and the development of their genetic maps, identification of new disease models for human genetics can be expected as well as additional information on potential candidate genes for disease loci (Figure 4).

A Systematic Approach to Gene Discovery

A key factor in the success of projects aimed at disease gene identification in the human genome is the availability of patient samples. Access to a large patient population brings commensurate benefits in the ability to map a disease locus as precisely as possible. Patients with gross, cytogenetically detectable chromosome abnormalities can be instrumental in pinpointing a specific chromosome or large chromosome region, whereas patients with submicroscopic rearrangements can be very useful in a subsequent phase of the project that narrows the critical region in which the gene is found. In the absence of patients with chromosomal rearrangements, linkage analysis can be performed on families with multiple affected cases in several generations.

In other organisms, such as the mouse or rat, the availability of homologous mutations may persuade investigators that this system may represent a more accessible system for disease gene isolation. In any event, the availability of an animal model is a bonus for investigators to follow up the function of the disease gene and the biology of the pathway involved.

The following are some key comparisons concerning disease gene isolation between human beings and other organisms:

- *Phenotypic characterization.* Clinical expertise in the analysis of pathophysiology is critical since a misdiagnosis could lead to a completely wrong mapping assignment. In the mouse or rat, for example, the ability to define and assess phenotype is just as important.
- *Genetic resolution.* The genetic resolution for mapping disease mutations in model organisms is limited only by recombination itself. In human beings, mapping accuracy will depend on the number of informative meioses in the patient population (for more information regarding these approaches, see Birren et al. 1998).
- *Genomic resources.* Human genomic resources are extremely well developed and well distributed, from genomic libraries (both large and small inserts) to a variety of cDNA libraries. In the mouse and rat, it is relatively straightforward to produce cDNA libraries from all tissues at all stages of development. Access is more limited in human beings.
- *Dissecting and identifying genetic modifiers of disease genes.* The identification and characterization of genes that affect the phenotypic expression of other disease genes are important aspects of identifying the genetic determinants of disease. The mouse and rat represent powerful systems for analyzing in a controlled manner the genetics of modifiers (Frankel 1995; Frankel and Schork 1996) and potentially for identifying the underlying modifying genes. The isolation of *Mom1*, the locus that modifies the effect of the dominant *Min* mutation in causing multiple intestinal adenomas in the mouse, provides one of the first examples of the cloning of a genetic modifier in mammals (MacPhee et al. 1995).

MAPPING STRATEGIES

The sections on pp. 10–17 describe the initial strategies appropriate for mammalian gene identification, emphasizing those critical differences between human beings and model organisms. The final strategies for disease gene identification—from a candidate gene to a transcript—are very similar in all organisms (see pp. 17–30), but the initial approaches depend heavily on the organism and the disease or phenotype being studied. Some of these strategies will be covered in detail in a future volume of this manual.

Genetic Mapping of the Disease Locus: Definition of the Critical Region

The important first step of disease gene identification is to determine the precise position of the target gene in the relevant genome. Associating a phenotype with a position on a given chromosome, and subsequently within a smaller region of that chromosome, is a crucial first step in gene discovery.

Typically, a gene is first assigned to a mapping interval by identifying two flanking genetic markers. The classic strategy in both human beings and other mammals is to use a genetic cross. Genetic linkage analysis in human pedigrees will be dealt with in detail in Chakravarti (1998). With a pedigree-based approach, mapping accuracy is highly dependent on the number of samples (i.e.,

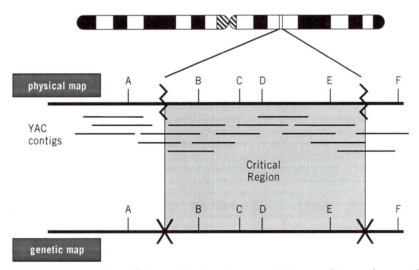

Figure 5 Definition of the critical region containing a disease locus. The boundaries are represented by chromosomal breakpoints in the physical map (*wavy lines*) and by meiotic crossing-over in the genetic map (Xs). Candidate genes within and surrounding the critical region are indicated by capital letters.

informative meioses) analyzed and on the distance between the two closest crossovers occurring at either side of the disease locus (see Figure 5). A major complication in this approach is the presence of locus heterogeneity for a particular disease, whereby a number of different genes may lead to very similar, if not identical, phenotypes. This problem can best be overcome by studying either very large families or inbred populations (e.g., mapping and dissecting the genes involved with nonsyndromic deafness; see, e.g., Guilford et al. 1994). In genetically homogeneous populations, the study of linkage disequilibrium (i.e., the identification of a specific haplotype segregating with the disease in affected individuals) sometimes provides a very useful guide to the location of the gene, as in the case of diastrophic dysplasia (Hästbacka et al. 1994).

The mapping interval in which the target gene (or at least part of the target gene) must lie is defined as the critical region (Figure 5). There are additional types of landmarks that might define a critical region besides crossovers, particularly in human beings. These include physical landmarks such as deletion breakpoints.

A short summary of how genetic mapping of the disease gene in general is carried out in the mouse and at what points mapping in the mouse differs from human genetic mapping is provided below.

In human beings, the mouse, and other model organisms, the ability to exploit genetic crosses, even those at high resolution, depends on the availability of correctly ascertained family samples and genomic resources such as well-mapped DNA markers (Weissenbach 1992).

LARGE FAMILIES AS A RESOURCE FOR MAPPING

Depending on the strain, Laboratory mice can produce several litters per year of between five and ten pups. Large numbers of mouse progeny can therefore be rapidly produced for a single disease-gene-mapping experiment, and the relevant locus can be mapped to a very high resolution within a limited time frame. For example, a mouse backcross with 1000 progeny provides a genetic resolution of 0.3 cM at a confidence level of 95% (Breen et al. 1994), a resolution comparable to the size of an average YAC clone. Producing high-resolution genetic maps of the mouse will assist in the establishment of both local and genome-wide physical maps. High-resolution maps consisting of large numbers of ordered markers enable the construction of robust physical maps using an STS content approach (see pp. 17–20). The separation and ordering of genetic markers in the high-resolution map of markers provide an extra degree of confidence during the construction of physical maps in any chromosome region; clearly, the order and orientation of markers on the genetic and physical maps should be consistent. It is equally clear that the generation of a large cross of 1000 or more mouse progeny segregating a disease mutation of interest provides a powerful resource for further narrowing the critical region containing the disease locus and defining a small interval on the physical map to search for the locus of interest (see pp. 17–20).

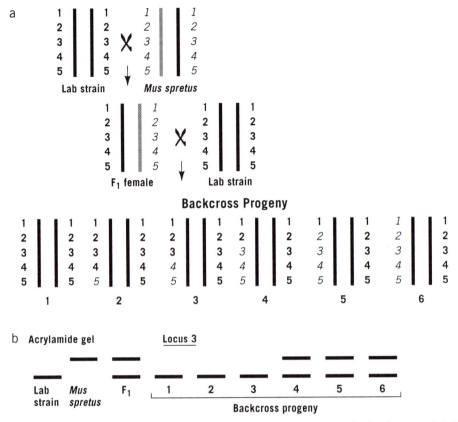

Figure 6 Genetic mapping in the mouse using an interspecific backcross. (*a*) Using multi-point analysis to construct a genetic map of five markers on one mouse chromosome. Laboratory strain marker variants are shown in bold type and *Mus spretus* marker variants are shown in italics. Backcross progeny are derived by mating F_1 females carrying both the laboratory strain chromosome (*bold*) and the *M. spretus* chromosome (*stippled*) to a laboratory strain mouse. Backcross progeny (e.g., 2, 3, 4, and 5) can carry recombinant chromosomes derived from crossing over between the laboratory strain chromosome and the *M. spretus* chromosome in the F_1 female. The genetic order of markers along the chromosome is determined by a straightforward haplotype analysis that minimizes the number of observed recombinations across the backcross progeny set. Altering the order of markers would require an overall increase in recombination among the backcross progeny and the appearance of triple recombinant chromosomes, which are very rare. (*b*) Analyzing the segregation of an SSLP (e.g., a variant microsatellite sequence) in an interspecific backcross. The SSLP can be detected by, for example, amplification using fluorescently labeled PCR primers and analysis on an automated sequencer. For locus 3, backcross progeny 4 has inherited an *M. spretus* variant from the F_1 female, but analysis of locus 2 for this mouse (not shown) reveals that it has not inherited the *M. spretus* variant (see *a*). Thus, backcross progeny 4 is recombinant between locus 2 and locus 3.

CONTROLLED MATINGS

Controlled crosses can be set up between two different *Mus* species to take advantage of the genetic divergence between parents and therefore to increase the informativeness of DNA markers in segregation analysis. For example, one of the most common mouse crosses used to map a disease locus of interest is known as the interspecific cross (Avner et al. 1988) in which the wild mouse *M. spretus* and laboratory strains are mated. Alternatively, the subspecies *M. castaneus* can be mated to laboratory strains in a cross known as an inter-subspecific cross (Deitrich et al. 1992). Fertile female offspring (F_1 animals) are produced that can be backcrossed to either parent to produce backcross progeny. (In the case of intersubspecific crosses, male F_1 progeny are also fertile.) The backcross progeny can be scored with a variety of DNA markers (e.g., micro-satellites and cDNAs) and segregation analysis of allelic variants from the paren-tal strains used to build a genetic map (see Figure 6). Most importantly, *M. spretus* (or *M. castaneus*) is highly divergent from the laboratory strains, and iden-tifying variants for any DNA marker between the parental strains is relatively straightforward. Thus, the genetic analysis is effectively multipoint since all markers can be scored through all progeny. Genetic order can be determined by a simple haplotype analysis in which the number of recombinants is minimized (see Figure 6).

RECOMBINANT INBRED, CONGENIC, AND RECOMBINANT CONGENIC STRAINS

A number of additional approaches are available for mapping mouse loci, in-cluding the use of RI strains and RC strains (for a description of the construction of these strains, see Frankel 1995). Each of these approaches depends on the in-terbreeding of laboratory strains, taking advantage of the genetic differences be-tween strains. RI strains are produced by interbreeding inbred strains—the F_1 hybrid is intercrossed, and brother-sister matings subsequently result in a new inbred strain carrying recombinant chromosomes from the two parental strains. In general, a set of approximately 20–30 strains is generated for any parental combination.

RC strains are created in similar fashion, except that the F_1 hybrid is back-crossed to one of the parents for two generations before brother-sister matings to establish a new inbred RC strain. Thus, whereas RI strains carry approximate-ly equal material from the two parental genomes, RC strains consist largely of one parental genome, with regions comprising approximately 10–15% of the genome from the donor parent. RC strains might more properly be termed mul-tiple congenic strains (Frankel 1995) in that they contain small regions of a donor strain on a recipient background. (Congenic strains are created by succes-sive rounds of backcrossing to achieve the transfer of one chromosome segment from the donor strain onto the recipient background.) RI and RC strains have been genotyped for a large number of markers across the mouse genome. Thus, both systems can be used to assess the map location of trait differences between the parental strains by comparison of the RI or RC segregation pattern for the trait's phenotype with the strain segregation pattern of genome-wide DNA markers.

The RI mapping approach does not immediately lend itself to the mapping of complex polygenic and quantitative trait loci (Frankel 1995; Melo et al. 1996).

Generally, too few RI strains develop from any two parental strains to ensure sufficient statistical power and identify unambiguous linkage. Only loci of major effect would be detected with any confidence, and it is for this reason that RI strains are used principally for mapping single locus traits (Taylor 1996). In addition, the available RI strains may not segregate a phenotype of interest. Nevertheless, there have been attempts to map complex traits (e.g., alcohol preference; Plomin and McClearn 1993) using RI strains. RC strains allow particular loci to be partly dissected in advance, and subsequent crosses between RC strains and the progenitor strains can be used for the further genetic dissection of complex traits and their interactions (Fijneman et al. 1996; Wezel et al. 1996). Ultimately, for the detailed genetic localization of individual loci, it is necessary to construct congenic strains consisting of one donor chromosome segment on a recipient background (see above). Subsequently, a congenic strain can be further dissected by successive rounds of backcrossing to identify additional recombinants that further delimit the locus region (see Frankel 1995).

Deletion Mapping of the Disease Locus: Definition of the Critical Region

In deletion mapping, the landmarks of the critical region are chromosome breakpoints, typically from patients carrying deletions or translocations (Ledbetter and Ballabio 1995). In some instances, more complex rearrangements, such as duplications and inversions, can also be used to define the critical region (Bardoni et al. 1994). In the best scenario, the study of a single affected individual carrying a significant chromosomal rearrangement (e.g., a balanced translocation) makes it possible to determine the precise location of the disease gene as in the case of Menkes disease (Chelly et al. 1993; Mercer et al. 1993; Vulpe et al. 1993). However, the definition of the critical region is usually achieved through the combined analysis of several patients carrying different but overlapping deletions. These deletions may involve more than one additional adjacent gene resulting in a complex phenotype, usually referred to as "contiguous gene syndrome" (Schmickel 1986; Ledbetter and Ballabio 1995). Deletion maps of large regions of the genome have been constructed (Vollrath et al. 1992; Schaefer et al. 1993; Leach et al. 1994) that are invaluable resources in the identification of disease genes by positional cloning.

Deletion mapping depends on the available patient resources and is not devoid of pitfalls. Summarized here are the key points for deletion mapping, including resources, methodologies, potential problems, and differences in approach for human and mouse disease gene identification projects.

It is very difficult to predict how large and how precise any given critical region will be, because the prediction will depend on fortuitous events, such as the finding of a particular chromosome rearrangement or the occurrence of a crossover in the vicinity of the disease gene. The use of deletion mapping is limited by the fact that many genetic diseases are not associated with a chromosome abnormality. In such cases, mapping must rely exclusively on the statistics of linkage analysis.

RESOURCES

Panels of cell lines from patients carrying deletions and translocations have been constructed for several chromosome regions and are available from cell banks upon request. An example of this type of panel is shown in Figure 7.

METHODOLOGY FOR HAPLOID VERSUS DIPLOID CHROMOSOMES

Deletions within haploid chromosomes (e.g., the X or Y chromosomes in males) can be directly tested in male individuals either by Southern blotting using various coding or noncoding probes or by PCR. A deletion of a particular locus results in the absence of a hybridization band or of a PCR product, respectively.

Deletions in diploid chromosomes (autosomes or X chromosomes in females) can be detected by gene dosage analysis using Southern blotting (Wolff and Gemmill 1997). However, the difficulty of interpreting differences in band intensities on a Southern blot can lead to erroneous interpretations. FISH analysis is a better method for deletion analysis, since it can distinguish between the detection of one or two signals in the deletion and normal chromosome karyotype, respectively (Trask 1998).

POTENTIAL PROBLEMS THAT CONFOUND ANALYSIS

Complex rearrangements such as double deletions or translocations/deletions can seriously confound analysis and lead to mistakes in deletion mapping disease gene loci (Page et al. 1990). A deletion of a chromosome region in an individual who does not manifest the disease does not automatically rule out the possibility that the gene involved in that particular disease lies within the deleted region (for review, see Green et al. 1995). The disease may be subject to incomplete penetrance, and therefore, it may not always manifest its phenotype.

A gene deletion in a patient with a genetic disease does not necessarily mean that the disease gene must lie within the deleted region. Deletions of regions located in the vicinity of the disease gene may have a position effect (perhaps due to the involvement of regulatory elements), thereby causing the disease without disrupting the disease gene itself. Notable examples are balanced translocations associated with campomelic dysplasia (Foster et al. 1994; Wagner et al. 1994) and with incontinentia pigmenti (Crolla et al. 1989; Gorski et al. 1992; Reed et al. 1994) and various chromosomal rearrangements causing aniridia (Fantes et al. 1995). Deletions in the Prader-Willi syndrome/Angelman syndrome region of chromosome 15 may have a position effect on the expression of genes located up to 200 kb away from the deleted regions (Sutcliffe et al. 1994; Buiting et al. 1995).

DELETION MAPPING IN THE MOUSE

Because genetic mapping is a powerful tool in the mouse that is capable of localizing mutations to a very small genetic interval (see above), mouse geneticists have tended to focus their efforts on using backcrosses to define the

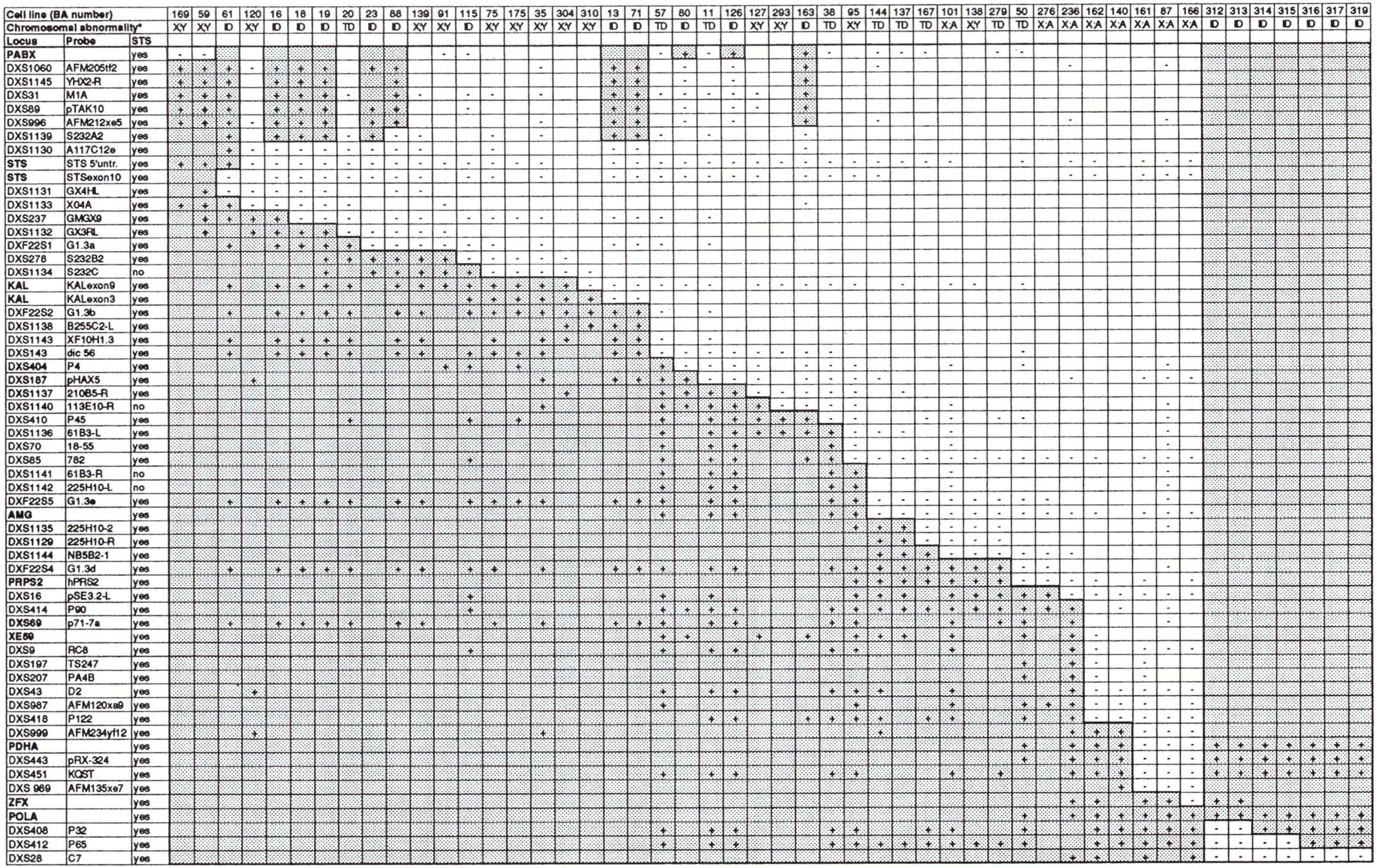

Figure 7 Characterization of patient breakpoints. Shown is a deletion map of the Xp22 region. Cell lines carrying X/autosome translocations (X/A), X/Y translocations (X/Y), interstitial deletions (ID), and terminal deletions (TD) are shown in the vertical columns. Loci tested are shown in the horizontal rows. Shaded boxes indicate regions that are present. Pluses and minuses indicate the experimentally demonstrated presence or absence of markers. (Reprinted, with permission, from Schaefer et al. 1993.)

position of a mutant gene instead of searching for deletions or translocations that might help localize the relevant locus. (This in any case would require some systematic trawling of mouse colonies to identify new mutations of the deletion/translocation type that demonstrate the appropriate phenotype.) Nevertheless, as in human beings, some mutations in the mouse are associated with deletions, translocations, or inversions that have been used to help define a critical chromosome region. For example, the *eed* (*e*mbryonic *e*ctoderm *d*evelopment) locus was localized to a 150-kb interval defined by deletion breakpoints in the albino region of mouse chromosome 7 (Schumacher et al. 1996).

Physical Mapping of the Critical Region

Physical mapping and isolation of transcribed sequences are the final stages in disease gene identification, and the approaches vary very little among the different organisms. Once the chromosomal localization of the gene locus has been established, the next steps are the characterization and cloning of the critical region into overlapping genomic clones. Mapping of a disease locus and cloning of the critical region often proceed in parallel, since markers generated by cloning efforts are useful tools to refine the critical region. The stages involved in this strategy are summarized in Figure 8. A few key points are discussed below. Various cloning systems will be covered in detail in a future volume of this manual.

CREATING A CLONE CONTIG

Construction of an overlapping series of clones (a clone contig) across the critical region is the first step in the production of a physical map encompassing the disease locus; the physical map, which will then provide access to candidate genes from any chromosome region. The advent of the YAC cloning system, which accommodates megabase-sized inserts, has revolutionized chromosome walking and large-size cloning strategies. YACs are clearly the best currently available tools for cloning large genomic regions, and their impact on positional cloning has been remarkable. In general, an STS content approach has been used for the construction of YAC clones across chromosome regions. STSs mapping to any chromosome, or chromosome region, are used to identify YAC clones, which are then assembled into a contig by STS content. For many human chromosomes, large uninterrupted YAC contigs with few gaps have been constructed, providing almost complete chromosome coverage in many instances (Chumakov et al. 1995; Collins et al. 1995; Doggett et al. 1995; Gemmill et al. 1995; Kruater et al. 1995). In addition, Hudson et al. (1995) report the establishment of a YAC-based physical map of the entire human genome with 94% coverage. The STS content approach to the construction of YAC maps also provides a detailed and ordered STS-based map of any chromosome region, sometimes at very high resolution (e.g., 75 kb on the human X chromosome; Nagajara et al. 1997 [http://genome.wustl.edu/cgm/cgm.html/]).

It follows that the physical mapping resources are already in place for the isolation and characterization of candidate genes from many regions. With the

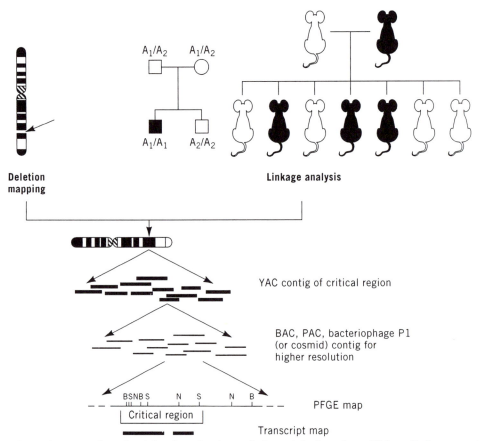

Figure 8 Steps involved in the cloning of the critical region. Either linkage analysis in human beings or the mouse or deletion analysis is the first step in defining a critical region. A physical map is subsequently produced (a YAC contig or contig from clones in a bacterium-based vector). From this, a transcript map can be derived containing potential candidate genes for the disease locus. Additional information may be derived from the long-range PFGE restriction map, where sites for infrequently cutting restriction enzymes within the critical region may indicate a putative CpG island associated with a gene or transcript. B, S, and N are restriction sites on the map.

benefit of a YAC-based map of the human genome, the tedious and time-consuming exercise of chromosome walking will no longer be needed, and YACs spanning critical regions of disease genes can be identified without going through the laborious process of screening YAC libraries. The YAC contig information for the whole-genome STS content map (Hudson et al. 1995) is available from:

    http://www-genome.wi.mit.edu

and the relevant YAC clones can be easily accessed from public sources (e.g., the Human Genome Mapping Program Resource Centre, UK):

    http://www.hgmp.mrc.ac.uk

or via commercial suppliers (e.g., Research Genetics). YAC clones provide an excellent tool for the construction of physical maps genome-wide, and YAC clone contigs are also an excellent framework on which to map other sequences. However, for a variety of reasons discussed on pp. 19–20, they are less than ideal as a primary tool for turning the physical map into a transcript map that identifies all of the genes across any disease locus region. Therefore, YAC maps are only the first step in the physical mapping of any chromosome region. Very often the YAC contigs are converted into PAC/BAC contigs which serve not only as a resource for good identification strategies, but also as a source for sequence-ready maps for the purpose of genome sequencing. Subsequently, more detailed physical maps are constructed across the region by using BAC, PAC, or bacteriophage P1 clones that contain inserts ranging in size from 70 kb to several hundred kilobases (see below). Smaller fragments (35–45 kb) of DNA can be further subcloned into cosmids to simplify analysis.

LONG-RANGE RESTRICTION MAPPING

A long-range restriction map constructed by PFGE (see Riethman et al. 1997) provides information on the physical distance between DNA markers and the size of the critical region and also helps in the identification of CpG islands (see pp. 22–23). Two major advantages for constructing a long-range restriction map starting from cloned material instead of uncloned genomic DNA are (1) it is technically easier because cloned material provides much stronger hybridization signals and (2) it circumvents technical problems related to genomic DNA methylation. However, it is important to confirm the cloned DNA map with the genomic map.

PREPARING TO ISOLATE TRANSCRIBED SEQUENCES

Although YACs are ideal tools for many applications, they are not ideal cloning vectors for gene identification. Two common problems are chimerism (Green et al. 1991)—the presence in a single YAC of DNA fragments from two or more noncontiguous regions of the genome—and internal deletions, which often produce gaps in the physical maps. Because of the similarities in size and composition between the YAC clone and endogenous yeast chromosomes, it is also not straightforward to obtain adequate amounts of YAC DNA free of the endogenous yeast DNA. The DNA thus obtained may not be suitable for performing some basic manipulations, such as restriction analysis, subcloning, labeling, and sequencing. To circumvent the problems of using yeast as a host, new bacterium-based vectors have been developed (for review, see Monaco and Larin 1994). These include the bacteriophage P1 system (cloning capacity of 70–100 kb) (Sternberg 1990, 1992; Sternberg et al. 1990), the PAC system (cloning capacity of 60–300 bp) (Ioannou et al. 1994), and the BAC system (cloning capacity of 200–300 kb) (Shizuya et al. 1992).

As discussed above, the first stage in preparing to isolate transcribed sequences from any chromosome region is to convert the YAC-based physical maps into maps based on BACs, PACs, or bacteriophage P1s. The conversion of YAC-based maps into cosmid-based maps is also an effective approach (Zuo et al. 1993; Nizetic et al. 1994; Wapenaar et al. 1994). The high density of STSs used to contruct the YAC maps (see above) provides an important framework for the construction of these higher-resolution physical maps. Overlapping sets of clones are derived by using the available STSs to screen BAC, PAC, bacteriophage P1, or cosmid libraries. In addition, hybridization approaches are very effective in turning the YAC maps into BAC, PAC, bacteriophage P1, or cosmid contig maps. Entire YACs and products of inter-*Alu*-PCR as well as primer sequences from appropriate STSs can be used to screen BAC, PAC, bacteriophage P1, or cosmid libraries. The clones isolated for any region are assembled into contigs using available STS content information and fingerprinting approaches to validate overlaps (Sulston et al. 1988). The contigs established using these methods represent excellent reagents for gene identification methods such as exon trapping and cDNA selection (see pp. 26–28). An alternative option is the construction of cosmid libraries directly from an entire YAC or group of YACs.

ISOLATION OF TRANSCRIBED SEQUENCES

Considerable effort is currently being expended on the generation and mapping of a huge collection of ESTs from the human and mouse genomes (Hillier et al. 1996; Schuler et al. 1996). The generation of dense EST maps covering the entire human and mouse genomes will provide a wealth of candidate genes in any and every chromosome region. Identifying those ESTs and genes that map to the critical disease region will ultimately be a trivial task involving querying the appropriate physical and genetic map information from the databases. Eventually, the complete sequence of the human genome will further improve the available gene map and enhance the ability to identify the critical candidate genes from the disease locus region. The development of EST maps and the genomic sequencing program are discussed on pp. 34–40. However, with the development of the EST map continuing and genome sequencing still in its infancy, investigators may still need to use a variety of direct approaches to identify expressed sequences from any chromosome region.

The identification of expressed sequences from a given region of genomic DNA is often achieved by using different, yet complementary, approaches. Some of these approaches, such as cloning CpG islands, testing for cross-species conservation, and hybridization to northern blots, are used to identify the presence of a gene in a genomic region without actually cloning the gene. Others, such as exon trapping, cDNA selection, genomic sequencing, and screening cDNA libraries, allow the isolation and at least partial sequence characterization of the gene and provide information regarding its subsequent protein product.

Once a set of overlapping clones spanning the entire critical region has been obtained, the investigator faces the dilemma of where and how to start looking for genes in that region. The choice of the "right" method varies a great deal, depending on factors such as the size of the critical region and the predicted pattern of expression of the target gene. A comparison of the features of several gene identification methods is shown in Figure 9. Detailed descriptions of most of these methods can be found elsewhere in this volume and Volume 1.

The sections on pp. 22–30 provide a brief guideline for when and how a particular method can be used, emphasizing the notable advantages or limitations of each. Figure 10 presents a flowchart of the steps involved in gene identification.

Methods for isolation of transcribed sequences are similar in all organisms, but investigation of early developmental defects may be more appropriate in the mouse because of tissue accessibility considerations. As the human and mouse EST and gene maps become denser, disease gene identification is increasingly likely to use these resources. Genomic sequencing will also become a more widely used approach for identifying genes in the critical disease region.

	Success rate	Informativeness	Expression independence	Cost/time efficiency	Suitability for		
					<50 kb	50–500 kb	>500 kb
Cloning CpG islands	●	•	●	•	●	●	●
Zoo blots	●	•	●	●	●	•	•
Northern blots	•	●	•	•	●	•	•
Exon trapping/amplification	●	●	●	●	●	•	●
cDNA selection	●	•	•	●	●	●	●
Genomic sequencing	•	●	●	•	●	•	•
Direct hybridization of cDNA libraries	●	●	•	●	●	●	•

Figure 9 Comparison of several gene identification methods. Rating is based on published data and on the author's (AB) personal experience. Relative value of the method: (*small circles*) low; (*medium circles*) medium; (*large circles*) high. Success rate is the number of times the method has been used successfully; informativeness is the potential to yield information on sequence/homology and expression pattern; expression independence is the ability of a method to work irrespective of the expression pattern of the target gene; cost/time efficiency is the efficiency of a method with respect to costs and time efforts.

Cloning CpG Islands

BASIS OF METHOD AND REQUIRED RESOURCES

The construction of a long-range restriction map (see also Reithman et al. 1997) either directly from genomic DNA or from cloned genomic DNA (typically maintained in cosmids or YACs) allows identification of clusters of rare-cutting restriction sites (RCRE sites) corresponding to regions with a high G+C content (Bird 1986). These GC-rich regions, referred to as CpG islands, are usually located at the 5′ end of genes, somewhere around the first exon (Bird 1987). RCRE sites could be methylated (and therefore resistant to cleavage) in genomic DNA (but not in yeast), whereas true CpG islands are typically unmethylated, allowing cleavage (Bird 1986; for further discussion, see Chapter 5). Most clusters of RCRE sites that have been identified in cloned genomic DNA (YACs and cosmids) actually correspond to true CpG islands. Once a putative CpG island is identified, this information can be used to subclone fragments of genomic DNA spanning or flanking the island. This can be achieved in several ways.

- If a YAC contig spanning the critical region is available and a CpG island has been mapped within one of the YAC clones, the YAC can be subcloned into more easily handled vectors, such as cosmids or bacteriophage λ, which can be subsequently digested with RCREs to identify and confirm which subclones contain the CpG island.

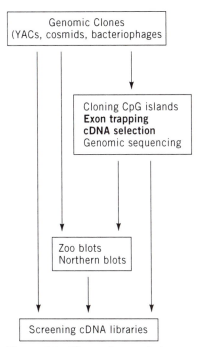

Figure 10 Flow chart of the steps involved in the identification of genes from cloned genomic DNA. Exon trapping and cDNA selection—two of the most commonly used methods—are highlighted.

- YACs and cosmids can be digested with an RCRE plus a restriction enzyme that cuts more frequently (e.g., *Eco*RI or *Hin*dIII), and the products obtained can be subcloned directly into a plasmid vector with compatible ends.
- The conversion of YAC-based maps into BAC, PAC, bacteriophage P1, or cosmid-based maps greatly facilitates the cloning of CpG islands. Cosmids or other large-insert clones can be readily digested with RCREs to identify which contains the CpG island (van Slegtenhorst et al. 1994). PCR-based approaches for the identification of CpG islands have also been developed, but their use has been limited (Patel et al. 1991).
- CpG islands can be isolated from genomic DNA by first using restriction enzymes that will cut within and on either side of a CpG island. The fragments produced are then separated based on their methylation (genomic DNA) and nonmethylation by using a methylated DNA-binding protein affinity column (see Chapter 5).

ADVANTAGES/DISADVANTAGES/LIMITATIONS

Cloning a CpG island can be helpful in isolating the corresponding gene, mainly by pointing to the precise position of the gene. However, the construction of a long-range restriction map is a laborious task, particularly for large regions (Evans 1991). Furthermore, a CpG island may not necessarily be associated with a gene and not all genes are associated with CpG islands (Bird 1995). To pinpoint the location of a given CpG island, a long-range restriction map will have to be constructed (Reithman et al. 1997).

Testing for Cross-species Conservation

BASIS OF METHOD AND REQUIRED RESOURCES

Protein-coding regions of genes are conserved to varying degrees across species, frequently resulting in detectable similarity at the DNA level. It is therefore possible to take advantage of sequence conservation to identify coding regions in genomic DNA. The method is simple and was used for the first time for the identification of the Duchenne muscular dystrophy gene (Monaco et al. 1986). Random subclones from the critical region of interest are derived and hybridized under relatively low stringency to a Southern blot containing DNA samples from several, more or less distantly related, species. Such blots have been named zoo blots or Noah's ark blots (if males and females of each species are included) because of the variety of species used. Because the endogenous repeat families (e.g., for human and *Alu* sequences) for most species show some conservation across species and will result in a smear after hybridization of a repeat-containing genomic probe, it is important to block repeats by adding denatured genomic DNA to the hybridization mixture. A random subclone that demonstrates cross-hybridizing bands in several distantly related species clearly contains conserved sequences and is usually an indication that the probe contains coding sequences. (Zoo blots and blocking repetitive DNA sequences are discussed in Wolff and Gemmill [1997].)

There is little value in applying this approach to whole YACs, BACs, PACs, bacteriophage P1s, or cosmids from the critical region. In most cases, such clones would by and large contain coding sequences. Thus, little information would be gained on the localization of potential gene sequences.

Another factor to consider in this method is the choice of the species. If too closely related species are used, there is a possibility that even noncoding sequences will be conserved, and cross-hybridization will not necessarily reveal the presence of a gene. When very distantly related species are used, coding sequences might diverge too greatly to allow for cross-hybridization. Species that are most commonly used to test for cross-species conservation of human DNA fragments include mouse, sheep, pig, bovine, rabbit, and chicken. Pig appears to have the advantage of not sharing common repeat sequences with the human genome, thus providing cleaner hybridization results.

ADVANTAGES/DISADVANTAGES/LIMITATIONS

The degree of conservation of any given gene depends on a variety of factors, including the phylogenetic distance between different species, the presence of sequence constraints to preserve a functional protein, the genetic code degeneracy, and the mutation rates in species. Therefore, this method will be successful only if the portion of the gene contained within the probe has been conserved across species (e.g., an exon coding a functional domain). However, the method is simple, and most importantly, it is completely independent of the expression pattern of the target gene. Because of its expression independence, this approach can be particularly useful in situations in which the gene has a tissue-specific and developmentally regulated expression pattern, as in the case of the identification of the *SRY* gene involved in sex determination (Sinclair et al. 1990). A major limitation of this method is that it is quite time-consuming, and it is therefore advisable to apply it to small genomic regions (<50 kb). Furthermore, many probes tend to give smeary hybridization results when low-stringency conditions are used, even when repeats are blocked.

Testing for cross-species conservation cannot be considered a robust method for gene identification, but it should be a useful complement to the other methods. For example, genomic clones spanning CpG islands, exon trapping products, and cDNAs isolated by cDNA selection can be hybridized to zoo blots to obtain additional evidence showing that they are indeed part of a gene before they are hybridized to cDNA libraries.

Hybridization of Genomic Clones to Northern Blots

BASIS OF METHOD AND REQUIRED RESOURCES

A feature shared by all genes is, by definition, that they are transcribed into RNA. One way to take advantage of this feature is to hybridize genomic probes to northern blots (Sambrook et al. 1989) containing RNAs from several human tissues (many of which are now commercially available). This allows the identification of mRNA species cross-hybridizing to the probe. The choice of the tissues to be used as RNA sources depends on the expected pattern of expression of the target gene, when this information is available. It is a good idea to use tissues from several developmental stages since the gene can be developmentally regulated. This method has been used successfully for the identification of a number of disease loci, including the cystic fibrosis locus (Rommens et al. 1989) and the Menkes disease gene (Vulpe et al. 1993).

ADVANTAGES/DISADVANTAGES/LIMITATIONS

The existence of cross-hybridizing bands is considered a good indication of the presence of a gene within the genomic fragment that was used as a probe. A notable exception is the presence of pseudogenes homologous to functional genes that are located elsewhere in the genome. A positive result on a northern blot will also provide information on tissue distribution of expression and on the transcript size. However, the method is highly dependent on the levels of expression of the target gene. Only abundant transcripts can be detected on northern blots, particularly when large genomic fragments are used as probes. Very large genomic probes may give hybridization results that are difficult to interpret.

This method provides the best results if applied to small genomic regions (<50 kb), but it is impractical for larger genomic regions. Furthermore, preparing northern blots with RNAs from different tissues is time-consuming and expensive, and availability of human samples is limited. Nevertheless, commercial (e.g., Clontech) northern blots are available for a range of mouse and human tissues, and although expensive, they can be probed multiple times.

Exon Trapping

BASIS OF METHOD AND REQUIRED RESOURCES

For a detailed description of exon trapping and its newer version, exon amplification, see Chapter 4. This method requires subcloning of the target DNA into a vector containing consensus splice sites (Duyk et al. 1990; Buckler et al. 1991; Krizman and Berget 1993; Church et al. 1994). Typically, transfection into eukaryotic cells followed by RNA preparation and RT-PCR with primers flanking the consensus splice sites will yield putative exon sequences. Any single genomic DNA clone, or pools of clones, can be scanned by exon trapping. BACs, PACs, bacteriophage P1s, and cosmids are all suitable substrates for exon trapping. Even small YACs of approximately 150 kb in size (Gibson et al. 1994) have been successfully used as targets for exon trapping. Bacteriophage λ clones (Nehls et al. 1994) have also been used successfully.

Putative exons identified by exon trapping must undergo a variety of verification steps. They can be first hybridized to ribosomal, mitochondrial, yeast, vector, and repeat sequences of various kinds, which are often found in exon-trapping products. Sequence analysis may reveal interesting homologies with previously identified genes and be instrumental in predicting the function of the target gene, as shown by the recent identification of a cluster of sulfatase genes by exon trapping (Franco et al. 1995). In addition, RT-PCR can be performed using oligonucleotide primers designed from exonic sequences to test for the expression of the putative gene in a variety of tissues. RT-PCR using primers from different exon-trap products can also be used to try to connect exons isolated from the same genomic region ("exon connection")—a PCR product should only be produced if the two exons are part of the same transcript. A final important step in gene discovery is to test whether the putative exons map to the critical region by hybridizing them to the original genomic clones and to genomic DNA. On the basis of these experiments, putative exons that are likely to be part of novel genes from the critical region are then hybridized to cDNA libraries. An efficient way to do this is to use exon pools (containing 10–50 clones with trapped exons) from the critical region for each cDNA library screening. Some exons within each pool may be part of the same gene. To determine which transcript corresponds to which exon, positive cDNA clones are subsequently hybridized to an array containing the entire panel of exon-trapping products.

ADVANTAGES/DISADVANTAGES/LIMITATIONS

This method provides a highly efficient, expression-independent way of cloning small exon-containing fragments of genomic DNA. However, a common problem in exon trapping is the presence of cryptic splice sites that can generate false positives. In addition, the method will only work with genes that have more than two exons (with the notable exception of 3′-terminal exon trapping), and it is highly dependent on the size of the exons. Best results are usually obtained with exons smaller than 200 bp. Exon trapping is particularly amenable to the construction of transcript maps of reasonably large (50–500 kb) genomic regions (Church et al. 1993).

cDNA Selection

BASIS OF METHOD AND REQUIRED RESOURCES

The principles and technical details of direct cDNA selection are described in Chapter 3. This method involves the capture of cDNA sequences on the basis of their hybridization to target genomic DNA (Lovett et al. 1991; Parimoo et al. 1991; Korn et al. 1992; Tagle et al. 1993; Lovett 1994). Entire YAC clones or pools of BACs, PACs, bacteriophage P1s, and cosmids are commonly used as sources of genomic DNA. Heterologous cDNA from another species can also be used to minimize problems related to common repeats such as *Alu* sequences (Sedlacek et al. 1993). Two major variants of this method have been developed. In the first, the target genomic DNA is immobilized on nylon membranes and the hybridization is performed on the membranes. In the second, the hybridization is performed in solution, and cDNAs hybridizing to biotinylated genomic DNAs are captured by using streptavidin-coated magnetic beads. In both cases, several rounds of selection will often be carried out.

The choice of cDNA sources is obviously very important in this approach and depends highly on the predicted pattern of expression of the target gene. Primary cDNA from reverse-transcribed RNA can be used after digestion, ligation of adapters, and amplification using primer-adapters. Alternatively, large pools of cDNA library inserts can be amplified using vector primers.

As with the putative exons from exon trapping, it is wise to verify the putative cDNAs identified by cDNA selection by several techniques before they are used as probes for the hybridization of cDNA libraries. This should include validating whether they arose from the target genomic region, sequencing, RT-PCR, and possibly hybridization to zoo blots and northern blots.

Because cDNA selection requires access to a specific tissue of interest, cDNA selection of mouse sequences may be more appropriate since the mouse affords access to all tissues at all developmental stages.

ADVANTAGES/DISADVANTAGES/LIMITATIONS

Although this method it is not completely expression-independent, it allows the isolation of genes with low-abundance expression patterns. Rare transcripts can be enriched up to 10,000-fold using cDNA selection. The advantage of cDNA selection over most other methods is that it has the ability to scan large (>500 kb) regions of genomic DNA relatively quickly because the size of the target DNA used has only a minimal effect on the selection process (Korn et al. 1992; Sedlacek et al. 1993; Lawrence et al. 1994). It is safer to use a YAC contig representing two to three times the coverage of the region, therefore minimizing the risk of not including in the analysis a genomic region that is deleted in a particular YAC clone. In addition, this method is not subject to the presence of introns (unlike exon trapping).

A disadvantage of cDNA selection compared to exon trapping is that 3'-untranslated regions (instead of coding regions) of genes are more commonly captured because of the longer cDNA/genomic DNA contiguity found in this region. Therefore, finding interesting protein homologies in the databases and cross-species conservation occurs less frequently with cDNA selection than with exon trapping. However, a growing number of 3′ESTs generated from the EST projects (Hillier et al. 1996) have been clustered with more 5′ESTs (the UniGene project; for further details, see pp. 29–30). It can thus be expected that where a cDNA selection product shows homology with a 3′EST, it will be increasingly possible to identify the associated protein-coding information. Common false positives are cDNA clones containing yeast and ribosomal DNA, repeat sequences, and cDNAs from heteronuclear RNA.

Genomic Sequencing

BASIS OF METHOD AND REQUIRED RESOURCES

Recent advances in bioinformatics now allow for faster and more accurate recognition of coding sequences in long sequences of genomic DNA (Uberbacher and Mural 1991; Claverie 1994; Solovyev et al. 1994; Boguski et al. 1996; see also Baxevanis et al. 1997). The number of new gene sequences entered in the databases is rapidly increasing. Therefore, the likelihood that a given gene sequence will hit an identical or closely related sequence in the database is growing exponentially. This, together with ongoing developments in automation and robotics for faster and cheaper long-range sequencing (Wilson and Mardis 1997), suggests the possibility that genomic sequencing may become the ultimate gene identification method.

A recent innovation that takes advantage of the human and mouse genome projects is comparative sequencing of homologous regions in each organism (Koop 1995; Oeltjen et al. 1997). This is likely to provide information on gene identification and on important regulatory sequences. The pufferfish, Fugu, is thought to contain the same number of genes as other vertebrates, including mice and human beings, but has a considerably smaller genome with a much lower complement of noncoding sequence (Brenner et al. 1994; Mileham and Brown 1994). Therefore, comparative sequencing of the pufferfish genome has the potential to be an efficient method of gene identification. Comparative sequencing of known genes in human beings and pufferfish has already assisted in the analysis of conserved functional domains (Maheshwar et al. 1996).

ADVANTAGES/DISADVANTAGES/LIMITATIONS

Advantages are the complete insensitivity of this method to factors such as expression pattern and size of exons. The major concern in using genomic sequencing as a gene identification method is the ability of most investigators to gain sufficient access to the necessary resources.

Screening cDNA Libraries

BASIS OF METHOD AND REQUIRED RESOURCES

As shown in Figure 10, the final step in most gene identification strategies is screening cDNA libraries. This method provides a great deal of information, allowing the cloning and sequence characterization of full-length cDNAs and, therefore, providing complete knowledge of the predicted protein product of a given gene. The isolation of the full-length cDNA also allows in vitro and in vivo expression of the gene and the development of functional assays for its protein product.

A variety of DNA fragments can be used as probes for screening cDNA libraries. They range from very large and complex fragments, such as YAC inserts or pools of cosmids, to small PCR products or plasmid inserts. As expected, smaller fragments give cleaner results. However, it is almost imperative to test the probe on a Southern blot to ensure that it gives clean results before it is hybridized to a cDNA library. Carrying out slightly more work in characterizing the DNA fragments that are going to be used as probes usually prevents wasting time chasing false positives at a later stage. Obviously, ideal probes include genomic fragments flanking CpG islands, sequences that are highly conserved across species, fragments detecting transcripts by northern analysis, and pooled exon trapping and/or cDNA selection products.

The large EST sets being developed from normalized human cDNA libraries by the IMAGE consortium (Hillier et al. 1996) can be accessed at:

```
http://www-bio.llnl.gov/bbrp/image/image.html
```

These will also provide a route to the identification of cDNAs homologous to coding sequences isolated by a variety of methods, including exon trapping and cDNA selection. The identification of a homologous EST by sequence comparison immediately provides a homologous cDNA from which the EST was derived that can subsequently be further characterized to abstract additional coding sequence. Much effort has been expended in organizing the available ESTs into clusters based on overlapping homologies identified between sequence tags. This so-called UniGene set provides additional cDNA sequence information beyond the initial observed homology as well as additional cDNAs from the relevant gene for further characterization. The UniGene set can be accessed at:

```
http://www.ncbi.nlm.nih.gov/UniGene/Home.html
```

For further discussion of the IMAGE consortium and the UniGene set, see pp. 34–35.

ADVANTAGES/DISADVANTAGES/LIMITATIONS

The major limitation of screening cDNA libraries by hybridization is that it is highly expression-dependent and time-consuming. Multiple purification steps are required to isolate a pure cDNA clone, and a large number of clones usually need to be screened (up to several million) to find the target gene since most of the clones found in cDNA libraries come from relatively few, yet highly abundant, transcripts.

The EST generation program by the IMAGE consortium has the advantage that a large number of normalized cDNA libraries have been constructed by skimming off cDNA clones from abundant transcripts (Soares et al. 1994; see also Chapter 2). Normalized cDNA libraries considerably reduce the number of cDNA clones that must be screened. The production of good normalized libraries is an essential prerequisite to generating ESTs representative of a given state. Protocols for these are provided in this manual, which have been used to generate such libraries for a significant proportion of the human and mouse EST collections. In addition, cDNA clones from the IMAGE consortium set have been arrayed on high-density filters (available from Research Genetics) and can be screened by hybridization. Use of arrayed cDNA libraries not only eliminated the tedious and time-consuming purification steps usually required for the screening of cDNA libraries, but also provided much stronger hybridization signals than those from conventional high-density platings. This latter factor is particularly important when complex or heterologous probes are used. Of course, as described above, an alternative to screening by hybridization is screening in silico (computer-based screening of databases) to discover homologous ESTs, which immediately identifies homologous cDNAs. Given the ease of this approach, it is increasingly superseding hybridization approaches to identifying cDNAs.

PROVING A CANDIDATE GENE IS CAUSALLY MUTATED

Once a candidate gene has been identified, it is still only a candidate until it can be empirically demonstrated that this gene, once mutated, underlies a mutant phenotype or disease state. A number of different approaches may be required to confirm that the candidate gene carries a causal mutation and is subsequently responsible for an aberrant phenotype or disease state.

Analyzing more than one allele is a particularly powerful approach to proving causality.

DNA Sequence Comparisons of Affected and Unaffected Individuals

Comparison of the DNA sequence of the gene carrying the mutation with the sequence obtained from an unaffected population may reveal an obvious deleterious mutation, such as a premature stop codon, which would suggest causation of the mutant phenotype (see Chapter 6). In most of the cases studied to date, however, a protein polymorphism is more likely to be found. In this case, statistically significant populations of affected and unaffected individuals must be sampled and compared to determine if the mutation is only found in phenotypically affected individuals and segregates as expected in families. Differences in penetrance (i.e., whether or not an individual carrying the mutation demonstrates the mutant phenotype) can clearly confound these analyses. In addition, if a protein different from the "unaffected" protein is produced, it must be demonstrated to be present in the mutated form in the tissue carrying the phenotype.

Gene Expression

If no differences in gene sequence are identified, a gene expression pattern must also be elucidated in affected and unaffected individuals to examine the possibility that a disruption in gene regulation is responsible for the phenotype. For example, the gene must first be expressed in the tissue(s) considered to be the most affected by the phenotype or disease. Both mRNA and protein levels should be quantified and checked for tissue specificity in affected individuals versus unaffected controls. In a truly polygenic disorder, this information may provide a clue to the development of a phenotype but not a direct answer.

Mutant Alleles

A very great aid in proving a candidate gene is causally mutated is the ability to analyze more than one allele of the mutation. The presence of multiple mutations in the same gene is a very strong indicator that the candidate is the correct gene. For many mutant loci in the mouse, multiple alleles exist and are well characterized. For example, at least 15 different mutations in the microphthal-mia-associated transcription factor cause microphthalmia in the mouse (Moore 1995). Mutations in this gene also cause Waardenburg's syndrome type 2 in human beings (Tasseabehji et al. 1995). In addition, many mouse mutations are co-isogenic with the background inbred strain on which they arose (i.e., the mutant and background strain are genetically identical except for the mutation itself). In these cases, discovery of a mutation identifies the causal gene. With the increased number of candidate genes available, mutation identification is likely to become the rate-limiting step. Mutation detection strategies have improved considerably in recent years (Grompe 1993). Several up to date approaches used for the detection of mutations in large stretches of DNA and the detection of mutations associated with human disease states are presented in Chapter 6. DNAs arrayed on a variety of surfaces (DNA microchips) promise to facilitate this process considerably (Pease et al. 1994).

Phylogenetic Comparisons

Cross-species comparisons can be particularly useful if a mutated gene causes a similar phenotype when mutated in more than one species. This is the case with mutation in the type VII myosin gene that gives rise to deafness in both human beings and mice (Gibson et al. 1995; Weil et al. 1995). However, more extensive phylogenetic comparisons may shed further light on the function of a potential candidate gene, for example, the gene encoding the ataxia-telangiectasia locus (*ATM*; Savitsky et al. 1995). The *ATM* gene appears to be part of the phosphatidyl inositol kinase gene family in which members share a highly conserved kinase domain close to the carboxyl terminus. Significant homologies across the carboxy-terminal kinase domain as well as in adjacent regions are shared by *ATM* and other members of this family, including the *rad3* gene from *S. pombe*, its homolog *ESR1* from *S. cerevisiae*, and the *ME1-41* gene in *D. melanogaster* (Lehmann and Carr 1995; Shiloh 1995). All of these homologous genes are involved in the cell-cycle-checkpoint response to DNA damage, which is consistent with one of the phenotypes observed for ataxia-telangiectasia in which cells fail to arrest at appropriate checkpoints after irradiation.

Further Analyses Using the Mouse as a System to Assess Candidate Genes

The mouse can be a powerful system for further assessment of a candidate genes using transgenic techniques or for further assessment of both temporal and spatial expression patterns.

DEVELOPMENTAL AND SPATIAL EXPRESSION PATTERNS

It may be instructive to examine the expression of the candidate gene in the mouse by using in situ hybridization or immunohistochemistry (Hogan et al.

1994). Detailed assessment of expression profiles in the embryo and adult tissues are likely to provide useful information to confirm or eliminate potential candidates.

TRANSGENIC APPROACHES

Transgenic approaches can be used to assist in the identification and confirmation of candidate genes in the mouse, especially when there are no multiple alleles to assist with gene identification (see above). It is possible to establish transgenic mice carrying BACs or bacteriophage P1s from contigs covering the critical region. Subsequently, individual transgenes can be transferred by breeding to a strain carrying the mutation of interest, and progeny mice can be examined for complementation of the defect. In this way, the critical region containing the gene can be further refined, and ultimately the identification of the correct candidate gene can be confirmed. A recent notable example is confirmation of the identification of the mouse Clock gene by transgenic rescues in BAC (Antoch et al. 1997; King et al. 1997).

Functional Analysis of Causal Mutations

All of the approaches above for proving that a gene is causally mutated provide strong evidence regarding the function of a given candidate gene. Additional dissection of the molecular and biochemical parts played by a gene product can be obtained by using a variety of experimental approaches in eukaryotic cell culture both to examine the mechanism of action of a candidate gene and to demonstrate a change in cellular function due to a given mutation. Although indepth discussion of such cell-culture-based assays and yeast two-hybrid analysis of candidate genes are beyond the scope of this chapter, a number of these approaches have recently been highlighted for the familial Alzheimer's disease genes, the presenilins (Haass 1997).

FUTURE DEVELOPMENTS

There are a number of areas where future developments are likely to have a significant impact on disease gene identification and the identification of gene function more broadly. Several of these areas are discussed below.

It is expected that more and more genes involved in disease will be identified in the near future. These genes will represent critical reagents for molecular diagnosis and for the assessment of genetic predisposition for human disease. The next challenge will be to use high-throughput cell biology approaches to translate this plethora of genetic information into knowledge of fundamental biological processes in mammalian organisms.

The Completion of a Dense Transcript Map of the Human Genome

Undoubtedly, the process of disease gene identification will be revolutionized by the construction of a detailed transcription map of the human genome. Several efforts are contributing to this goal. They can be divided into two main types: cDNA-based and genomic-based approaches (Figure 11).

cDNA-BASED APPROACHES

Among the cDNA-based approaches, mapping ESTs (Adams et al. 1992, 1995) promises to make a major contribution to gene identification and analysis (Polymeropoulos et al. 1992, 1993; Berry et al. 1995; Boguski and Schuler 1995;

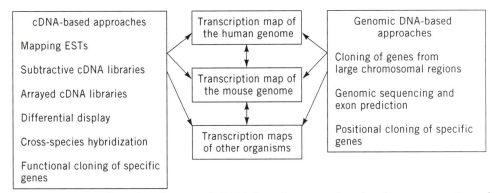

Figure 11 cDNA-based and genomic DNA-based approaches for the construction of transcription maps of the human genome. It is important to recognize that transcript maps of different mammals are reciprocal because of synteny and that the map position of genes from one organism can be related to the map position in another organism.

Banfi et al. 1996). The main effort is currently devoted to the production of ESTs from normalized human and mouse cDNA libraries (Soares et al. 1994). The IMAGE consortium, which is involved with the development of sequence, map, and expression information, has currently arrayed more than 800,000 clones from normalized human cDNA libraries and more than 300,000 clones from normalized mouse cDNA libraries. The consortium set can be accessed at:

```
http://www-bio.llnl.gov/bbrp/image/image.html
```

More than 1,000,000 EST sequences (Hillier et al. 1996) from these arrayed clones have been deposited in dbEST, which can be accessed at:

```
http://www.ncbi.nlm.nih.gov
```

and it is estimated that more than 50,000 distinct human genes are available within the IMAGE consortium set. In addition, the IMAGE consortium will soon provide a re-arrayed set of clones, including full-length clones representing known genes as well as clones representing unique ESTs that may or may not be full length.

ESTs will continue to be produced at a very high rate from a variety of tissues at different developmental stages. At the same time, the UniGene project will be developing a large collection of distinct gene sequences from the available EST information by identifying clusters of homologous ESTs. To date, the UniGene collection comprises more than 40,000 distinct cDNA clusters, which represents a valuable tool for developing a transcript map of the human genome. (For further information, see pp. 28 and 29.)

ESTs are mapped most efficiently by assigning them to a chromosome region using radiation hybrid panels (Cox et al. 1990; Stewart et al. 1997). Recently, the UniGene set has been used to develop a transcript map of the human genome comprising more than 16,000 ESTs (Schuler et al. 1996). In general, the construction of this gene map proceeded by the use of radiation hybrid panels, but in some cases, ESTs were also more finely mapped by assignment to YAC contigs available in the appropriate map region (Berry et al. 1995; Boguski and Schuler 1995; Hudson et al. 1995). Linking an EST to a specific YAC clone allows immediate integration of the EST locus into the physical map and provides its relationship to the critical disease locus region as defined by both genetic and physical mapping studies (see pp. 10–20).

Additional sources of cDNAs will come from subtractive cDNA libraries (Jones et al. 1991) and the use of the differential display technique (Liang and Pardee 1992; Welsh et al. 1992; Liang et al. 1993; Fanning and Gibbs 1997). Functional cloning of specific genes by traditional methods (e.g., immunoscreening of cDNA libraries [Young and Davis 1983] and cross-species hybridization) or new techniques (e.g., the two-hybrid system; Fields and Sternglanz 1994) will continue to yield new genes, even though at a lower rate than the other approaches.

GENOMIC DNA-BASED APPROACHES

Among the genomic DNA-based approaches, gene identification in large chromosome regions by exon trapping and cDNA selection will represent additional sources of mapped genes (see Chapters 3 and 4). More focused positional cloning efforts will yield additional genes. Of particular interest is the strategy of reciprocal probing of arrayed cDNA and cosmid libraries in which pools of amplified and arrayed inserts of cDNA libraries are hybridized to chromosome-specific arrayed cosmid libraries. Subsequently, positive cosmids are hybridized to the arrayed cDNA clones to identify transcripts associated with specific cosmids (Lee et al. 1995). A similar approach can be applied to small subsets of cosmids from specific chromosome regions (Guo et al. 1993). Genomic sequencing (p. 28) and exon prediction are set to play a major part in genomic DNA-based approaches to the development of transcript maps across genomes.

Comparative Maps: Reciprocal Framework Maps of Human and Mouse Genes

Concurrent with the development of a human EST map, gene mapping in other organisms will play a key part in establishing the gene–phenotype relationship. Work is well under way to identify approximately 400,000 ESTs from the mouse genome (see above). Nearly 200,000 ESTs have already been added to the databases at:

```
http://www.ncbi.nlm.nih.gov
```

The mapping of ESTs in both the mouse and human genomes will lead to the development of reciprocal framework maps of genes that will ultimately enhance the process of gene identification in both organisms (see Figures 2 and 4). The presence of conserved ordered segments (see Figure 4) in which both gene content and gene order are conserved between the two genomes allows the position of an EST in one genome to be related to its likely position in the other genome. It is unlikely that the EST maps in either the mouse or human genomes will be totally complete, and they will be nonoverlapping to some extent. ESTs identified in one species but not the other can be related in a reciprocal fashion, and new ESTs can be subsequently assigned to a species map according to their map position in the related species.

The end result of all of these efforts will be a detailed catalog of all mammalian genes associated with their precise map position in each species. This catalog will be extremely valuable in the positional candidate approach for the subsequent identification of genes involved in disease. When a disease gene is assigned to a specific map position, it is now possible, via a combination of experimental and database approaches, to examine the available gene information both in the human database (see above address) and in the homologous chromosome region in the mouse at:

```
http://www.mgd.jax.org
```

Genes in the correct region should be considered candidates for that particular disease, and their features should be compared to the features of the disease to find which candidate gene is the strongest contender for that particular disease. The types of data that may help in this process of "partner" identification are

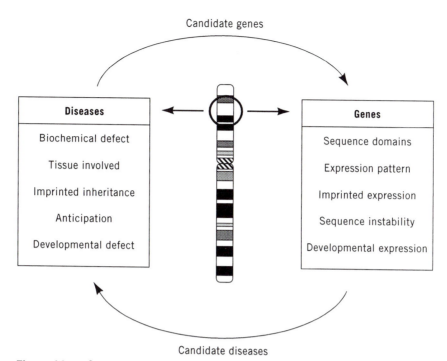

Candidate genes

Diseases	Genes
Biochemical defect	Sequence domains
Tissue involved	Expression pattern
Imprinted inheritance	Imprinted expression
Anticipation	Sequence instability
Developmental defect	Developmental expression

Candidate diseases

Figure 12 Schematic representation of the positional candidate approach. When a new disease locus is assigned to a chromosome region (indicated in the middle), candidate genes from the same region are analyzed. The features of these candidate genes are compared with the features of that particular disease. Several types of features are listed here in a specific order so that they can be matched with the disease features (for specific examples of each feature, see text). When a new gene is assigned to a chromosome region, candidate diseases are analyzed by an analogous process. (Reprinted, with permission, from Ballabio 1993.)

listed in Figure 12 in a specific order so that disease features can be directly matched with the gene features. The following are specific examples of features that could be matched:

- The presence of a previously recognized functional domain in the predicted protein product of the candidate gene could be matched with a biochemical defect identified (or predicted) in patients with a specific disease (e.g., homology of an EST of the *GK* gene with the bacterial glycerol kinase gene could be matched with deficiency of glycerol kinase activity in the patients; Sargent et al. 1993).
- The expression pattern of a gene could be matched with the tissue affected in the disease (e.g., expression of the *PMP22* gene in peripheral nerves could be matched with peripheral neuropathy in Charcot-Marie-Tooth disease type 1A; Matsunami et al. 1992; Patel et al. 1992; Timmerman et al. 1992; Valentijn et al. 1992).
- Diseases showing imprinted inheritance patterns will be candidates for genes showing imprinted expression patterns (e.g., imprinted expression of the *SNRPN* gene could be matched with imprinted inheritance of Prader-Willi syndrome; Leff et al. 1992; Ozcelik et al. 1992).

- Genes showing instability caused by triplet repeat expansion will be candidates for diseases showing anticipation, i.e., a progressive increase of the severity of the phenotype from one generation to the next (e.g., triplet repeat expansion in the *SCA1* gene could be matched with anticipation in the spinal cerebellar ataxia phenotype; Orr et al. 1993; Banfi et al. 1994).
- Developmental defects could be matched with developmentally regulated genes (e.g., developmentally regulated expression of the *PAX3* gene could be matched with the developmental defect in Waardenburg syndrome; Tassabehji et al. 1992).

Comparative Sequence Reference Databases

Cross-species informatics has an increasingly important role in gene identification, particularly in assigning potential functions to newly characterized genes and in assessing their potential as candidates for a particular disease phenotype. In many cases, a related gene or sequence motif may have already been characterized and associated with a phenotype in another organism. The comparative sequence reference database can be accessed at:

```
www.ncbi.nlm.nlh.gov/XREFdb/index-html
```

XREFdb is an important tool for genome cross-referencing (Bassett et al. 1995; Bassett 1997). XREFdb is systematically placing on the mouse and human maps novel ESTs that are highly related to genes of known function in other model organisms. XREFdb now includes a map-position-based query tool for identifying and assessing potential candidate genes for disease loci. Users can also query XREFdb with a protein sequence of interest from any organism. Each sequence is searched against dbEST on a monthly basis, and XREFdb provides all relevant positional data in mouse and human phenotypes as well as disease phenotypes from the same region.

Another publicly available bioinformatic resource that is exclusively focused on comparison between the sequence of *D. melanogaster*, one of the most valuable model organisms, and the human and mouse sequences is represented by the *Drosophila*-related expressed sequence (DRES) database. This bioinformatic tool provides access to a variety of information (sequence, mapping, as well as expression data) on human and murine homologs of *Drosophila* mutant genes:

```
http://www.tigem.it/LOCAL/drosophila/html/drestable.html
```

This cross-species comparative approach has brought important benefits to a number of investigations either by assisting with identification of the underlying gene for a disease locus or by providing clues to the function of a disease locus gene (see p. 32). A recent example involves the identification of a novel protein, HIP1, interacting with huntingtin (the product of the Huntington's disease locus). HIP1 shares sequence homology and biochemical characteristics with Sla2p from *S. cerevisiae* (Kalchman et al. 1997). Genetic analysis of Sla2p in the yeast indicates a role in the regulation of membrane events through interaction with the underlying cytoskeleton. Huntingtin–HIP1 interaction may therefore be crucial for the normal function of the membrane cytoskeleton of neurons, and disruption of this interaction in Huntington's disease may contribute to the phenotype observed in Huntington's disease patients.

Identification of Genes Underlying Polygenic Diseases

The construction of high-density genetic maps of highly polymorphic markers in a variety of mammalian organisms is an important step for the identification of loci involved not only in monogenic diseases, but also in genetic predisposition for polygenic diseases (Lander and Schork 1994). This latter category will represent the real challenge in disease gene identification. In the human genome, much progress has been made in localizing polygenic loci for some diseases through linkage analyses and allele association/linkage disequilibrium studies in outbred populations (for further references and an in-depth treatment, see Cordell and Todd 1995; Tomlinson and Bodmer 1995). It is potentially possible to refine the critical region at any trait locus to a small enough chromosome segment to apply the methods for gene identification presented in this manual. If the appropriate animal facilities are accessible for the mouse, the development of congenic strains (see pp. 13–14) allows the genetic dissection of the loci involved (Frankel 1995) and also allows refinement of the map position to a fraction of a centimorgan on the genetic map, thus providing methods for a suitably small physical region to apply the gene identification.

An Expansion in Classical Mutagenesis Programs in the Mouse

There is a growing awareness that the available animal models of disease processes can be limited. For example, although there are a large number of mouse mutants in which a wide variety of biochemical, physiological, and developmental pathways are affected (Lyon et al. 1996), access to all of the available phenotypes is not available—i.e., a phenotype gap exists (Brown and Peters 1996). Only approximately 1000 mutations are available in the mouse, clearly representing only a small fraction of the likely total number of mammalian genes. The breadth of the mouse mutation resource must be increased. The expansion of efficient chemical (particularly N-ethyl-N-nitrosourea) mutagenesis programs in the mouse is likely to be seen. When this expansion is coupled with sophisticated screens, it will identify novel interesting phenotypes that may be useful models for disease processes. At the same time, the development of high-resolution genetic and physical maps in the mouse, as well as the development of transcript (EST) maps, will lead to more efficient routes for the identification of the genes underlying novel mutations recovered from these mutagenesis programs.

Genomic Sequencing

Ultimately, investigators can expect to be able to use the complete sequence of the human genome to assist in gene identification. Current efforts to sequence the entire human genome are focused on the concept of a "sequence" map in which the sequence of most of the genome (~95%) is recovered at high fidelity (99.99%). Regions that are not amenable to sequencing (e.g., long repeats) are not necessarily recovered in their entirety, but their position on the sequence map is known. Current target dates for finishing the sequence map of the human genome lie in the first decade of the next century. Although the available sequence map will be produced largely from one mammal—the human being—it will be of immense assistance to gene identification in all other mammals and will effectively act as the pivotal comparison for comparative gene mapping in the mammalian order. Despite the identification of a candidate gene, however, investigators will still be dependent on the manipulation of model organisms to understand fully the mechanism whereby any given mutation can affect function and render a characteristic phenotype.

REFERENCES

Adams, M.D., M. Dubnick, A.R. Kerlavage, R. Moreno, J.M. Kelley, T.R. Utterback, J.W. Nagle, C. Fields, and J.C. Venter. 1992. Sequence identification of 2,375 human brain genes. *Nature* **355**: 632–634.

Adams, M.D., A.R. Kerlavage, R.D. Fleischmann, R.A. Fuldner, C.J. Bult, N.H. Lee, E.F. Kirkness, K.G. Weinstock, J.D. Gocayne, O. White, G. Sutton, J.A. Blake, R.C. Brandon, M.W. Chiu, R.A. Clayton, R.T. Cline, M.D. Cotton, J. Earle Hughes, L.D. Fine, L.M. Fitzgerald, W.M. Fitzhugh, J.L. Fritchman, N.S.M. Geoghagen, A. Glodek, C.L. Gnehm, M.C. Hanna, E. Hedblom, P.S. Hinkle, J.M. Kelley, K.M. Klimek, J.C. Kelley, L. Liu, S.M. Marmaros, J.M. Merrick, R.F. Moreno Palanques, L.A. McDonald, D.T. Nguyen, S.M. Pellegrino, C.A. Phillips, S.E. Ryder, J.L. Scott, D.M. Saudek, R. Shirley, D.V. Small, T.A. Spriggs, T.R. Utterback, J.F. Wildman, Y. Li, R. Barthlow, D.P. Bednarik, L. Cao, M.L. Cepeda, T.A. Coleman, E.J. Collins, D. Dimke, P. Feng, A. Ferrie, C. Fischer, G.A. Hastings, W.W. He, J.S. Hu, K.A. Huddleston, J.M. Greene, J. Gruber, P. Hudson, A. Kim, D.L. Kozak, C. Kunsch, H. Hi, H. Li, P.S. Meissner, H. Olsen, L. Raymond, Y.F. Wei, J. Wing, C. Zu, G.L. Yu, S.M. Ruben, P.J. Dillon, M.R. Rannon, C.A. Rosen, W.A. Haseltine, C. Fields, C.M. Fraser, and J.C. Venter. 1995. Initial assessment of human gene diversity and expression patterns based upon 83 million nucleotides of cDNA sequence. *Nature* (suppl.) **377**: 3–176.

Andersson, L., A. Archibald, M. Ashburner, S. Audun, W. Barendse, J. Bitgood et al. 1996. The First International Workshop on Comparative Genome Organisation. Comparative genome organisation of vertebrates. *Mamm. Genome* **7**: 717–734.

Antoch, M.P., E.-J. Song, A.-M. Chang, M.H. Vitaterna, Y. Zhao, L.D. Wilsbacher, A.M. Sangoram, D.P. King, L.H. Pinto, and J.S. Takahashi. 1997. Functional identification of the mouse circadian Clock gene by transgenic BAC rescue. *Cell* **89**: 655–667.

Avner, P., L. Amar, L. Dandalo, and J.L. Guenet. 1988. Genetic analysis of the mouse using interspecific crosses. *Trends Genet.* **4**: 18–23.

Ballabio, A. 1993. The rise and fall of positional cloning? *Nat. Genet.* **3**: 277–279.

Banfi, S., A. Servadio, M.-Y. Chung, T.J. Kwiatkowski, Jr., A.E. McCall, L.A. Duvick, Y. Shen, E.J. Roth, H.T. Orr, and H.Y. Zoghbi. 1994. Identification and characterization of the gene causing type 1 spinocerebellar ataxia. *Nat. Genet.* **7**: 513–520.

Banfi, S., G. Borsani, E. Rossi, L. Bernard, F. Rubboli, A. Marchitiello, A. Guffanti, S. Giglio, E. Coluccia, M. Zollo, O. Zuffardi, and A. Ballabio. 1996. Identification and mapping of *Drosophila* mutant genes through EST database searching. *Nat. Genet.* **13**: 167–174.

Bardoni, B., E. Zanaria, S. Guioli, G. Floridia, K.C. Wor-ley, G. Tonini, E. Ferrante, G. Chiumello, and E.R. McCabe. 1994. A dosage sensitive locus at chromosome Xp21 is involved in male to female sex reversal. *Nat. Genet.* **7**: 497–501.

Bassett, D.E. M.S. Boguski, F. Spencer, R. Reeves, S. Kim, T. Weaver, and P. Hieter. 1997. Genome cross-referencing and XREFdb: Implications for the identification and analysis of genes mutated in human disease. *Nat. Genet.* **15**: 339–344.

Bassett, D.E., M.S. Boguski, and P. Hieter. 1996. Yeast genes and human disease. *Nature* **379**: 589–590.

Bassett, D.E., M.S. Boguski, F. Spencer, R. Reeves, M. Goebl, and P. Hieter. 1995. Comparative genomics, genome cross-referencing and XREFdb. *Trends Genet.* **11**: 372–373.

Baxevanis, A.D., M.S. Boguski, and B.F.F. Ouellette. 1997. Computational analysis of DNA and protein sequences. In *Genome analysis: A laboratory manual.* Vol. 1 *Analyzing DNA* (ed. B. Birren et al.), pp. 533–586. Cold Spring Harbor Laboratory Press, Cold Spring Harbor, New York.

Beaudet, A.L. and A. Ballabio. 1994. Molecular genetics and medicine. In *Harrison's principles of internal medicine* (ed. K.J. Isselbacher), pp. 349–365. McGraw-Hill, New York.

Beaudet, A.L., C.R. Scriver, W.S. Sly, and D. Valle. 1995. Genetics, biochemistry, and molecular basis of variant human phenotypes. In *The metabolic and molecular bases of inherited disease* (ed. C.R. Scriver et al.), pp. 53–228. McGraw-Hill, New York.

Berry, R., T.J. Stevens, N.A.R. Walter, A.S. Wilcox, T. Rubano, J.A. Hopkins, J. Weber, R. Goold, B.M. Soares, and J.M. Sikela. 1995. Gene-based sequence-tagged-sites (STSs) as the basis for a human gene map. *Nat. Genet.* **10**: 415–423.

Bellen, H.J. and R.F. Smith. 1995. FlyBase: A virtual *Drosophila* cornucopia. *Trends Genet.* **11**: 456–457.

Bird, A.P. 1986. CpG-rich islands and the function of DNA methylation. *Nature* **321**: 209–213.

———. 1987. CpG islands as gene markers in the vertebrate nucleus. *Trends Genet.* **3**: 342347.

———. 1995. Gene number; noise reduction and biological complexity. *Trends Genet.* **11**: 94–100.

Birren, B., E.D. Green, P. Hieter, S. Klapholz, R.M. Myers, H. Riethman, and J. Roskams, eds. 1998. *Genome analysis: A laboratory manual.* Vol. 4 *Mapping genomes.* Cold Spring Harbor Laboratory Press, Cold Spring Harbor, New York. (In press.)

Boguski, M.S. and G.D. Schuler. 1995. Establishing a human transcript map. *Nat. Genet.* **10**: 369–371.

Boguski, M., A. Chakrvarti, R. Gibbs, E. Green, and R. Myers. 1996. The end of the beginning: The race to begin human genome sequencing. *Genome Res.* **6**: 771–772.

Breen, M., L. Deakin, B. Macdonald, S. Miller, R. Sibson, E. Tarttelin et al. 1994. Towards high resolution maps of the mouse and human genomes—A facility for ordering markers to 0.1cM resolution. *Hum. Mol. Genet.* **3**: 621–627.

Brenner, S., G. Elgar, R. Sandford, A. Macrae, B. Venkatesh, and S. Aparacio. 1993. Characterisation of the pufferfish (Fugu) genome as a compact model vertebrate genome. *Nature* **366**: 265–268.

Brown, S.D.M. 1994. Integrating maps of the mouse genome. *Curr. Opin. Genet. Dev.* **4**: 389–394.

———. 1996. Mouse genome. *Encyclopedia of molecular biology and medicine* (ed. R.A. Meyers), vol. 4, pp. 120–128. VCH-Weinheim, New York.

Brown, S.D.M. and J. Peters. 1996. Combining mutagenesis and genomics in the mouse—Closing the phenotype gap. *Trends Genet.* **12**: 433–435.

Buckler, A.J., D.D. Chang, S.L. Graw, J.D. Brook, D.A. Haber, P.A. Sharp, and D.E. Housman. 1991. Exon amplification: A strategy to isolate mammalian genes based on RNA splicing. *Proc. Natl. Acad. Sci.* **88**: 4005–4009.

Buiting, K., S. Saitoh, S. Gross, B. Dittrich, S. Schwartz, R.D. Nicholls, and B. Horsthemke. 1995. Inherited microdeletions in the Angelman and Prader-Willi syndromes define an imprinting centre on human chromosome 15. *Nat. Genet.* **9**: 395–400.

Chakravarti, A. 1998. Meiotic mapping of human chromosomes. In *Genome analysis: A laboratory manual.* Vol. 4 *Mapping genomes* (ed. B. Birren et al.). Cold Spring Harbor Laboratory Press, Cold Spring Harbor, New York. (In press.)

Chelly, J., X. Tümer, T. Tonnensen, A. Petterson, Y. Ishikawa-Brush, F.N. Tommerup, N. Horn, and A.P. Monaco. 1993. Isolation of a candidate gene for Menkes disease that encodes a potential heavy metal binding protein. *Nat. Genet.* **3**: 14–19.

Church, D.M., L.T. Banks, A.C. Rogers, S.L. Graw, D.E. Housman, J.F. Gusella, and A.J. Buckler. 1993. Identification of human chromosome 9 specific genes using exon amplification. *Hum. Mol. Genet.* **2**: 1915–1920.

Church, D.M., C.J. Stotler, J.L. Rutter, J.R. Murrell, J.A. Trofatter, and A.J. Buckler. 1994. Isolation of genes from complex sources of mammalian genomic DNA using exon amplification. *Nat. Genet.* **6**: 98–105.

Chumakov, I.M., P. Rigault, I. Le Gall, C. Bellanne-Chantelot, A. Billault, S. Guillou, P. Soularue, G. Guasconi, E. Poullier, I. Gros et al. 1995. A YAC contig map of the human genome. *Nature* (suppl.) **377**: 175–298.

Clayton, R.A., O. White, K.A. Ketchum, and J.C. Venter. 1997. The first genome from the third domain of life. *Nature* **387**: 459–462.

Claverie, J.-M. 1994. A streamlined random sequencing strategy for finding coding exons. *Genomics* **23**: 575–581.

Collins, F.S. 1992. Positional cloning: Let's not call it reverse anymore. *Nat. Genet.* **1**: 3–6.

———. 1995. Positional cloning moves from perditional to traditional. *Nat. Genet.* **9**: 347–350.

Collins, J.E., C.G. Cole, L.J. Smink, C.L. Garrett, M.A. Leversha, C.A. Soderlund, G.L. Maslen, L.A. Everett, K.M. Rice, A.J. Coffey et al. 1995. A high-density YAC contig map of human chromosome 22. *Nature* (suppl.) **377**: 367–379.

Cordell, H.J. and J.A. Todd. 1995. Multifactorial inheritance in type 1 diabetes. *Trends Genet.* **11**: 499–504.

Cox, D.R., M. Burmeister, E.R. Price, S. Kim, and R.M. Myers. 1990. Radiation hybrid mapping: A somatic cell genetic method for constructing high-resolution maps of mammalian chromosomes. *Science* **250**: 245–250.

Crolla, J.A., S. Gilgenkrantz, J. de Grouchy, T. Kajii, and M. Bobrow. 1989. Incontinentia pigmenti and X-autosome translocations. Non-isotopic in situ hybridization with an X-centromere-specific probe (pSV2X5) reveals a possible X-centromeric breakpoint in one of five published cases. *Hum. Genet.* **81**: 269–272.

Deitrich, W.F., G. Copeland, D.J. Gilbert, J.C. Miller, N.A. Jenkins, and E.S. Lander. 1995. Mapping the mouse genome: Current status and future prospects. *Proc. Natl. Acad. Sci.* **92**: 10849–10853.

Dietrich, W.F., H. Katz, S.E. Lincoln, H.S. Shin, J. Friedman, N. Dracopoli, and E.S. Lander. 1992. A genetic map of the mouse suitable for typing intraspecific crosses. *Genetics* **131**: 423–447.

Doggett, N.A., L.A. Goodwin, J.G. Tesmer, L.J. Meincke, D.C. Bruce, L.M. Clark, M.R. Altherr, A.A. Ford, H.C. Chi, B.L. Marrone et al. 1995. An integrated physical map of human chromosome 16. *Nature* (suppl.) **377**: 335–366.

Dujon, B. 1996. The yeast genome project: What did we learn? *Trends Genet.* **12**: 263–270.

Duyk, G.M., S. Kim, R.M. Myers, and D.R. Cox. 1990. Exon trapping: A genetic screen to identify candidate transcribed sequences in cloned mammalian genomic DNA. *Proc. Natl. Acad. Sci.* **87**: 8995–8999.

Evans, G.A. 1991. Physical mapping of the human genome by pulsed field gel analysis. *Curr. Opin. Genet. Dev.* **1**: 75–81.

Fanning, S. and R.A. Gibbs. 1997. PCR in genome analysis. In *Genome analysis: A laboratory manual.* Vol. 1 *Analysing DNA* (ed. B. Birren et al.), pp. 249–299. Cold Spring Harbor Laboratory Press, Cold Spring Harbor, New York.

Fantes, J., B. Redeker, M. Breen, S. Boyle, J. Brown, J. Fletcher, S. Jones, W. Bickmore, Y. Fukushima, M. Mannens, S. Danes, and I.H. van Heyningen. 1995. Aniridia-associated cytogenetic rearrangements suggest that a position effect may cause the mutant phenotype. *Hum. Mol. Genet.* **4**: 415–422.

Fields, S. and R. Sternglanz. 1994. The two-hybrid system: An assay for protein-protein interactions. *Trends Genet.* **10**: 286–292.

Fijneman, R.J.A., S.S. de Vries, R.C. Jansen, and R. Demant. 1996. Complex interactions of new quantitative trait loci, Sluc1, Sluc2, Sluc3 and Sluc4 that influence the susceptibility to lung cancer in the mouse. *Nat. Genet.* **14**: 465–467.

Foster, J.W., M.A. Dominguez-Steglich, S. Guioli, C. Kwot, P.A. Weller, M. Stevanovic, J. Weissenbach, S. Mansour, I.D. Young, P.N. Goodfellow, J.D. Brook, and A.J. Schafer. 1994. Campomelic dysplasia and autosomal sex reversal caused by mutations in an SRY-related gene. *Nature* **372**: 525–530.

Franco, B., G. Meroni, G. Parenti, J. Levilliers, L. Bernard, M. Gebbia, L. Cox, P. Maroteaux, L. Sheffield, G.A. Rappold, G. Andria, C. Petit, and A. Ballabio. 1995. A cluster of sulfatase genes on Xp22.3: Mutations in chondrodysplasia punctata (CDPX) and implications for Warfarin embryopathy. *Cell* **81**: 15–25.

Frankel, W.N. 1995. Taking stock of complex trait genetics in mice. *Trends Genet.* **11**: 471–477.

Frankel, W.N. and N.J. Schork. 1996. Who's afraid of epistasis. *Nat. Genet.* **14**: 371–373.

Gemmill, R.M., I. Chumakov, P. Scott, B. Waggoner, P. Rigault, J. Cypser, Q. Chen, J. Weissenbach, K. Gardiner, H. Wang et al. 1995. A second generation YAC contig map of human chromosome 3. *Nature* (suppl.) **377**: 299–320.

Gibson, F., H. Lehrach, A. Buckler, S.D.M. Brown, and N. North. 1994. Isolation of conserved sequences from YACs by exon amplification. *BioTechniques* **16**: 453–458.

Gibson, F., J. Walsh, P. Mburu, A. Varela, K.A. Brown, M. Antonio, K.W. Beisel, K.P. Steel, and S.D.M. Brown. 1995. A type VII myosin encoded by the mouse deafness gene shaker-1. *Nature* **374**: 62–64.

Gorski, J.L., M. Boehnke, E.L. Reyner, and E.N. Burright. 1992. A radiation hybrid map of the proximal short arm of the human X chromosome spanning incontinentia pigmenti 1 (IP1) translocation breakpoints. *Genomics* **14**: 657–665.

Green, E.D., D.R. Cox, and R.M. Myers. 1995. The human genome project and its impact on the study of human disease. In *The metabolic and molecular bases of inherited disease* (ed. C.R. Scriver et al.), pp. 401–436. McGraw-Hill, New York.

Green, E.D., H.C. Riethman, J.E. Dutchik, and M.V. Olson. 1991. Detection and characterization of chimeric yeast artificial-chromosome clones. *Genomics* **11**: 658–669.

Grompe, M. 1993. The rapid detection of unknown mutations in nucleic acids. *Nat. Genet.* **5**: 111–117.

Guilford, P., H. Ayadi, S. Blanchard, H. Chaib, D. Le Paslier, J. Weissenbach, M. Drira, and C. Petit. 1994. A human gene responsible for neurosensory, non-syndromic recessive deafness is a candidate homologue of the mouse sh-1 gene. *Hum. Mol. Genet.* **3**: 989–993.

Guo, W., K. Worley, V. Adams, J. Mason, D. Sylvester-Jackson, Y.-H. Zhang, J.A. Towbin, D.D. Fogt, S. Madu, D.A. Wheeler, and E.R.B. McCabe. 1993. Genomic scanning for expressed sequences in Xp21 identifies the glycerol kinase gene. *Nat. Genet.* **4**: 367–372.

Haass, C. 1997. Presenilins: Genes for life and death. *Neuron* **18**: 687–690.

Haldi, M.L., P. Lim, K. Kaphingst, U. Akella, J. Whang, and E.S. Lander. 1997. Construction of a large-insert yeast library of the rat genome. *Mamm. Genome* **8**: 284.

Hästbacka, J., A. de la Chapelle, M.M. Mahtani, G. Clines, M.P. Reeve-Daly, M. Daly, B.A. Hamilton, K. Kusumi, B. Trivedi, A. Weaver, A. Coloma, M. Lovett, A. Buckler, I. Kaitila, and E.S. Lander. 1994. The diastrophic dysplasia gene encodes a novel sulfate transporter: Positional cloning by fine-structure linkage disequilibrium mapping. *Cell* **78**: 1073–1087.

Hillier, L., G. Lennon, B. Becker, M.F. Bonaldo, B. Ciapelli, G. Chissoe et al. 1996. Generation and analysis of 280,000 human expressed sequence tags. *Genome Res.* **6**: 807–828.

Hogan, B., R. Beddington, F. Costantini, and E. Lacey. 1994. *Manipulating the mouse embryo: A laboratory manual*, 2nd. ed. Cold Spring Harbor Laboratory Press, Cold Spring Harbor Laboratory, New York.

Hudson, T.J., L.K. Stein, S.S. Gerety, J. Ma, A.B. Castle, J. Silva, D.K. Slonim, R. Baptista, L. Kruglyak, S.H. Xu, X. Hu, A.M.E. Colbert, C. Rosenberg, M.P. Reeve-Daly, S. Rozen, L. Hui, X. Wu, C. Vestergaard, K.M. Wilson, J.S. Bae, S. Maitra, S. Ganitsas, C.A. Evans, M.M. DeAngelis, K.A. Ingalls, R.W. Nahf, L.T. Horton, M.O. Anderson, A.H. Collymore, W. Ye, V. Kouyoumijian, I.S. Zemsteva, J. Tam, R. Devine, D.F. Courtney, M.T. Renaud, H. Nguyen, R.J. O'Connor, C. Fizames, S. Faure, G. Gyapay, C. Dib, J. Morissette, B.J. Orlin, B.W. Birren, N. Goodman, J. Wissenbach, T.L. Hawkins, S. Foote, C.D. Page, and E.S. Lander. 1995. An STS-based map of the human genome. *Science* **270**: 1945–1954.

Ioannou, P.A., C.T. Amemiya, J. Garnes, P.M. Kroisel, H. Shizuya, C. Chen, M.A. Batzer, and P.J. de Jong. 1994. A new bacteriophage P1-derived vector for the propagation of large human DNA fragments. *Nat. Genet.* **6**: 84–89.

Jacob, H.J., D.M. Brown, R.K. Bunker, M.J. Daly, V.J. Dzau, A. Goodman, G. Koike, V. Kren, T. Kurtz, A. Lernmark, G. Levan, Y. Mao, A. Pettersson, M. Pravenec, J.S. Simon, C. Szpirer, J. Szpirer, M.R. Trolliet, E.S. Winer, and E.S. Lander. 1995. A genetic linkage map of the laboratory rat, Rattus norvegicus. *Nat. Genet.* **9**: 63–69.

James, M.R. and K. Lindpainter. 1997. Why map the rat? *Trends Genet.* **13**: 171–173.

Jones, K.W., M.H. Shapero, M. Chevrette, and R.E. Fournier. 1991. Subtractive hybridization cloning of a tissue-specific extinguisher: TSE1 encodes a regulatory subunit of protein kinase A. *Cell* **66:** 861–872.

Kalchman, M.A., H.B. Koide, K. McCutcheon, R.K. Graham, K. Nichol, N. Kazutoshi, P. Kazemi-Esfarjani, F.C. Lynn, C. Wellington, M. Metzler, Y.P. Goldberg, I. Kanazawa, R.D. Gietz, and M.R. Hayden. 1997. HIP1, a human homologue of *S. cerevisiae* Sla2p, interacts with membrane-associated huntingtin in the brain. *Nat. Genet.* **16:** 44–53.

King, D.P., Y. Zhao, A.M. Sangoram, L.D. Wilsbacher, M. Tanaka, M.P. Antoch, T.D.L. Steeves, M.H. Vitaterna, J.M. Kornhauser, P.L. Lowery, F.W. Turek, and J.S. Takahashi. 1997. Positonal cloning of the mouse circadian Clock gene. *Cell* **89:** 641–653.

Koonin, E.V. 1997. Big time for small genomes. *Genome Res.* **7:** 418–421.

Koop, B.F. 1995. Human and rodent DNA sequence comparisons: A mosaic model of genomic evolution. *Trends Genet.* **11:** 367–371.

Korn, B., Z. Sedlacek, A. Manca, P. Kioschis, D. Konecki, H. Lehrach, and A. Poustka. 1992. A strategy for the selection of transcribed sequences in the Xq28 region. *Hum. Mol. Genet.* **1:** 235–242.

Krauter, K., K. Montgomery, S.J. Yoon, J. LeBlanc-Straceski, B. Renault, I. Marondel, V. Herdman, L. Cupelli, A. Banks, J. Lieman et al. 1995. A second-generation YAC contig map of human chromosome 12. *Nature* (suppl.) **377:** 321–334.

Krizman, D.B. and S.M. Berget. 1993. Efficient selection of 3′-terminal exons from vertebrate DNA. *Nucleic Acids Res.* **21:** 5198–5202.

Lander, E.S. and N.J. Schork. 1994. Genetic dissection of complex traits. *Science* **265:** 2037–2048.

Lawrence, B.J., W. Schwabe, P. Kioschis, J.F. Coy, A. Poustka, M.B. Brennan, and U. Hochgeschwender. 1994. Rapid identification of gene sequences for transcriptional map assembly by direct cDNA screening of genomic reference libraries. *Hum. Mol. Genet.* **3:** 2019–2023.

Leach, R.J., R. Chinn, B.E. Reus, S. Hayes, L. Schantz, B. Dubois, J. Overhauser, A. Ballabio, H. Drabkin, T.B. Lewis, G. Mendgen, and S.L. Naylor. 1994. Regional localization of 188 sequence tagged sites on a somatic cell hybrid mapping panel for human chromosome 3. *Genomics* **24:** 549–556.

Ledbetter, D.H. and A. Ballabio. 1995. Molecular cytogenetics of contiguous gene syndromes: Mechanisms and consequences of gene dosage imbalance. In *The metabolic and molecular bases of inherited disease* (ed. C.R. Scriver et al.), pp. 811–839. McGraw-Hill, New York.

Lee, C.C., A. Yazdani, M. Wehnert, Z.Y. Zhao, E.A. Lindsay, J.Bailey, M.I. Coolbaugh, L. Couch, M. Xiong, A.C. Chinault, A. Baldini, and C.T. Caskey. 1995. Isolation of chromosome-specific genes by reciprocal probing of arrayed cDNA and cosmid libraries. *Hum.*

Mol. Genet. **4:** 1373–1380.

Lee, G.H., R. Proenca, J.M. Montez, K.M. Carroll, J.G. Darvishzadeh, J.I. Lee, and J.M. Friedman. 1996. Abnormal splicing of the leptin receptor in diabetic mice. *Nature* **379:** 632–635.

Leff, S.E., C.I. Brannan, M.L. Reed, T. Ozcelik, U. Francke, N.G. Copeland, and N.A. Jenkins. 1992. Maternal imprinting of the mouse Snrpn gene and conserved linkage homology with the human Prader-Willi syndrome region. *Nat. Genet.* **2:** 259–264.

Lehman, A.R. and A.M. Carr. 1995. The ataxia-telangiectasia gene: A link between checkpoint controls, neurodegeneration and cancer. *Trends Genet.* **375:** 375–376.

Liang, P. and A.B. Pardee. 1992. Differential display of eukaryotic messenger RNA by means of the polymerase chain reaction. *Science* **257:** 967–971.

Liang, P., L. Averboukh, and A. Pardee. 1993. Distribution and cloning of eukaryotic mRNAs by means of differential display: Refinements and optimization. *Nucleic Acids Res.* **21:** 3269–3275.

Lovett, M. 1994. Fishing for complements: Finding genes by direct selection. *Trends Genet* **10:** 352–357.

Lovett, M., J. Kere, and L.M. Hinton. 1991. Direct selection: A method for the isolation of cDNAs encoded by large genomic regions. *Proc. Natl. Acad. Sci.* **88:** 9628–9632.

Lyon, M.F., S. Rastan, and S.D.M. Brown. 1996. *Genetic variants and strains of the laboratory mouse.* Oxford University Press, United Kingdom.

MacPhee, M., K.P. Chepenik, R.A. Liddell, K.K. Nelson, L.D. Siracusa, and A.M. Buchberg. 1995. The secretory phospholipase A2 gene is a candiate for the Mom1 locus, a major modifier of APcMin-induced intestinal neoplasia. *Cell* **81:** 957–966.

Madueno, E., G. Papagiannakis, G. Rimmington, R.D. Saunders, C. Savakis, I. Siden-Kiamos, G. Skavdis, L. Spanos, J. Trenear, P. Adam et al. 1995. A physical map of the X chromosome of *Drosophila melanogaster:* Cosmid contigs and sequence tagged sites. *Genetics* **139:** 1631–1647.

Maheshwar, M.M., R. Sandford, N. Nellist, J.P. Cheadle, B. Sgotto, M. Vaudin, and J.R. Sampson. 1996. Comparative analysis and genomic structure of the tuberous sclerosis 2 (TSC2) gene in human and pufferfish. *Hum. Mol. Genet.* **5:** 131–137.

Matsunami, N., B. Smith, L. Ballard, M.W. Lensch, M. Robertson, H. Albertsen, C.O. Hanemann, H.W. Müller, T.D. Bird, R. White, and P.F. Chance. 1992. Peripheral myelin protein-22 gene maps in the duplication in chromosome 17p11.2 associated with Charcot-Marie-Tooth 1A. *Nat. Genet.* **1:** 176–179.

Melo, J.A., J. Shendure, K. Pciask, and L.M. Silver. 1996. Indentification of sex-spepcific quantitative trait loci controlling alcohol preference in C57BL/6 mice. *Nat. Genet.* **13:** 147–153.

Mercer, J.F.B., J. Livingston, B. Hall, J.A. Paynter, C.

Begy, S. Chandrasekharappa, P. Lockhart, A. Grimes, M. Bhave, D. Siemieniak, and T.W. Glover. 1993. Isolation of a partial candidate gene for Menkes disease by positional cloning. *Nat. Genet.* **3:** 20–25.

Mewes, H.W., K. Albermann, M. Bahr, D. Frishman, A. Gleissner, J. Hani, K. Heumann, K. Kleine, A. Maierl, S.G. Oliver, F. Pfeiffer, and A. Zollner. 1997. Overview of the yeast genome. *Nature* (suppl.) **387:** 7–8.

Mileham, P. and S.D.M. Brown. 1994. The pufferfish genome: Small is beautiful? *BioEssays* **16:** 153–154.

Monaco, A.P. and Z. Larin. 1994. YACs, BACs, PACs and MACs: Artificial chromosomes as research tools. *Trends Biotechnol.* **12:** 280–286.

Monaco, A.P., R.L. Neve, C. Colletti-Feener, C.J. Bertelson, D.M. Kurnit, and L.M. Kunkel. 1986. Isolation of candidate cDNAs for portions of the Duchenne muscular dystrophy gene. *Nature* **323:** 646–650.

Moore, K.J. 1995. Insight into the micropthththalmia gene. *Trends Genet.* **11:** 442–448.

Ngaraja, R., S. MacMillan, J. Kere, C. Jones, S. Griffin, M. Schmatz, J. Terrell, M. Shomaker, C. Jermak, C. Hott, M. Masisi, S. Mumm, A. Srivastava, G. Pilia, T. Featherstone, R. Mazzarella, S. Kesterson, B. McCauley, B. Railey, F. Burough, V. Nowotny, M. D'Urso, D. States, B. Brownstein, and D. Schlessinger. 1997. X chromosome map at 75 kb STS resolution, revealing extremes of recombination and GC content. *Genome Res.* **7:** 210–222.

Nehls, M., D. Pfiefer, M. Schorpp, H. Hedrich, and T. Boehm. 1994. New member of the winged-helix protein family in mouse and rat nude mutations. *Nature* **372:** 103–107.

Nizetic, D., L. Gellen, R.M.J. Hamvas, R. Mott, A. Grigoriev, R. Vatcheva, G. Zehetner, M.-L. Yaspo, A. Dutriaux, C. Lopes, J.-M. Delabar, C. van Broeckhoven, M.-C. Potier, and H. Lehrach. 1994. An integrated YAC-overlap and ''cosmid-pocket'' map of the human chromosome 21. *Hum. Mol. Genet.* **3:** 759–770.

Oeltjen, J.C., T.M. Malley, D.M. Muzny, W. Miller, R.A. Gibbs, and J.W. Belmont. 1997. Large-scale comparative sequence analysis of the human and murine Bruton's tyrosine kinase loci reveals conserved regulatory domains. *Genome Res.* **7:** 315–329.

Orkin, S.H. 1986. Reverse genetics and human disease. *Cell* **47:** 845–850.

Orr, H.T., M.-Y. Chung, S. Banfi, T.J. Kwiatkowski, Jr., A. Servadio, A.L. Beaudet, A.E. McCall, L.A. Duvick, L.P.W. Ranum, and H.Y. Zoghbi. 1993. Expansion of an unstable trinucleotide (CAG) repeat in spinocerebellar ataxia type 1. *Nat. Genet.* **4:** 221–226.

Özcelik, T., S. Leff, W. Robinson, T. Donlon, M. Lalande, E. Sanjines, A. Schinzel, and U. Francke. 1992. Small nuclear ribonucleoprotein polypeptide N (SNRPN), an expressed gene in the Prader-Willi syndrome critical region. *Nat. Genet.* **2:** 265–269.

Page, D.C., E.M.C. Fisher, B. McGillivray, and L.G.

Brown. 1990. Additional deletion in sex-determining region of human Y chromosome resolves paradox of X,t(Y;22) female. *Nature* **346:** 279–281.

Pandolfo, M. 1992. A rapid method to isolate $(GT)_n$ repeats from yeast artificial chromosomes. *Nucleic Acids Res.* **20:** 1154.

Parimoo, S., S.R. Patanjali, H. Shukla, D.D. Chaplin, and S.M. Weissman. 1991. cDNA selection: Efficient PCR approach for the selection of cDNAs encoded in large chromosomal DNA fragments. *Proc. Natl. Acad. Sci.* **88:** 9623–9627.

Patel, K., R. Cox, J. Shipley, F. Kiely, K. Frazer, D.R. Cox, H. Lehrach, and D. Sheer. 1991. A novel and rapid method for isolating sequences adjacent to rare cutting sites and their use in physical mapping. *Nucleic Acids Res.* **19:** 4371–4375.

Patel, P.I., B.B. Roa, A.A. Welcher, R. Schoener-Scott, B.J. Trask, L. Pentao, G.J. Snipes, C.A. Garcia, U. Francke, E.M. Shooter, J.R. Lupski, and U. Suter. 1992. The gene for the peripheral myelin protein PMP-22 is a candidate for Charcot-Marie-Tooth disease type 1A. *Nat. Genet.* **1:** 159–165.

Pease, A.C., D. Solas, E.J. Sullivan, M.T. Cronin, C.P. Holmes, and S.P. Fodor. 1994. Light-generated oligonucleotide arrays for rapid DNA sequence analysis. *Proc. Natl. Acad. Sci.* **91:** 5022–5026.

Plomin, R. and G.E. McClearn. 1993. Quantitative trait loci (QTL) analyses and alcohol-related behaviours. *Behav. Genet.* **23:** 197–211.

Polymeropoulos, M.H., H. Xiao, J.M. Sikela, M. Adams, J.C. Venter, and C.R. Merril. 1993. Chromosomal distribution of 320 genes from a brain cDNA library. *Nat. Genet.* **4:** 381–386.

Polymeropoulos, M.H., H. Xiao, A. Glodek, M. Gorski, M.D. Adams, R.F. Moreno, M.G. Fitzgerald, J.C. Venter, and C.R. Merril. 1992. Chromosomal assignment of 46 brain cDNAs. *Genomics* **12:** 492–496.

Reed, V., S. Rider, G.L. Maslen, E. Hatchwell, H.J. Blair, I.C. Uwechue, I.W. Craig, S.H. Laval, A.P. Monaco, and Y. Boyd. 1994. A 2-Mb YAC contig encompassing three loci (DXF34, DXS14, and DXS390) that lie between Xp11.2 translocation breakpoints associated with incontinentia pigmenti type. *Genomics* **20:** 341–346.

Riethman, H., B. Birren, and A. Gnirke. 1997. Preparation, manipulation, and mapping of HMW DNA. In *Genome analysis: A laboratory manual.* Vol. 1 *Analyzing DNA* (ed. B. Birren et al.), pp. 83–248. Cold Spring Harbor Laboratory Press, Cold Spring Harbor, New York.

Rommens, J.M., M.C. Iannuzzi, B.S. Kerem, M.L. Drumm, G. Melmer, M. Dean, R. Rozmahel, J.L. Cole, D. Kennedy, N. Hidaka, M. Zsiga, M. Buchwald, J.R. Riordan, L.C. Tsui, and F.S. Collins. 1989. Identification of the cystic fibrosis gene: Chromosome walking and jumping. *Science* **245:** 1059–1065.

Rubin, G.M. 1988. *Drosophila melanogaster* as an experimental organism. *Science* 240: 1453–1459.

Ruddle, F.H. 1984. Reverse genetics and beyond. *Am. J. Hum. Genet.* 6: 944–953.

Sargent, C.A., N.A. Affara, E. Bentley, A. Pelmear, D.M.D. Bailey, P. Davey, D. Dow, M. Leversha, H. Aplin, G.T.N. Besley, and M.A. Ferguson-Smith. 1993. Cloning of the X-linked glycerol kinase deficiency gene and its identification by sequence comparison to the *Bacillus subtilis* homologue. *Hum. Mol. Genet.* 2: 97–106.

Sambrook, J., E.F. Fritsch, and T. Maniatis. 1989. *Molecular cloning: A laboratory manual*, 2nd ed. Cold Spring Harbor Laboratory Press, Cold Spring Harbor, New York.

Savitsky, K., A. Bar-Shira, S. Gilad, G. Rotman, Y. Ziv, L. Vanagaite, D.A. Tagle, S. Smith, T. Uziel, S. Sfez et al. 1995. A single ataxia telangiectasia gene with a product similar to PI-3 kinase. *Science* 268: 1749–1753.

Schaefer, L., G.B. Ferrero, A. Grillo, M.T. Bassi, E.J. Roth, M.C. Wapenaar, G.J.B. van Ommen, T.K. Mohandas, M. Rocchi, H.Y. Zoghbi, and A. Ballabio. 1993. A high resolution deletion map of human chromosome Xp22. *Nat. Genet.* 4: 272–279.

Schmickel, R.D. 1986. Contiguous gene syndromes: A component of recognizable syndromes. *J. Pediatr.* 109: 231–241.

Schuler, G.D., M.S. Boguski, E.A. Stewart, L.D. Stein, G. Gyapay, K. Rice, R.E. White, P. Todriguez-Tome, A. Aggarwal, E. Bajorek, S. Bentolila, B.B. Birren, A. Butler, A.B. Castle, N. Chiannilkulchai, A. Chu, C. Clee, S. Cowles, P.J. Day, T. Dibling, N. Drouot, I. Dunham, S. Duprat, C. East, T.J. Hudson et al. 1996. A gene map of the human genome. *Science* 274: 540–546.

Schumacher, A., C. Faust, and T. Magnusson. 1996. Positional cloning of a global regulator of anterior-posterior patterning in mice. *Nature* 383: 250–253.

Sedlacek, Z., B. Korn, D.S. Konecki, R. Siebenhaar, J.F. Coy, P. Kioschis, and A. Poustka. 1993. Construction of a transcription map of a 300 kb region around the human G6PD locus by direct cDNA selection. *Hum. Mol. Genet.* 2: 1865–1869.

Shiloh, Y. 1995. Relationship of the ataxia-tenagiectasia protein ATM to phosphoinositide 3-kinase. *Trends Biochem. Sci.* 20: 382–383.

Shizuya, H., B. Birren, U. Kim, V. Mancino, T. Slepack, Y. Tachiiri, and M. Simon. 1992. Cloning and stable maintenance of 300-kilobase-pair fragments of human DNA in *Escherichia coli* using an F-factor-based vector. *Proc. Natl. Acad. Sci.* 89: 8794–8797.

Sinclair, A.H., P. Berta, M.S. Palmer, J.R. Hawkins, B.L. Griffiths, M.J. Smith, J.W. Foster, A.M. Frischauf, R. Lovell-Badge, and P.N. Goodfellow. 1990. A gene from the human sex-determining region encodes a protein with homology to a conserved DNA-binding motif. *Nature* 346: 240–244.

Soares, M.B., M.F. Bonaldo, P. Jelene, L. Su, L. Lawton,

and A. Efstratiadis. 1994. Construction and characterization of a normalized cDNA library. *Proc. Natl. Acad. Sci.* 91: 9228–9232.

Solovyev, V.V., A.A. Salamov, and C.B. Lawrence. 1994. Predicting internal exons by oligonucleotide composition and discriminant analysis of spliceable open reading frames. *Nucleic Acids Res.* 22: 5156–5163.

Sternberg, N. 1990. Bacteriophage P1 cloning system for the isolation, amplification, and recovery of DNA fragments as large as 100 kilobase pairs. *Proc. Natl. Acad. Sci.* 87: 103–107.

———. 1992. Cloning high molecular weight DNA fragments by the bacteriophage P1 system. *Trends Genet* 8: 11–16.

Sternberg, N., J. Ruether, and K. deRiel. 1990. Generation of a 50,000-member human DNA library with an average DNA insert size of 75-100kbp in a bacteriophage P1 cloning vector. *New Biol.* 2: 151–162.

Stewart, E.A., K.B. McKusick, A. Aggarwal, E. Bajorek, S. Brady, A. Chu, N. Fang, D. Hadley, M. Harris, S. Hussain, R. Lee, A. Maratukulam, K. O'Connor, S. Perkins, M. Piercy, F. Qin, T. Reif, C. Sanders, X. She, W.L. Sun, P. Tabar, S. Voyticky, S. Cowles, J.B. Fan, D.R. Cox et al. 1997. An STS-based radiation hybrid map of the human genome. *Genome Res.* 7: 422–433.

Sulston, J., F. Mallett, R. Staden, R. Durbin, T. Hornsell, and A. Coulson. 1988. Software for genome mapping by fingerprinting techniques. *Comput. Appl. Biosci.* 4: 125–132.

Sutcliffe, J.S., M. Nakao, S. Christian, K.H. Orstavik, N. Tommerup, D.H. Ledbetter, and A.L. Beaudet. 1994. Deletions of a differentially methylated CpG island at the SNRPN gene define a putative imprinting control region. *Nat. Genet.* 8: 52–58.

Tagle, D.A., M. Swaroop, M. Lovett, and F.S. Collins. 1993. Magnetic bead capture of expressed sequences encoded within large genomic segments. *Nature* 361: 751–753.

Tassabehji, M., A.P. Read, V.E. Newton, R. Harris, R. Balling, P. Gruss, and T. Strachan. 1992. Waardenburg's syndrome patients have mutations in the human homologue of the Pax-3 paired box gene. *Nature* 355: 635–636.

Tassabehji, M., V.E. Newton, X.Z. Liu, A. Brady, D. Donnai, M. Krajewska-Walasek, V. Murday, A. Norma, E. Oberszytn, W. Reardon, J.C. Rice, R. Trembath, P. Wieacker, M. Whiteford, R. Winter, and A.P. Read. 1995. The mutational spectrum in Waardenburg syndrome. *Hum. Mol. Genet.* 4: 2131–2137.

Taylor, B. 1996. Recombinant inbred strains. In *Genetic variants and strains of the laboratory mouse* (ed. M. Lyon et al.), pp. 1597–1659. Oxford University Press, United Kingdom.

Timmerman, V., E. Nelis, W. van Hul, B.W. Nieuwenhuijsen, K.L. Chen, S. Wang, K. Ben Othman, B. Cullen, R.J. Leach, C.O. Hanemann, P. de Jonghe, P. Raeymaekers, G.-J.B. van Ommen, J.-J. Martin, H.W.

Müller, J.M. Vance, K.H. Fischbeck, and C. van Broeckhoven. 1992. The peripheral myelin protein gene PMP-22 is contained within the Charcot-Marie-Tooth disease type 1A duplication. *Nat. Genet.* **1**: 171–175.

Tomlinson, I.P.M. and W.F. Bodmer. 1995. The HLA system and the analysis of multifactorial genetic disease. *Trends Genet.* **11**: 493–498.

Trask, B. 1998. Fluorescence in situ hybridization. In *Genome analysis: A laboratory manual.* Vol. 4 *Mapping genomes* (ed. B. Birren et al.). Cold Spring Harbor Laboratory Press, Cold Spring Harbor, New York. (In press.)

Tugendreich, S., M.S. Boguski, M.S. Seldin, and P. Hieter. 1993. Linking yeast genetics to mammalian genomes: Identification and mapping of the human homolog of CDC27 via the expressed sequence tag (EST) data base. *Proc. Natl. Acad. Sci.* **90**: 10031–10035.

Uberbacher, E.C. and R.J. Mural. 1991. Locating protein-coding regions in human DNA sequences by a multiple sensor-neural network approach. *Proc. Natl. Acad. Sci.* **88**: 11261–11265.

Valentijn, L.J., P.A. Bolhuis, I. Zorn, J.E. Hoogendijk, N. van den Bosch, G.W. Hensels, V.P. Stanton, Jr., D.E. Housman, K.H. Fischbeck, D.A. Ross, G.A. Nicholson, E.J. Meershoek, H.G. Dauwerse, G.J.B. van Ommen, and F. Baas. 1992. The peripheral myelin gene PMP-22/GAS-3 is duplicated in Charcot-Marie-Tooth disease type 1A. *Nat. Genet.* **1**: 166–170.

van Slegtenhorst, M.A., M.T. Bassi, G. Borsani, M.C. Wapenaar, G.B. Ferrero, L. de Conciliis, E. Rugarli, A. Grillo, B. Franco, H.Y. Zoghbi, and A. Ballabio. 1994. A gene from the Xp22.3 region shares homology with voltage-gated chloride channels. *Hum. Mol. Genet.* **3**: 547–552.

van Wezel, T., A.P.M. Stassen, C.J.A. Moen, A.A.M. Hart, M.A. van der Valk, and P. Demant. 1996. Gene interaction and single gene effects in colon tumour susceptibility in mice. *Nat. Genet.* **14**: 468–470.

Vollrath, D., S. Foote, A. Hilton, L.G. Brown, P. Beer-Romero, J.S. Bogan, and D.C. Page. 1992. The human Y chromosome: A 43-interval map based on naturally occurring deletions. *Science* **258**: 52–59.

Vulpe, C., B. Levinson, S. Whitney, S. Packman, and J. Gitschier. 1993. Isolation of a candidate gene for Menkes disease and evidence that it encodes a copper-transporting ATPase. *Nat. Genet.* **3**: 7–13.

Wagner, T., J. Wirth, J. Meyer, B. Zabel, M. Held, M. Zimmer, J. Pasantes, F.D. Bricarelli, J. Keutel, E.

Huster, U. Wolf, N. Tommerup, W. Schempp, and G. Scherer. 1994. Autosomal sex reversal and campomelic dysplasia are caused by mutations in and around the SRY-related gene SOX9. *Cell* **79**: 1111–1120.

Wapenaar, M.C., M.V. Schiaffino, M.T. Bassi, L. Schaefer, A.C. Chinault, H.Y. Zoghbi, and A. Ballabio. 1994. A YAC-based binning strategy facilitating the rapid assembly of cosmid contigs: 1.6 Mb of overlapping cosmids in Xp22. *Hum. Mol. Genet.* **3**: 1155–1161.

Waterston, R. and J. Sulston. 1995. The genome of *Caenorhabditis elegans. Proc. Natl. Acad. Sci.* **92**: 10836–10840.

Weil, D., S. Blanchard, J. Kaplan, P. Guilford, F. Gibson, J. Walsh, P. Mburu, A. Varela, J. Levilliers, M.D. Weston, P.M. Kelley, W.J. Kimberling, M. Wagenaar, F. Levi-Ascobas, D. Parget-Piet, A. Munnich, K.P. Steel, S.D.M. Brown, and C. Petit. 1995. Defective myosin VIIA gene responsible for Usher syndrome type 1B. *Nature* **374**: 60–61.

Weissenbach, J., G. Gyapay, C. Dib, A. Vignal, J. Morissette, P. Millasseau, G. Vaysseix, and M. Lathrop. 1992. A second-generation linkage map of the human genome. *Nature* **359**: 794–801.

Welsh, J., K. Chada, S. Dalal, R. Cheng, D. Ralph, and M. McClelland. 1992. Arbitrarily primed PCR fingerprinting of RNA. *Nucleic Acids Res.* **20**: 4965–4970.

Wilson, R.K. and E.R. Mardis. 1997. Fluorescence-based DNA sequencing. In *Genome analysis: A laboratory manual.* Vol. 1 *Analyzing DNA* (ed. B. Birren et al.), pp. 301–395. Cold Spring Harbor Laboratory Press, Cold Spring Harbor, New York.

Wolff, R. and R. Gemmill. 1997. Purifying and analyzing genomic DNA. In *Genome analysis: A laboratory manual.* Vol. 1 *Analyzing DNA* (ed. B. Birren et al.), pp. 1–81. Cold Spring Harbor Laboratory Press, Cold Spring Harbor, New York.

Yamamoto, H., S. Takeuchi, T. Kudo, K. Makino, A. Nakata, T. Shinoda, and T. Takeuchi. 1987. Cloning and sequencing of mouse tyrosinase cDNA. *Jpn. J. Genet.* **62**: 271–274.

Young, R.A. and R.W. Davis. 1983. Efficient isolation of genes by using antibody probes. *Proc. Natl. Acad. Sci.* **80**: 1194–1198.

Zuo, J., C. Robbins, S. Baharloo, D.R. Cox, and R.M. Myers. 1993. Construction of cosmid contigs and high-resolution restriction mapping of the Huntington disease region of human chromosome 4. *Hum. Mol. Genet.* **2**: 889–899.

2

Constructing and Screening Normalized cDNA Libraries

MARCELO BENTO SOARES AND MARIA DE FATIMA BONALDO

A cDNA library is considered normalized when the frequency of each cDNA in the library is equally represented regardless of whether the cDNA is derived from a rare or a frequently occurring mRNA species. The frequencies of all clones are therefore within the same order of magnitude range in a typical normalized library (see, e.g., Soares et al. 1994; Bonaldo et al. 1996). This chapter provides detailed protocols for construction and normalization of oligo(dT)-primed, directionally cloned cDNA libraries in phagemid vectors. Modifications in the existing methods for library construction (Okayama and Berg 1982; Gubler and Hoffman 1983; D'Alessio et al. 1987) and solutions for the problems that are most commonly observed in cDNA libraries are discussed. These problems are (1) the high frequency of clones with small inserts, some of which contain nothing except the poly(A) tails of mRNAs, (2) the presence of a long poly(A) tail at the 3′ end of cDNAs, which makes it difficult to obtain 3′-terminal sequences, and (3) the occurrence of chimeric cDNA clones. In addition, a protocol for screening cDNA libraries by colony hybridization is presented.

Primary Uses and Applications for cDNA Libraries

STUDY OF GENE STRUCTURE AND EXPRESSION

One of the first steps toward cloning a gene of interest typically involves constructing a cDNA library from a tissue or cell type likely to express it and screening the library either by hybridization (Young and Davis 1983a,b; Huynh et al. 1985; Sambrook et al. 1989) or by PCR (Frohman et al. 1988; Saiki et al. 1988; Compton 1990; Frohman 1990). After a cDNA clone is isolated and sequenced for verification that it corresponds to the desired gene, it is then used as a probe to screen a genomic library and thereby to isolate its cognate gene. It is also used for northern analysis to study its pattern of expression in different cell types, tissues, and stages of development.

The comparative analysis of the cDNA and its corresponding genomic clone reveals, at least in part, the structure of the gene. However, since cDNA clones often represent incomplete products of reverse transcription, characterization of the entire transcript frequently involves additional screenings of cDNA libraries to isolate overlapping cDNAs. Alternatively, the missing cDNA fragments can be obtained by RT-PCR approaches (5′ and 3′ RACE) with specific oligonucleotides designed on the basis of sequence information derived from the truncated cDNA (Compton 1990; Frohman 1990; see also Fanning and Gibbs 1997).

RAPID IDENTIFICATION OF GENES BY PARTIAL SEQUENCING OF cDNAs

An alternative approach for gene discovery has emerged in the past years that is expected to facilitate quite dramatically the work involved in cloning a gene (Adams et al. 1991, 1992, 1993a,b; Khan et al. 1992; Okubo et al. 1992; Matsubara and Okubo 1993; Matoba et al. 1994; Okubo 1994). In this approach, large numbers of directionally cloned cDNAs are randomly chosen from libraries and sequenced from one or both (3′ and 5′) ends. Their nucleotide sequences, and translations thereof, are compared with those available in public databases for identification of homologies that occasionally either reveal or suggest their function. It is anticipated that once most genes are identified and their clones become available, the process of isolating a cDNA clone may no longer require constructing and screening a cDNA library but instead may simply involve searching the public databases for sequences homologous to the one of interest.

Overview of the Methods for Synthesizing Double-stranded cDNA

Much progress has been made since the first method for cloning cDNAs was reported (Efstratiadis et al. 1976; Rougeon and Mach 1976). In these early protocols, the hairpin-loop structure naturally formed at the 3′ end of first-strand cDNA is used to self-prime the synthesis of a second strand after hydrolysis of the RNA template. The end product is a hairpin double-stranded DNA with the 5′ end of the mRNA sequence in the form of a single-stranded loop of variable size and location. The resulting cDNAs are then treated with nuclease S1 to digest the single-stranded DNA loop, a step that invariably removes portions of the cDNA corresponding to coding or 5′-noncoding regions of the mRNA.

To circumvent the need for digestion with nuclease S1, Land et al. (1981) developed an alternative protocol in which a homopolymeric tail is added to the 3′ end of the first-strand cDNA and a complementary oligonucleotide is used to prime second-strand synthesis.

In their classic paper, Okayama and Berg (1982) described an elegant method that introduced the idea of synthesis of second-strand cDNA by RNA-primed nick translation. In this method, (1) the mRNA is reverse transcribed into cDNA by using T-tailed cloning vector as primer; (2) a linker fragment is ligated to the DNA:RNA hybrid molecules to ensure their circularization; and (3) the second strand is synthesized by using a combination of RNase H, *E. coli* DNA polymerase I, and *E. coli* DNA ligase to replace the RNA in the hybrids with DNA. RNase H is an endoribonuclease specific for the RNA strand of a DNA:RNA hybrid (Leis et al. 1973). This endoribonuclease introduces nicks and small gaps on the RNA strand, which are extended by the 5′→3′ exonucleolytic activity of *E. coli* DNA polymerase I (nick translation), whereas a DNA strand is synthesized by the polymerase.

Gubler and Hoffman (1983) described a simple modification of the Okayama and Berg (1982) protocol that combines the classic oligo(dT)-primed first-strand synthesis with the novel RNase H–DNA polymerase I-mediated second-strand synthesis. Homopolymeric tails are added to the ends of both the cloning vector (dCTP tail) and the cDNAs (dGTP tail) and then ligated. The advantage of this procedure is that neither the elaborate Okayama and Berg vector–primer system nor the classic hairpin-loop cleavage by nuclease S1 is used.

A limitation of the RNase H–DNA polymerase I-mediated method for second-strand synthesis is that it results in the loss of sequence information from the 5′ end of the mRNA template (D'Alessio and Gerard 1988). This can be explained as follows. Removal of the 5′ most RNA primer by the combination of *E. coli* RNase H activity and the 5′→3′ exonucleolytic activity of *E. coli* DNA polymerase I makes the complementary dNTPs, which are located at the 3′ end of a full-length first-strand cDNA, single-stranded. Single-stranded DNA, in turn, is hydrolyzed by the 3′→5′ exonucleolytic activity of *E. coli* DNA polymerase I, leaving the end of the double-stranded cDNA product blunt or nearly so. Therefore, during subsequent cloning, sequence information corresponding to the 5′ most RNA oligonucleotide that primed second-strand synthesis is lost. However, this loss can be minimized by adding RNase H after, instead of with, *E. coli* DNA polymerase I in the second-strand synthesis reaction (D'Alessio and Gerard 1988) since RNase H is not capable of cleaving hybrid RNA at the junction between a ribonucleotide and a deoxyribonucleotide (Omer and Faras 1982; Resnick et al. 1984).

For further discussion and figures illustrating these methods for synthesizing double-stranded cDNA, see Sambrook et al. (1989, Chapter 8).

Cloning Vectors for Construction of cDNA Libraries

There are two main choices of cloning vector for construction of cDNA libraries: bacteriophage λ vectors (Murray et al. 1977; Young and Davis 1983a,b; Huynh et al. 1985; Short et al. 1988; Palazzolo et al. 1990) and plasmid vectors (Vieira and Messing 1982, 1987, 1991; Mead et al. 1986; Dente and Cortese 1987; Heidecker and Messing 1987; Mead and Kemper 1988). Both cloning systems have advantages and disadvantages, and the best choice will greatly depend on the specific application.

GENERAL PRINCIPLES IN CHOOSING A VECTOR

Cloning efficiencies (as defined by the number of pfu or cfu per microgram of bacteriophage λ or plasmid vector DNA, respectively) are no longer a factor in deciding which cloning vector to use. With the advent of electroporation (Dower et al. 1988), cloning efficiencies of recombinant plasmid DNA became comparable to those attained by in vitro packaging of recombinant bacteriophage λ DNA. Efficiencies of in vitro packaging (using Gigapack II Gold [Stratagene 200214]) and electroporation (using Electromax *E. coli* DH10B [Life Technologies 18290-015]) are very similar for control DNA (2 x 10^9 pfu/µg of wild-type bacteriophage λ DNA and 1 x 10^{10} cfu/µg of CsCl-purified supercoiled pUC19 plasmid DNA). Thus, a typical ligation containing 50 ng of cDNA (with an average size of 1 kb) and a twofold molar excess of vector DNA (5 µg of bacteriophage λ DNA or 0.3 µg of pUC19 DNA) would be expected to yield approximately three times more recombinants with the bacteriophage vector than with the plasmid vector.

Bacteriophage λ libraries are easier to screen than plasmid libraries. More clones can be screened per plate, replica filters can be more rapidly prepared, and the hybridization background tends to be lower. However, plasmid vectors offer greater versatility than bacteriophage λ vectors for subsequent manipulation. For example, a plasmid DNA library can easily be purified by agarose gel electrophoresis to eliminate all of the nonrecombinant vector molecules as well as any clones with small inserts that may have escaped size selection (see pp. 121–123). This type of purification cannot be as easily accomplished in bacteriophage λ libraries. From a technical standpoint, it is therefore far easier to obtain greater yields of individually selected plasmid DNA than bacteriophage λ DNA for subsequent subcloning and manipulation of the insert.

BACTERIOPHAGE λ VECTORS

The most widely used bacteriophage λ vectors for cDNA cloning are λgt10, λgt11, and Lambda ZAP. These bacteriophage λ insertion vectors typically accept inserts of 0–10 kb. Lambda ZAP (Stratagene) has some advantages over the other bacteriophage λ cloning vectors because it contains a plasmid component (pBluescript, which has an f1 origin of replication and a polylinker with multiple cloning sites) that can be excised in vivo after superinfection with a helper bacteriophage and then recircularized, thus eliminating the time involved in subcloning. Lambda ZAP is therefore an easier vector with which to work.

PHAGEMID VECTORS

The observation by Dotto et al. (1981) that a plasmid carrying the intergenic region of the filamentous bacteriophage f1 can be packaged as single-stranded DNA into a viral particle by a helper bacteriophage led to the construction of vectors (phagemids) that combine the advantages of both plasmid and bacteriophage systems (Zinder and Boeke 1982; Mead and Kemper 1986). The problem originally encountered in the use of these phagemids was the significant reduction in the amount of single-stranded DNA produced as compared to bacteriophage vectors such as bacteriophage M13.

This reduction was due to interference by the phagemid with the replication of the helper bacteriophage (Enea and Zinder 1982). This problem has now been solved. With the use of bacteriophage mutants such as bacteriophage M13KO7, which preferentially packages plasmid DNA over bacteriophage DNA (Vieira and Messing 1987), sufficient amounts of single-stranded plasmid can be routinely obtained from small-volume cultures (1 μg of single-stranded plasmid DNA per milliliter of culture). A detailed protocol for construction of cDNA libraries in phagemid vectors is provided on pp. 86–123.

Overview of the Procedure for Constructing Directionally Cloned cDNA Libraries

An outline of the steps involved in the construction of directionally cloned cDNA libraries in phagemid vectors is depicted in Figure 1. A more detailed description of the process of cloning a bona fide cDNA is presented in Figure 2 and in the protocol on pp. 86–123.

Briefly, a *Not*I-(dT)$_{18}$ oligonucleotide (i.e., an oligo(dT)$_{18}$ flanked by a *Not*I restriction site at its 5′ end) is used to prime the synthesis of first-strand cDNA with RNase H⁻ reverse transcriptase from MMLV (Life Technologies). (This enzyme is recommended for its ability to generate longer reverse transcripts than most commercially available forms of reverse transcriptase.) "One-tube" first- and second-strand cDNA syntheses are performed essentially as described by D'Alessio et al. (1987). Double-stranded cDNAs thus generated are treated with bacteriophage T4 DNA polymerase to create blunt ends, size selected by gel filtration on a Bio-Gel A-50m (100–200 mesh; Bio-Rad) column, and ligated to a large excess of adapter molecules (e.g., *Eco*RI adapters). Adapter-ligated cDNAs are then digested with *Not*I and purified on a Sepharose CL-4B (Pharmacia) column to remove the excess adapter molecules. Size-selected cDNAs (>350 bp) are treated with bacteriophage T4 polynucleotide kinase to phosphorylate the adapter ends (one of the two oligonucleotides of the adapter molecule has a 5′-hydroxyl group to prevent concatemerization of adapters during ligation) and ligated directionally to the *Not*I- and *Eco*RI-digested ends of a phagemid vector (pT7T3-Pac; for vector description, see p. 86). Bacteria are transformed with the purified ligated material by electroporation and propagated under selection with an appropriate antibiotic. Finally, plasmid DNA is prepared.

A recommended option at this point is to eliminate from the library all of the non-recombinant ("empty vector") molecules (see pp. 121–123) as follows: A plasmid DNA preparation of the library can be linearized with *Not*I and the recombinant DNA separated from the empty vector by agarose gel electrophoresis. The recombinant DNA is then purified using β-agarase (New England Biolabs) and recircularized in a large-volume ligation. The recircularized DNA is precipitated, bacteria are transformed by electroporation and propagated under selection with an appropriate antibiotic, and plasmid DNA is prepared.

Problems Commonly Encountered in cDNA Libraries

Large-scale sequencing of cDNAs has uncovered some of the problems commonly encountered in oligo(dT)-primed, directionally cloned cDNA libraries (Adams et al. 1991, 1993b; Soares 1994), such as the high frequency of clones with a long poly(A)

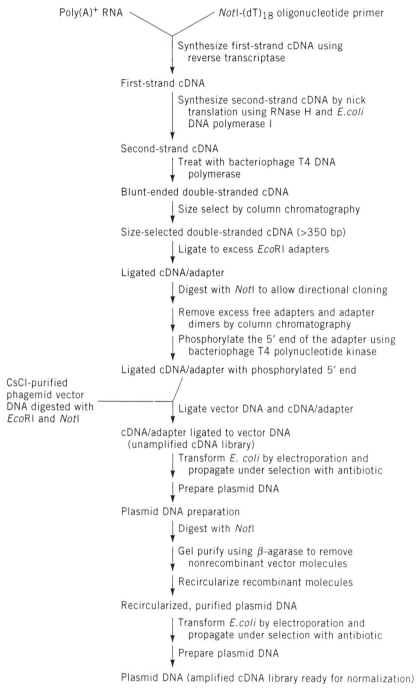

Poly(A)$^+$ RNA   *Not*I-(dT)$_{18}$ oligonucleotide primer

Synthesize first-strand cDNA using
reverse transcriptase

First-strand cDNA

Synthesize second-strand cDNA by nick
translation using RNase H and *E.coli*
DNA polymerase I

Second-strand cDNA

Treat with bacteriophage T4 DNA
polymerase

Blunt-ended double-stranded cDNA

Size select by column chromatography

Size-selected double-stranded cDNA (>350 bp)

Ligate to excess *Eco*RI adapters

Ligated cDNA/adapter

Digest with *Not*I to allow directional cloning

Remove excess free adapters and adapter
dimers by column chromatography

Phosphorylate the 5' end of the adapter using
bacteriophage T4 polynucleotide kinase

Ligated cDNA/adapter with phosphorylated 5' end

CsCl-purified
phagemid vector
DNA digested with
*Eco*RI and *Not*I

Ligate vector DNA and cDNA/adapter

cDNA/adapter ligated to vector DNA
(unamplified cDNA library)

Transform *E. coli* by electroporation and
propagate under selection with antibiotic

Prepare plasmid DNA

Plasmid DNA preparation

Digest with *Not*I

Gel purify using β-agarase to remove
nonrecombinant vector molecules

Recircularize recombinant molecules

Recircularized, purified plasmid DNA

Transform *E.coli* by electroporation and
propagate under selection with antibiotic

Prepare plasmid DNA

Plasmid DNA (amplified cDNA library ready for normalization)

Figure 1 Flow chart of the steps involved in the construction of directionally cloned cDNA libraries in phagemid vectors. See protocol on pp. 86–123 for details.

tail and the occurrence of chimeric clones (i.e., cDNA fragments derived from different mRNA molecules that are joined by ligation in a single cDNA clone). Although the former poses limitations for generation of sequence from the 3' end of a cDNA, the latter is even more serious since it often cannot be identified and may therefore result in misleading information. In addition, cDNA clones with small inserts, sometimes consisting exclusively of pieces of mRNA tails, prevail in many libraries (Adams et al. 1991, 1992). Together, these undesirable features represent major impediments to the process of obtaining the complete nucleotide sequence of a transcript.

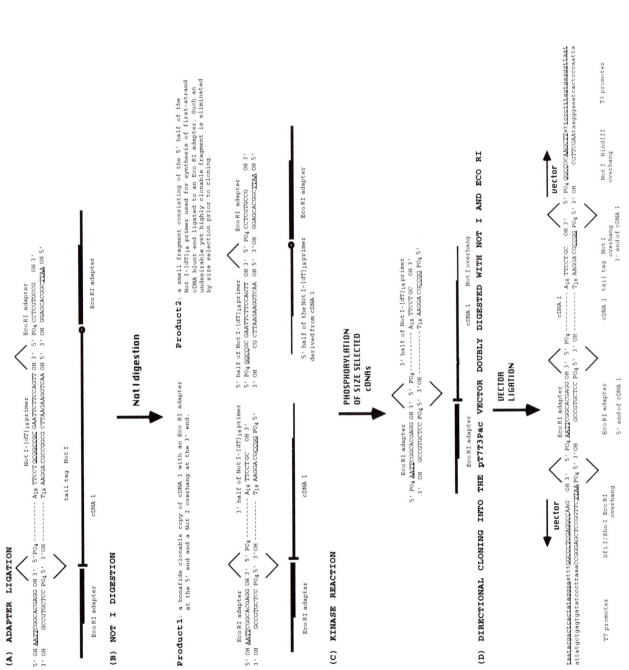

Figure 2 (*See facing page for legend.*)

The occurrence of clones with long tails can be minimized by taking advantage of the fact that the enzyme reverse transcriptase is incapable of displacing a complementary strand of DNA if its synthesis has begun at a site downstream from its direction of synthesis (Kornberg and Baker 1992). By using the NotI-(dT)$_{18}$ oligonucleotide in great excess over the mRNA and by controlling the annealing conditions very carefully to achieve complete saturation of the poly(A) tails, only the most proximal primer molecules can be successfully extended to reverse-transcribe the mRNA. Primers that anneal to the poly(A) tract further upstream (respective to the orientation of the first-strand DNA) are extended for a very short distance until they encounter the primer located immediately downstream. The resulting small fragments of tail can be effectively eliminated by a strict size-selection procedure.

The occurrence of clones with small inserts can be resolved by using a reliable system for size-selection of cDNAs. Spin columns often perform rather poorly for this purpose. cDNAs can be very efficiently size-selected by gel filtration on a Bio-Gel A-50m (100–200 mesh) column (see pp. 104–107).

The occurrence of chimeric clones is potentially the most serious of the three problems commonly encountered in cDNA libraries because such artifacts are often undetectable by sequencing analysis and consequently may result in misleading information. As depicted in Figure 3, there are two steps during which chimeric clones can be generated: The first is during the ligation of the adapter molecules to the cDNAs, and the second is during the final ligation of the cDNAs to the cloning vector. However, since the cDNAs have two different ends to allow for directional cloning and since vector is always present in slight excess over cDNA in that ligation, the latter event is not likely to occur. In fact, this type of chimeric clone could only be generated if three cDNA molecules are ligated to one another and the resulting chimeric molecule is then ligated to the vector. The former event, however, can happen if the adapter molecules are not present in great excess over cDNAs, which may occur, for example, if the amount of cDNA is underestimated. It is noteworthy (as mentioned above) that a significant fraction of the synthesized double-stranded cDNA corresponds to small fragments that represent exclusively poly(A) tails of mRNAs. Accordingly, if [^{32}P]dCTP is the radioisotope that is incorporated during synthesis to allow for quantitation of the cDNA, all of these molecules will remain unaccounted for. Consequently, two blunt-ended cDNA molecules can be joined by ligation and the resulting chimera can then be ligated to adapter molecules. As depicted in Figure 3, two cDNAs can be blunt-end ligated either in a head-to-tail (A) or in a head-to-head (B) orientation. It should be noted, however, that most of such chimeric cDNA molecules are likely to be eliminated at a later step (i.e., digestion with NotI, which precedes the final ligation to the vector). The exceptions would be those cases in which NotI fails to cleave at one or both of the NotI sites present in the chimeric molecule (for details, see Figure 3). In any event, the generation of chimeric molecules during ligation to the adapter can be substantially minimized if small molecules are quantitatively eliminated by using a reliable system for size selection of cDNAs. Furthermore, it is noteworthy that both types of chimeric clones are eliminated during the subsequent purification steps involving linearization of a plasmid DNA preparation of the library with NotI followed by gel purification, recircularization of the linearized recombinant molecules, and then transformation of bacteria by electroporation and propagation under selection with an appropriate antibiotic.

Cloning Efficiencies Required to Ensure That a Representative cDNA Library Is Obtained

How many clones are needed to obtain a representative cDNA library? The probability that a given

Figure 2 Process of cloning a bona fide cDNA molecule. This figure depicts the normal process involved in cloning a cDNA molecule. In the first step, EcoRI adapters are blunt-end-ligated to the cDNAs. Note that only one of the strands of the adapter molecules is phosphorylated (this prevents avoid multiple concatemerization). Adapter-ligated cDNAs are then digested with NotI. Note that one of the restriction fragments is a small fragment generated by NotI plus EcoRI that consists of part of the NotI–(dT)$_{18}$ oligonucleotide used to prime the synthesis of first-strand cDNA ligated to an EcoRI adapter molecule. This highly clonable, yet undesirable, fragment must be eliminated during size selection. The cDNAs are then phosphorylated and ligated to a phagemid vector digested with both NotI and EcoRI.

I. BLUNT END LIGATION OF TWO cDNAs DURING ADAPTER ADDITION

A. BLUNT END LIGATION OF cDNAs 1 AND 2 IN THE HEAD TO TAIL ORIENTATION

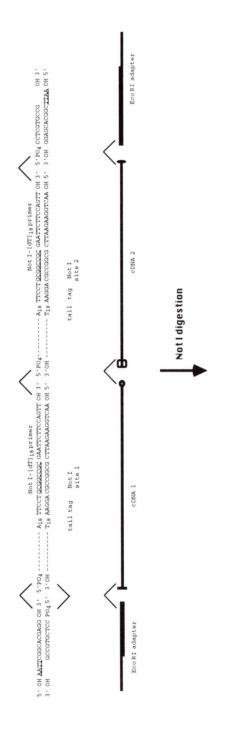

A.1. Products that are generated if Not I digestion is complete

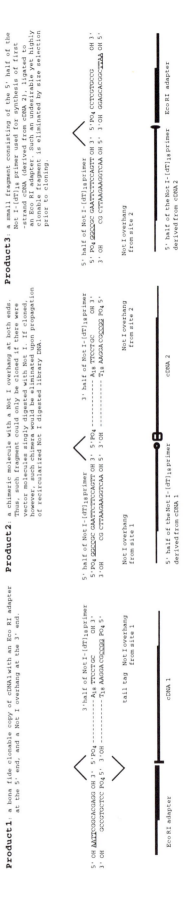

Product 1: a bona fide clonable copy of cDNA1 with an Eco RI adapter at the 5' end, and a Not I overhang at the 3' end.

Product 2: a chimeric molecule with a Not I overhang at both ends. Thus, such fragment could only be cloned if there were vector molecules singly digested with Not I. If cloned, however, such chimera would be eliminated upon propagation of recircularized Not I digested library DNA.

Product 3: a small fragment consisting of the 5' half of the Not I-[dT]18 primer used for synthesis of first-strand cDNA (derived from cDNA 2), ligated to an Eco RI adapter. Such an undesirable yet highly clonable fragment is eliminated by size selection prior to cloning.

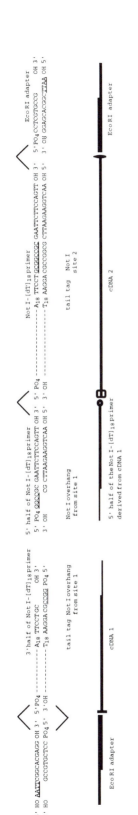

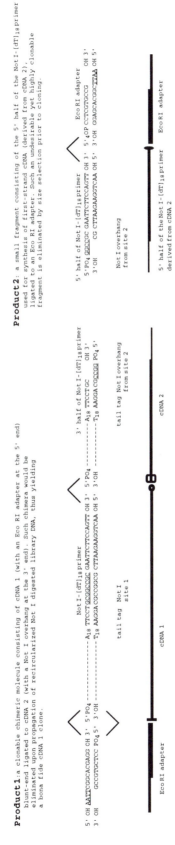

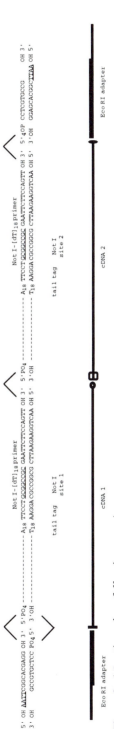

Figure 3 *(Continued on following page.)*

B. BLUNT END LIGATION OF CDNAS 1 AND 2 IN THE HEAD TO HEAD ORIENTATION

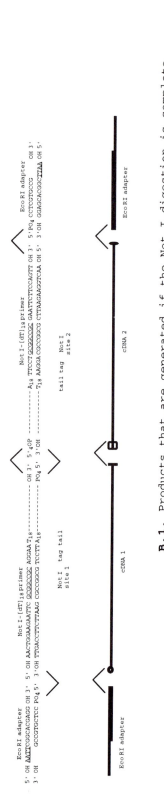

B.1. Products that are generated if the Not I digestion is complete

Product 1: a small fragment consisting of the 5' half of the Not I-[dT]18 primer used for synthesis of first-strand cDNA (derived from cDNA 1) blunt-end ligated to an Eco RI adapter. Such an undesirable yet highly clonable fragment can be eliminated by size selection prior to cloning.

Product 2: a chimeric Not I fragment containing blunt end ligated cDNAs 1 and 2. Thus, such fragment could only get cloned if there were vector molecules singly digested with Not I. If cloned, however, such chimera would be eliminated upon propagation of recircularized Not I digested-library DNA.

B.2. Products that are generated if the Not I digestion is incomplete and only site 1 is cleaved

Product 1: a small fragment consisting of the 5' half of the Not I-[dT]18 primer used for synthesis of first-strand cDNA (derived from cDNA 1) blunt-end ligated to an Eco RI adapter. Such an undesirable yet highly clonable fragment can be eliminated by size selection prior to cloning.

Product 2: a clonable chimeric molecule containing cDNAs 1 and 2 (blunt end ligated). Such chimera would be eliminated upon propagation of recircularized Not I digested library DNA, thus yielding a bona fide cDNA 2 clone.

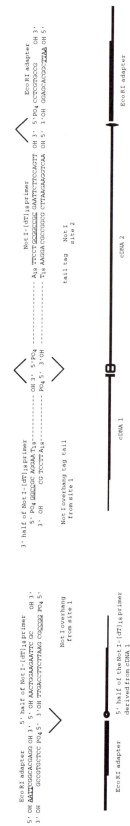

Product 3: a small fragment consisting of the 5' half of the Not I-[dT]18 primer used for synthesis of first-strand cDNA (derived from cDNA 2), blunt end ligated to an Eco RI adapter. Such an undesirable yet highly clonable fragment can be eliminated by size selection prior to cloning.

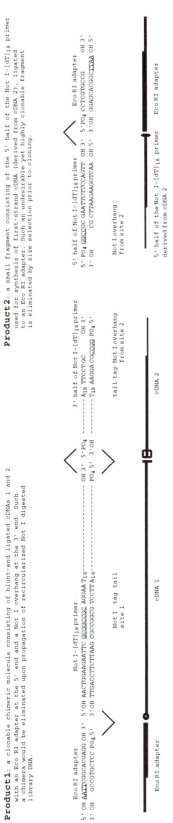
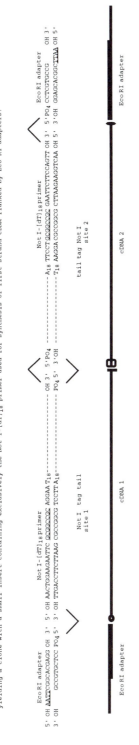
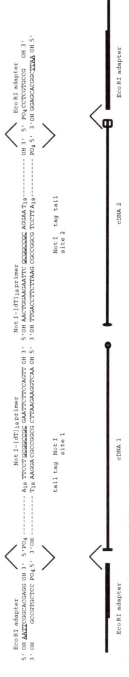

Figure 3 (*Continued on following page.*)

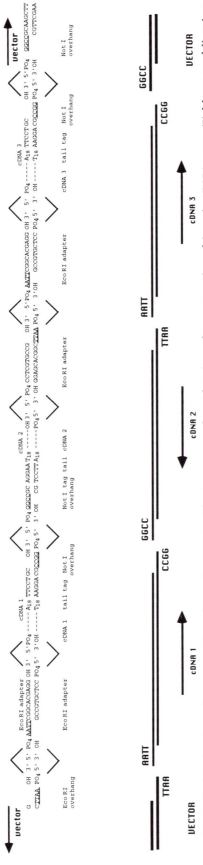

II. GENERATION OF A CHIMERIC CLONE DURING LIGATION OF cDNAS TO THE CLONING VECTOR

Product: a chimeric molecule containing three different cDNAs (1, 2 and 3). Such a chimera would be eliminated upon propagation or recircularized Not I digested library DNA, thus yielding a bona fide copy of one of the cDNAs (cDNA 1 in the example below).

Figure 3 Mechanisms for generating chimeric cDNAs. The two general mechanisms for generating chimeric cDNAs are: (I) blunt-end ligation of two cDNAs during the addition of adapters and (II) co-cloning of cDNAs during ligation to the cloning vector. As depicted in this figure, cDNAs can be blunt-end ligated in the head-to-tail (*A*) or in the head-to-head (*B*) orientation. No ligation is possible in the tail-to-tail orientation (*C*). A complicating factor is the fact that digestion with *Not*I may be complete (A.1.; B.1.) or incomplete (A.2., A.3., A.4.; B.2., B.3., B.4.). The various products that may be generated in each case are described in the figure.

mRNA will be represented in a cDNA library can be calculated by the following formula:

$$p(x) = (1 - [1 - f]_n)$$

where n is the number of recombinant clones and f is the frequency of a particular mRNA. Since a typical somatic cell contains approximately 0.6 pg of mRNA, it follows that there are approximately 500,000 mRNA molecules per cell (0.6 pg x 1 molecule/11 x 10^{-7} pg = 545,454), assuming that the average size of an mRNA is 2 kb (11 x 10^{-7} pg). Accordingly, there is a probability of 99.9% that the rarest mRNA (1 copy per cell; f = 1/500,000) will be represented in a library of 5,000,000 recombinants from such a somatic cell (note that this probability drops to 86.5% for a library of 1,000,000 recombinants). Therefore, it is imperative that the highest cloning efficiencies be achieved.

Transformation by electroporation is the method that yields the highest cloning efficiencies for cDNAs ligated to plasmid vectors. Typically, a total of 1,000,000–5,000,000 recombinant clones can be obtained from 50 ng of cDNA by electroporation. High cloning efficiencies can be attained by electroporation using commercially available competent bacteria (Electromax [Life Technologies]; electroporation efficiency is 1 x 10^{10} cfu/µg of CsCl-purified supercoiled pUC19 control DNA). However, existing protocols for the preparation of competent bacteria (Dower et al. 1988) have had some modifications introduced that reproducibly yield cells with 1 x 10^{10} to 3 x 10^{10} cfu/µg of CsCl-purified supercoiled pUC19 control DNA (see pp. 118–119).

Theoretical Considerations in Normalizing cDNA Populations

The mRNAs of a typical somatic cell are distributed into three frequency classes (Bishop et al. 1974; Davidson and Britten 1979): (I) prevalent, (II) intermediate, and (III) rare (complex). The classes at the two extremes (10% and 40–45% of the total mRNA species, respectively) include members occurring at vastly different relative frequencies. On average, the most prevalent class consists of approximately ten mRNA species, each represented by 5000 copies per cell, whereas the class of high complexity comprises 15,000 different species, each represented by only 1–15 copies. Even though the rarest mRNA sequence from any tissue is likely to be represented in a cDNA library of 5,000,000–10,000,000 recombinants, its identification is very difficult (its frequency of occurrence may be as low as 2 x 10^{-6} on average or even 10^{-7} for complex tissues such as the brain). Thus, for a variety of purposes, it is advantageous to apply a normalization procedure and bring the frequency of each clone in a cDNA library within a narrow range.

It has been estimated (Soares et al. 1994) that in a complex organ such as the brain, there are approximately 36 prevalent mRNAs, 2150 intermediate mRNAs, and as many as 45,000 different rare (complex) mRNAs, which comprise 16%, 46%, and 38% of the total mRNA mass, respectively. Thus, 62% of the clones in a library correspond to either a class I or a class II transcript. In contrast, only 4.6% of the clones in a normalized library correspond to transcripts of these two frequency classes ([36 + 2150]/[36 + 2150 + 45,000]), whereas 95.4% of the clones correspond to class III transcripts. Therefore, on average, normalization results in a 13.5-fold reduction in the total frequency of class I plus class II transcripts and in a simultaneous 2.5-fold increase in the total frequency of class III transcripts.

Two general approaches have been proposed to normalize cDNA libraries (Weissman 1987). One approach is based on hybridization to genomic DNA. The frequency of each hybridized cDNA in the resulting normalized library would be proportional to that of each corresponding gene in the genomic DNA. The feasibility of such an approach has not yet been demonstrated. The other approach is based on reassociation kinetics (i.e., the time course of the reassociation process). If a population of identical DNA duplexes is dissociated by heating and then allowed to reassociate, the rate of the reaction follows second-order kinetics (with respect to the concentration of single-stranded DNA), that is,

$$dC/dt = -kC^2$$

where C is the concentration (moles of dNTPs/liter) of DNA remaining single stranded at time t, t is time (seconds) spent under reassociation conditions, k is the reassociation rate constant (liter/

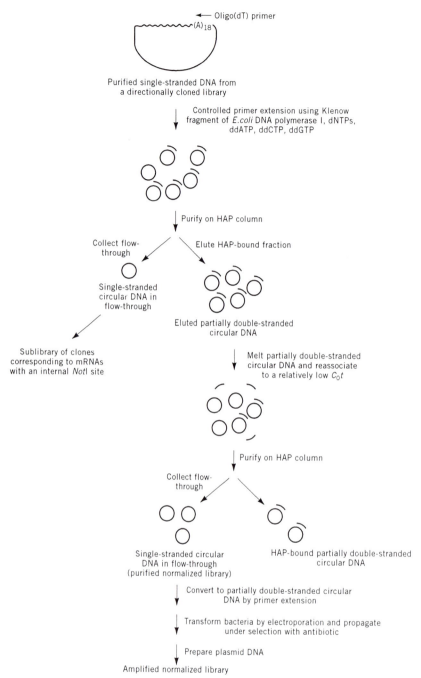

Figure 4 Normalization of directionally cloned cDNA libraries constructed in phagemid vectors (Soares et al. 1994). A preparation of single-stranded DNA from a directionally cloned library is annealed to an oligo(dT)$_{12-18}$ primer (represented by an arrow), and controlled extensions are performed with the Klenow fragment of *E. coli* DNA polymerase I in the presence of dNTPs (dATP, dTTP, dCTP, and dGTP) and ddNTPs (ddATP, ddCTP, and ddGTP). The resulting partially double-stranded circular DNA is purified on a HAP column to separate it from any remaining single-stranded circular DNA and is then melted and allowed to reassociate to a relatively low C_0t (~5–20 seconds-moles/liter). The fraction that remains single-stranded (the normalized library) is purified from the reassociated material on a HAP column, partially converted into double strands by primer extension, and amplified using transformation of bacteria by electroporation and propagation under selection with ampicillin. See protocol on pp. 124–148 for details.

a relatively low C_0t; (5) the fraction that remains single-stranded is purified from the reassociated molecules by chromatography on HAP columns; and (6) the single-stranded flow-through DNA from the HAP column is partially converted into double strands for improvement of electroporation efficiencies and bacteria are then transformed by electroporation for propagating a normalized library without subcloning steps. Accordingly, although only 3′-noncoding sequences participate in the reassociation reaction, the resulting normal-ized library consists of clones with large inserts encompassing both coding and noncoding sequences. This method can be applied to any library constructed in a phagemid vector or in any other vector with an f1 origin of replication such that library cDNA can be obtained in the form of single-stranded circular DNA after superinfection of a growing culture with a helper bacteriophage. A detailed protocol and further discussion of this method are provided on pp. 124–148.

METHODS FOR ISOLATING RNA

The preparation of a good cDNA library, representative of all transcripts present in a given tissue or cell type, is entirely dependent on the quality of the RNA that is originally isolated and purified. This section contains several RNA purification and isolation protocols. Although one of the many kits available commercially for synthesizing poly(A)$^+$ RNA is suitable for many applications (e.g., northern blotting or RT-PCR), using one of the RNA isolation protocols in this chapter is recommended for library construction.

Choosing a Method for Isolating RNA

Total cellular RNA can be extracted from tissues and cells in culture by using a variety of standard procedures. The protocol on pp. 69–71 is probably the most classic of all methods of RNA isolation. It uses guanidinium thiocyanate and ultracentrifugation in a CsCl solution essentially as described by Glisin et al. (1974), Ullrich et al. (1977), and Chirgwin et al. (1979). This method yields very clean preparations of RNA that are suitable for all applications. However, because of its greater simplicity and the fact that it is particularly well suited to extracting RNA from small amounts of tissue, the method originally described by Chomczynski and Sacchi (1987) (see pp. 73–76) has become the procedure of choice for RNA isolation.

The protocol on p. 72 is for the preparation of total cellular RNA from cell monolayers (either cell lines or primary cultures) grown to confluence on 10-cm dishes. Once the cells are homogenized, RNA is isolated as described on pp. 69–71. Each dish (~10^7 cells) is expected to yield 30–100 μg of total cellular RNA.

The method on pp. 73–76, originally developed by Chomczynski and Sacchi (1987) and subsequently modified by Chomczynski and Mackey (1995), is the simplest and most advantageous of all procedures for isolation of total cellular RNA. It is fast, it does not require ultracentrifugation, and yet it yields RNA of very high quality. Chomczynski and Mackey (1995) modified the original procedure to include a precipitation with only 0.5 volume of isopropanol and 0.5 volume of high-salt solution. This modification eliminates contaminating polysaccharides from the final RNA preparation.

The protocol on pp. 77–79, which is a modification of the procedure described by Favaloro et al. (1980), is designed for isolation of RNA from cultured cells. Because it is possible to lyse cell membranes yet maintain intact nuclei, both cytoplasmic and nuclear RNA can be isolated with this protocol. Whenever possible, cytoplasmic RNA should used instead of total cellular RNA for the construction of cDNA libraries. However, the conditions used for cell lysis in this protocol are too gentle to allow efficient lysis of cells from most tissues. The methods on pp. 69–76 must therefore be used for RNA isolation in most cases. It should be noted, however, that the RNA obtained with these other methods also contains heterogeneous nuclear transcripts (i.e., unprocessed and semiprocessed transcripts).

Mammalian mRNAs have tracts of poly(A) at their 3′ ends. This property has been exploited in the purification and separation of mRNA from the majority of cellular rRNAs and tRNAs (which do not have the tracts). An oligo(dT)-cellulose matrix can be used under fairly stringent conditions to enrich for RNAs with poly(A)$^+$ tracts (a protocol for standard chromatography on oligo[dT]-cellulose columns is provided on pp. 83–85). Usually, only a small percentage (1–5%) of total cellular RNA actually represents the mRNA population. Before poly(A)$^+$ RNA is purified, thorough examination of the total cellular RNA is recommended to evaluate its quality and estimate the quantity (see below). Typically, a representative cDNA library ($\geq 10^6$ recombinants) can be prepared from 0.5–1 μg of poly(A)$^+$ RNA (which can be obtained from ~100 μg of total cellular RNA).

Evaluating the RNA Preparation

Regardless of the preparation method used, the quality (intactness) of the total cellular RNA should be evaluated before poly(A)$^+$ RNA selection. A sample of the RNA (~0.5–1 μg) should be analyzed on a formaldehyde-agarose minigel (see pp. 80–82), and the relative intensities of the ethidium bromide fluorescence of both the 28S and 18S mammalian rRNAs should be visually inspected. Since mammalian 28S and 18S rRNAs are approximately 4718 and 1874 bases long, respectively (Lewin 1994), the intensity of the 28S rRNA band should be approximately 2.5-fold higher than that of the 18S rRNA if the RNA is intact (for an example, see Figure 4.2.2 [p. 4.2.8] in Ausubel et al. 1995). The amount of total cellular RNA can be estimated by measuring the OD$_{260}$ of a small aliquot (see Appendix).

Special Considerations for Working with RNA

RNA is especially susceptible to breakdown by the enzyme RNase (which is ubiquitous) and should therefore be handled with special care at all stages of preparation and analysis. RNase is released during the lysis of cells (a critical step in mRNA isolation) and is also present on skin and hair (and subsequently everything with which they come into contact). Working with RNA requires the use of the following precautions to avoid breakdown by RNase:

- Wear gloves at all times during work with RNA and change them frequently.
- Wash all glassware (tubes, bottles, beakers, and conical flasks) used in the preparation or analysis of RNA, cover the rims with foil, and then bake at 180ºC for 1 hour. The glassware should only be handled with gloves and should be stored separately from other laboratory glassware.
- Prepare all solutions and stock solutions in baked glassware or in sterile disposable (RNase-free) plasticware that has been rinsed with chloroform (see

Appendix for Caution). Designate these solutions "for RNA work only" and store them separately from stock solutions used for DNA preparation.

- For weighing reagents for the preparation of stock solutions, use spatulas that have been sterilized by autoclaving and then baked at 180°C for 1 hour and weighing boats that have only been handled with gloves.

- Prepare Tris-based buffers with RNase-free H_2O, which has been treated with DEPC (for DEPC treatment of H_2O and Caution, see Appendix). Other buffers and stock solutions should be treated directly with DEPC. DEPC inactivates RNase activity by covalent modification.

- Autoclave all pipette tips used in the preparation and analysis of RNA. Store sterile (RNase-free) tips stored separately from other pipette tips. Sterile disposable pipettes should be used for pipetting large volumes of solutions.

RNase Inhibitors

Several different RNase inhibitors are available commercially. Where appropriate, these inhibitors are recommended in the protocols in this chapter. Guanidinium thiocyanate and β-mercaptoethanol (both potent inhibitors of RNases) are used throughout cell lysis and RNA isolation in this chapter.

PROTOCOL

Isolation of Total Cellular RNA from Tissues Using Guanidinium Thiocyanate and CsCl

This method was first developed by Glisin et al. (1974) and then modified by Ullrich et al. (1977) and Chirgwin et al. (1979). Be sure to observe the special considerations for working with RNA throughout this procedure (see pp. 67–68).

1. Remove tissue from the organism and quick-freeze it in **liquid nitrogen**. If the sample is not small (≥2 g), rapidly split it into smaller pieces before quick-freezing it. Alternatively, remove tissue from the organism and immediately homogenize the unfrozen tissue in the guanidinium thiocyanate solution as described in step 2.

   Notes: Wide-mouth plastic containers (e.g., 50-ml conical tubes) can be used to store samples at –80°C for later processing. These containers facilitate removal of the tissue for subsequent processing.

   For tissue that may be resistant to homogenization (e.g., bone or liver), freeze fracturing the tissue to a powdered form while it is still deep frozen (e.g., mechanically breaking the tissue by smashing it with a hammer immediately after it is removed from storage at –80°C or immediately after it is removed from the liquid nitrogen) is recommended to maximize the yield and to minimize breakdown by RNase.

   Procedures for the humane treatment of animals must be observed at all times. Check with the local animal facility for guidelines.

   liquid nitrogen (see Appendix for Caution)

2. Add approximately 5 volumes (or 10 ml per gram of tissue) of guanidinium thiocyanate solution to the tissue fragment. Immediately homogenize by using either a Polytron PT3000 (Brinkmann) or a dounce homogenizer.

   Notes: Guanidinium thiocyanate is a strong denaturant that is used to disrupt the cells.

   Homogenized tissue can be stored in the guanidinium thiocyanate solution at –80°C for later processing.

Guanidinium thiocyanate solution for RNA isolation

Component and final concentration	Amount to add per 250 ml
4 M guanidinium thiocyanate (ICN 820991)	118.16 g
50 mM sodium acetate	4.15 ml of 3 M (pH 7.0)
10 mM EDTA	5 ml of 0.5 M (pH 8.0)
DEPC-treated H$_2$O	to make 250 ml
100 mM β-**mercaptoethanol**	1.78 ml of 14 M (Fisher Scientific BP176-100)

Combine all of the components except the β-mercaptoethanol in the bottle the guanidinium thiocyanate comes in. (This solution can be stored in a dark bottle at room temperature for at least 6 months.) Add the β-mercaptoethanol just before use.

DEPC, β-**mercaptoethanol** (see Appendix for Caution)

3. Place the homogenized tissue in a 13-ml Sarstedt tube (55.518 and 65.816). Centrifuge in a Sorvall SS34 rotor at 10,000 rpm (12,000g) at 4°C for 10 minutes to remove cellular and nuclear debris.

4. Transfer the supernatant into a new tube and discard the pellet. Add 0.1 volume of 20% Sarkosyl and heat at 65°C for 2–5 minutes.

5. Add 1 g of solid ultrapure CsCl for each 2.5 ml of homogenate and mix by vortexing until all of the CsCl has dissolved.

6. Prepare step gradients from the entire volume of CsCl/homogenate and centrifuge as follows: Layer 3.5 ml of CsCl/homogenate onto 1.5 ml of CsCl gradient solution in a silanized cellulose acetate or Beckman ultraclear SW50.1 ultracentrifuge tube. Centrifuge in a Beckman SW50.1 rotor at 35,000 rpm (146,816g) at 20°C for 18 hours. Alternatively, layer 8 ml of the CsCl/homogenate onto 3 ml of CsCl gradient solution in a silanized Beckman ultraclear SW41 ultracentrifuge tube. Centrifuge in a Beckman SW41 rotor at 30,000 rpm (153,900g) at 20°C for 24 hours.

Notes: Silanizing the tube decreases the amount of RNA lost because of the stickiness of the sides of the tube. To silanize a tube, fill it with silanizing solution (**Sigmacote**; Sigma SL-2), allow it to stand for 5 minutes, empty it, rinse it thoroughly with deionized H$_2$O, and then invert the tube and allow it to air dry. Silanized tubes can also be purchased.

To avoid mixing the layers of the step gradient, it is sometimes easier to pipette the CsCl/homogenate into the tube first and then underlay the CsCl gradient solution. To do this, place a glass pipette or a plastic transfer pipette through the CsCl/homogenate layer and pipette the CsCl gradient solution directly into the bottom of the tube.

Be sure to weigh the tubes carefully and balance the load in the swinging-bucket rotor before centrifugation.

CsCl gradient solution for RNA isolation

Component and final concentration	Amount to add per 100 ml
5.7 M ultrapure CsCl	96 g
25 mM sodium acetate	0.83 ml of 3 M (pH 5.2)
10 mM EDTA	2 ml of 0.5 M (pH 8.0)
0.1% DEPC	100 µl
H$_2$O	to make 100 ml

Dissolve the CsCl in 90 ml of H$_2$O in a 250-ml bottle. Add the sodium acetate and EDTA, and then add the DEPC. Adjust the volume to 100 ml with H$_2$O. Store at 4°C for up to 1 month.

Sigmacote (see Appendix for Caution)

7. Remove the tubes from the rotor, being careful not to disturb the gradient. Carefully remove the supernatant from each tube with a pasteur pipette, starting at the top of the gradient and moving down. To remove the final 0.5 ml, invert the tube to decant the remaining liquid and then place the inverted tube on a paper towel to drain.

Notes: The cellular DNA (which may appear white) forms a band at the interface and may be smeared on the side of the tube; the RNA precipitates. The small, round RNA pellet will be clear and glassy and may not be readily visible until the supernatant has been removed and the tube is held up to the light and viewed from the side.

8. Add 200–300 µl of DEPC-treated H_2O to each RNA pellet and dissolve by pipetting up and down with RNase-free pipette tips. Transfer each RNA sample into a silanized 1.5-ml microcentrifuge tube. Add another 100 µl of DEPC-treated H_2O to the tube and wash the tube to collect any residual RNA.

9. Extract each sample with 1 volume of **phenol:chloroform**:isoamyl alcohol (25:24:1) and then with 1 volume of chloroform:isoamyl alcohol (24:1) (see Appendix). Use ultrapure buffer-saturated phenol (Life Technologies 15513-047).

   Notes: Before use, phenol must be equilibrated to a pH greater than 7.6 because DNA (but not RNA) will partition into the organic phase at acidic pHs. The ultrapure-saturated buffer specified here works well.
       The second extraction should remove any residual phenol, which could interfere with quantitation of the RNA.
       If desired, quantitate the RNA by measuring the OD_{260} of a 1-µl aliquot (see Appendix). A 40-µg/ml solution of single-stranded RNA corresponds to an OD_{260} of 1 in a quartz cuvette with a 1-cm path length.

   phenol, chloroform (see Appendix for Caution)

10. Precipitate the RNA in the aqueous phase as follows:

    a. Add 0.1 volume of 3 M sodium acetate (pH 5.2) and mix thoroughly.

    b. Add 2.5 volumes of ice-cold absolute ethanol and mix by vigorous vortexing. Centrifuge briefly to collect the liquid at the bottom of the tube.

    c. Place at –20°C overnight or in a dry-ice/ethanol bath for 30 minutes.

       Note: If the RNA is not to be used immediately, it can be stored in ethanol at –20°C indefinitely.

    d. Centrifuge in a microcentrifuge at 12,000g at 4°C for 20 minutes. Carefully remove the supernatant with a pasteur pipette and discard.

       Note: It is not necessary to wash the pellet with 70% ethanol.

    e. Dry the RNA pellet under vacuum in a SpeedVac Concentrator.

11. Dissolve the RNA in DEPC-treated H_2O at a final concentration of approximately 0.5 mg/ml.

    Note: The expected yield per gram of tissue differs among tissues (e.g., 1 g of liver might yield 8 mg of total RNA, whereas 1 g of muscle might only yield 1.25 mg).

12. Assess the integrity of the RNA by electrophoresis on a formaldehyde–agarose gel as described on pp. 80–82. Keep the remainder of the RNA on ice while the gel is running. If the RNA is intact, proceed with the purification of poly(A)$^+$ RNA (pp. 83–85).

Isolation of Total Cellular RNA from Cultured Cells

Be sure to observe the special considerations for working with RNA throughout this procedure (see pp. 67–68).

1. Add 7–10 ml of ice-cold 1x PBS for RNA isolation (pH 7.4) to each 10-cm dish and wash the cell monolayers for 1–5 minutes to remove any residual growth medium. Repeat the wash and place the cells on ice.

   Note: For preparing total cellular RNA, cell monolayers (either cell lines or primary cultures) should be grown to confluence on 20 10-cm petri dishes. Each dish (~10^7 cells) is expected to yield 30–100 μg of total cellular RNA.

   10x PBS for RNA isolation (pH 7.4)

Component and final concentration	Amount to add per 1 liter
27 mM KCl	2 g
14 mM KH$_2$PO$_4$ (monobasic, anhydrous)	2 g
1.37 M NaCl	80 g
43 mM Na$_2$HPO$_4$ (dibasic, anhydrous)	6.1 g
0.1% **DEPC**	1 ml
H$_2$O	to make 1 liter

   Mix the components in a 2-liter flask. Sterilize by autoclaving. Store at 4°C for up to 6 months.

   DEPC (see Appendix for Caution)

2. Add 0.5 ml of guanidinium thiocyanate solution for RNA isolation (for preparation, see p. 69) to each dish (~10 ml of guanidinium thiocyanate solution for 20 dishes) and scrape off the cells with a rubber policeman. Transfer the cells into a 50-ml conical tube and homogenize by vortexing.

3. Follow steps 5–12 on pp. 70–71 for isolation and further processing of the RNA.

PROTOCOL

Isolation of Total Cellular RNA from Tissues by Extraction with Acid Guanidinium Thiocyanate and Phenol:Chloroform

This method was developed by Chomczynski and Sacchi (1987) and modified by Chomczynski and Mackey (1995). Be sure to observe the special considerations for working with RNA throughout this procedure (see pp. 67–68).

1. Remove tissue from the organism and quick-freeze it in **liquid nitrogen**. If the sample is not small (≥2 g), rapidly split it into smaller pieces before quick-freezing it. Alternatively, remove tissue from the organism and immediately homogenize the unfrozen tissue in the guanidinium thiocyanate/Sarkosyl solution as described in step 2.

   Notes: Wide-mouth plastic containers (e.g., 50-ml conical tubes) can be used to store samples at –80°C for later processing. These containers facilitate removal of the tissue for subsequent processing.

   For tissue that may be resistant to homogenization (e.g., bone or liver), freeze fracturing the tissue to a powdered form while it is still deep frozen (e.g., mechanically breaking the tissue by smashing it with a hammer immediately after it is removed from storage at –80°C or immediately after it is removed from the liquid nitrogen) is recommended to maximize the yield and to minimize breakdown by RNase.

   Procedures for the humane treatment of animals must be observed at all times. Check with the local animal facility for guidelines.

   liquid nitrogen (see Appendix for Caution)

2. Add 10 ml of guanidinium thiocyanate/Sarkosyl solution to 1 g of fresh or frozen tissue in a 50-ml conical tube. Immediately homogenize at 4°C by using a Polytron PT3000 homogenizer (Brinkmann).

   Guanidinium thiocyanate/Sarkosyl solution for RNA isolation

Component and final concentration	Amount to add per 50 ml
4 M guanidinium thiocyanate (ICN 820991)	23.63 g
25 mM sodium citrate	1.67 ml of 0.75 M (pH 7.0)
0.5% Sarkosyl	2.5 ml of 10%
DEPC-treated H_2O	to make 50 ml
100 mM β-mercaptoethanol	360 µl of 14 M (Fisher Scientific BP176-100)

   Combine all of the components except the β-mercaptoethanol in the bottle the guanidinium thiocyanate comes in. (This solution can be stored in a dark bottle at room temperature for at least 6 months.) Add the β-mercaptoethanol just before use.

   DEPC, β-mercaptoethanol (see Appendix for Caution)

0.75 M Sodium citrate (pH 7.0)
In a 50-ml polypropylene tube, dissolve 11.03 g of sodium citrate (dihydrate; m.w. = 294.1) in sufficient DEPC-treated H$_2$O to make a final volume of 50 ml. Mix thoroughly. Store at room temperature for up to 6 months.

3. Divide the homogenate into two 13-ml Sarstedt tubes (55.518 and 65.816). Add 0.5 ml of 2 M sodium acetate to each tube (pH 4.0) and mix by vigorous vortexing.

2 M Sodium acetate (pH 4.0)

Component and final concentration	Amount to add per 50 ml
2 M sodium acetate anhydrous	8.2 g
glacial acetic acid	28.5 ml
DEPC-treated H$_2$O	to make 50 ml

Mix the components in a 50-ml polypropylene tube. Check the pH. Store at 4°C for up to 6 months.

glacial acetic acid (see Appendix for Caution)

4. Extract each sample with **phenol** and **chloroform** as follows:

phenol, chloroform (see Appendix for Caution)

a. Add 1 volume of phenol and mix by vigorous vortexing. Use ultrapure buffer-saturated phenol (Life Technologies 15513-047).

Note: Before use, phenol must be equilibrated to a pH greater than 7.6 because the DNA (but not RNA) will partition into the organic phase at acidic pHs. The ultrapure-saturated buffer specified here works well.

b. Add 0.2 volume of chloroform and mix by vigorous vortexing.

c. Place on ice for 15 minutes.

d. Centrifuge in a Sorvall SS34 rotor at 10,000 rpm (12,000*g*) at 4°C for 10 minutes to separate the phases.

e. Transfer the aqueous (upper) phases into new tubes and discard the organic phases.

5. Precipitate the RNA in the aqueous phases as follows:

a. Add 0.5 volume of high-salt precipitation solution and 0.5 volume of isopropanol and mix thoroughly by vortexing.

High-salt precipitation solution

Component and final concentration	Amount to add per 100 ml
1.2 M NaCl	7.01 g
0.8 M sodium citrate (dihydrate)	23.52 g
DEPC-treated H$_2$O	to make 100 ml

Sterilize the solution by passing it through a 0.22-μm filter. Store at 4°C for 6 months.

b. Centrifuge in a Sorvall SS34 rotor at 10,000 rpm (12,000g) at 4°C for 20 minutes. Carefully remove the supernatant with a pasteur pipette and discard. Invert the tube to drain the remaining supernatant.

c. Dissolve each pellet in 5 ml of guanidinium thiocyanate/Sarkosyl solution by gently pipetting up and down.

d. Add 1 volume of isopropanol and mix. Place at –20°C for 1 hour.

e. Repeat step b.

f. Add 5 ml of 70% ethanol and repeat step b.

6. Dissolve each pellet in 1 ml of Tris-Cl/EDTA/SDS. Mix and combine the pellets in one tube.

Tris-Cl/EDTA/SDS

Component and final concentration	Amount to add per 1 liter
10 mM Tris-Cl	10 ml of 1 M (pH 7.5 at 25°C)
1 mM EDTA	2 ml of 0.5 M (pH 8.0)
0.5% **SDS**	50 ml of 10%
DEPC-treated H$_2$O	938 ml

To prepare this solution, use Tris buffer prepared with DEPC-treated H$_2$O and stock solutions treated directly with DEPC. Store at room temperature for up to 6 months.

SDS (see Appendix for Caution)

7. Extract with chloroform as follows:

a. Add 1 volume of chloroform and mix by vigorous vortexing.

b. Centrifuge in a Sorvall SS34 rotor at 10,000 rpm (12,000g) at 4°C for 10 minutes to separate the phases.

c. Transfer the aqueous (upper) phase into a new tube and discard the organic phase.

Note: If desired, quantitate the RNA by measuring the OD$_{260}$ of a 1-μl aliquot (see Appendix). A 40-μg/ml solution of single-stranded RNA corresponds to an OD$_{260}$ of 1 in a quartz cuvette with a 1-cm path length.

8. Precipitate the RNA in the aqueous phase as follows:

a. Add 0.1 volume of 3 M sodium acetate (pH 5.2) and 1 volume of isopropanol and mix thoroughly by vortexing. Place at –20°C for 1 hour.

b. Centrifuge in a Sorvall SS34 rotor at 10,000 rpm (12,000g) at 4°C for 30 minutes. Carefully remove the isopropanol by aspiration.

c. Dry the RNA pellet under vacuum in a SpeedVac Concentrator.

9. Dissolve the RNA in DEPC-treated H$_2$O at a final concentration of approximately 0.5 mg/ml.

Notes: The RNA can be stored at −80°C indefinitely.

Expect an average of approximately 4.5 mg of total RNA per gram of tissue. However, yields differ among tissues (e.g., 1 g of liver might yield 8 mg of RNA, whereas 1 g of muscle might only yield 1.25 mg).

10. Assess the integrity of the RNA by electrophoresis on a formaldehyde–agarose gel as described on pp. 80–82. Keep the remainder of the RNA on ice while the gel is running. If the RNA is intact, proceed with the purification of poly(A)$^+$ RNA (pp. 83–85).

PROTOCOL

Isolation of Cytoplasmic and Nuclear RNA from Cultured Cells

This method was developed by Favaloro et al. (1980). Be sure to observe the special considerations for working with RNA throughout this procedure (see pp. 67–68).

1. Add 7–10 ml of ice-cold 1x PBS for RNA isolation (pH 7.4; for preparation, see p. 72) to each 10-cm dish and wash the cell monolayers for 1–5 minutes to remove any residual growth medium. Repeat the wash and place the cells on ice.

   Note: Cell monolayers (either cell lines or primary cultures) should be grown to confluence on 20 10-cm petri dishes. Each dish ($\sim 10^7$ cells) is expected to yield approximately 67.5 μg ($\sim 95\%$) of cytoplasmic RNA or 3.75 μg ($\sim 5\%$) of nuclear RNA.

2. Add 2–5 ml of PBS to each dish and scrape off the cells with a rubber policeman. Transfer the cells into a 50-ml conical tube and homogenize by vortexing.

3. Centrifuge in a clinical centrifuge at 3500 rpm (2000g) at 4°C for 5 minutes to recover the cells. Discard the supernatant.

4. Resuspend the cell pellet in ice-cold RNA extraction buffer (~ 250 μl per 10-cm dish). Mix by vortexing for 10 seconds.

   Note: RNA extraction buffer contains a nonionic detergent (NP-40), which lyses the plasma membrane but leaves the nuclei intact.

RNA extraction buffer

Component and final concentration	Amount to add per 20 ml
140 mM NaCl	0.56 ml of 5 M
1.5 mM MgCl$_2$	30 μl of 1 M
10 mM Tris-Cl	200 μl of 1 M (pH 8.6 at room temperature)
0.5% NP-40	1 ml of 10%
10 mM vanadyl–ribonucleoside complexes (Life Technologies 15522-014)	1 ml of 200 mM
DEPC-treated H$_2$O	17.21 ml

Mix the components in a 50-ml polypropylene tube. Store at room temperature for up to 1 month.

DEPC (see Appendix for Caution)

5. Use a glass pipette or a plastic transfer pipette to overlay each 10 ml of lysed cell suspension onto 10 ml of sucrose-containing lysis buffer in a 30-ml Corex tube. Place on ice for 5 minutes.

Sucrose-containing lysis buffer

Component and final concentration	Amount to add per 20 ml
140 mM NaCl	0.56 ml of 5 M
1.5 mM MgCl$_2$	30 µl of 1 M
10 mM Tris-Cl	200 µl of 1 M (pH 8.6 at room temperature)
1% NP-40	2 ml of 10%
10 mM vanadyl–ribonucleoside complexes (Life Technologies 15522-014)	1 ml of 200 mM
0.7 M sucrose	4.8 g
DEPC-treated H$_2$O	to make 20 ml

Mix the components in a 50-ml polypropylene tube. Store at room temperature for up to 1 month.

6. Centrifuge in a Sorvall SS34 rotor at 10,000 rpm (12,000*g*) at 4°C for 20 minutes.

Note: As a result of this centrifugation, the cytoplasmic RNA will constitute an upper layer on top of the sucrose-containing lysis buffer, and the nuclei will precipitate.

7. Isolate cytoplasmic RNA from the upper layer or nuclear RNA from the pellet.

For isolation of cytoplasmic RNA:

a. Use a pipette to transfer the upper layer (the cytoplasmic RNA) into a 13-ml Sarstedt tube (55.518 and 65.816). Add 1 volume of 2x proteinase K buffer and add proteinase K solution (13.6 mg/ml in 10 mM Tris-Cl [pH 7.5]; Boehringer Mannheim 1413783) to a final concentration of 200 µg/ml.

2x Proteinase K buffer

Component and final concentration	Amount to add per 20 ml
20 mM Tris-Cl	400 µl of 1 M (pH 7.8 at 25°C)
10 mM EDTA	400 µl of 0.5 M (pH 8.0)
1% **SDS**	2 ml of 10%
DEPC-treated H$_2$O	17.2 ml

Mix the components in a 50-ml polypropylene tube. Store at room temperature for up to 1 month. (This buffer can also be purchased from Boehringer Mannheim.)

SDS (see Appendix for Caution)

b. Incubate at 37°C for 30 minutes.

c. Extract with 1 volume of **phenol:chloroform**:isoamyl alcohol (25:24:1) (see Appendix). Use ultrapure buffer-saturated phenol (Life Technologies 15513-047).

   Notes: Before use, phenol must be equilibrated to a pH greater than 7.6 becuase the DNA (but not RNA) will partition into the organic phase at acidic pHs. The ultrapure-saturated buffer specified here works well.

   If desired, quantitate the RNA by measuring the OD_{260} of a 1-µl aliquot (see Appendix). A 40-µg/ml solution of single-stranded RNA corresponds to an OD_{260} of 1 in a quartz cuvette with a 1-cm path length.

   phenol, chloroform (see Appendix for Caution)

d. Follow steps 10–12 on p. 71, but use a Sorvall SS34 rotor at 10,000 rpm ($12,000g$) at 4°C for 30 minutes.

For isolation of nuclear RNA:

a. Dissolve the nuclear pellet in 20 ml of guanidinium thiocyanate solution for RNA isolation (for preparation, see p. 69) by vortexing.

b. Follow steps 5–11 on pp. 70–71 for isolation and further processing of the RNA.

   Notes: If the cytoplasmic or nuclear RNA is not to be used immediately, it can be stored in ethanol at –20°C indefinitely.

   The nuclear RNA produced as a byproduct here is not used for library construction.

PROTOCOL

Formaldehyde-Agarose Gels for Electrophoresis of RNA

The conditions for electrophoresis of RNA through gels containing formaldehyde are based on the procedures described by Lehrach et al. (1977), Goldberg (1980), and Thomas (1983), with some modifications (Kroczek 1989; Rosen and Villa-Komaroff 1990). Be sure to observe the special considerations for working with RNA throughout this procedure (see pp. 67–68).

1. Prepare and, if necessary, seal the ends of the gel-casting mold according to the manufacturer's instructions. Place the mold on a level surface in a chemical fume hood.

2. Prepare a gel containing 1.2% (w/v) agarose and 6.7% (w/v) formaldehyde.

   Note: A 1.2% agarose gel will give good size separation and resolution in the 1–5-kb size range. For separation of larger RNAs (2–12 kb), a 0.7% agarose gel is recommended. For separation of smaller RNAs (<2 kb), a 2% agarose gel can be used.

   a. Combine 0.6 g of electrophoresis-grade agarose and 36 ml of **DEPC**-treated H_2O in a conical flask.

      DEPC (see Appendix for Caution)

   b. Heat in a microwave oven set on high for approximately 1 minute to melt the agarose.

   c. Swirl the flask and closely inspect the solution to make sure that all small particles of agarose have been dissolved. If the solution is not uniform, heat for an additional 30 seconds. Repeat as necessary.

   d. Allow the melted agarose solution to cool to 55°C in a water bath.

   e. Add 5 ml of 10x formaldehyde-agarose electrophoresis running buffer to the melted agarose and swirl gently to mix.

      10x Formaldehyde–agarose electrophoresis running buffer

Component and final concentration	Amount to add per 50 ml
200 mM MOPS	5 ml of 100x MOPS (pH 7.3)
50 mM sodium acetate 10 mM EDTA DEPC-treated H_2O	5 ml of 100x sodium acetate/EDTA (pH 7.0) 40 ml

      Prepare this buffer from RNase-free stock solutions just before use.

      Note: This running buffer can be stored at 4°C for short periods of time, but it will become straw-colored as it ages. When this occurs, fresh buffer should be prepared.

100x MOPS (pH 7.3)

Component and final concentration	Amount to add per 100 ml
2 M MOPS (Sigma)	41.86 g (F.W. = 209.3)
0.165 N **NaOH**	8.25 ml of 2 N
DEPC-treated H_2O	to make 100 ml

Sterilize the solution by passing it through a 0.22-μm filter. Store in a dark or foil-wrapped bottle at 4°C for up to 6 months.

100x Sodium acetate/EDTA (pH 7.0)

Component and final concentration	Amount to add per 100 ml
0.5 M sodium acetate (trihydrate)	6.8 g (m.w. = 136.08)
100 mM EDTA	2.92 g of EDTA (m.w. = 292.25) or 3.72 g of Na_2 EDTA·$2H_2O$ (m.w. = 372.24)
0.029 N NaOH	2.9 ml of 1 N
DEPC-treated H_2O	to make 100 ml

Sterilize the solution by passing it through a 0.22-μm filter. Store at 4°C for up to 6 months.

NaOH (see Appendix for Caution)

f. In a chemical fume hood, add 9 ml of 37% (w/v) **formaldehyde** GR solution (EM Science FX0410-5) and swirl gently to mix thoroughly. Immediately pour the gel into the prepared mold.

Note: Immediately after the gel has been poured, small bubbles may collect on the surface of the agarose. Sweep these bubbles to the sides of the gel-casting mold by gently stroking the surface of the agarose with the gel comb before positioning the comb in the apparatus and allowing it to set.

formaldehyde (see Appendix for Caution)

g. Allow the gel to solidify at room temperature for 30–45 minutes.

3. Gently transfer the solidified gel into an electrophoresis chamber and submerge it in 1x formaldehyde–agarose electrophoresis running buffer. Carefully remove the comb.

Note: Electrophoresis chambers that have previously been used for DNA should be washed thoroughly with detergent and then rinsed with DEPC-treated H_2O and ethanol before they are used in the analysis of RNA.

4. Prepare the RNA samples and markers for loading on the gel.

a. Dissolve (or dilute) approximately 0.5–1 μg of RNA (and each molecular-weight marker) in 5 μl of deionized **formamide**.

Notes: Use RNA molecular-weight markers (e.g., 240 bp to 9.5 kb; Life Technologies 15620016) that have been precipitated with ethanol (see p. 71), dried under vacuum in a SpeedVac Concentrator, and dissolved in 5 μl of deionized formamide. If RNA markers

6. Wash the column with 1 ml of loading buffer. Collect the eluate.

Note: To minimize losses of RNA, this 1 ml of loading buffer can be used to wash the tube that contained the RNA before it is used to wash the column.

7. Combine the eluates from steps 5 and 6 and reload on the column. Collect and save this eluate until the end of the procedure.

Notes: At this point, the RNA is in a buffer containing SDS, which probably helps minimize the formation of secondary structures. Therefore, the eluate does not need to be reheated before it is loaded.

This eluate can be discarded at the end of the procedure if an adequate yield of poly(A)$^+$ RNA is obtained.

8. Wash the column with 2 ml of medium-salt washing buffer for oligo(dT)-cellulose columns. Discard the eluate.

Note: The purpose of washing the column with a buffer of lower salt concentration is to make conditions more stringent and thus minimize retention of RNAs (other than mRNAs) that may have bound to the column fortuitously (e.g., RNAs that happen to contain an internal stretch of A residues).

Medium-salt washing buffer for oligo(dT)-cellulose columns

Component and final concentration	Amount to add per 50 ml
150 mM LiCl	0.75 ml of 10 M
10 mM Tris-Cl	250 µl of 2 M (pH 7.5 at room temperature)
1 mM EDTA	100 µl of 0.5 M (pH 8.0)
0.1% SDS	0.5 ml of 10%
DEPC-treated H$_2$O	48.4 ml

Mix the components in a 50-ml polypropylene tube. Store at room temperature for up to 1 month.

9. Elute the poly(A)$^+$ mRNA with 2 ml of elution buffer for oligo(dT)-cellulose columns. Collect the eluate (~2 ml) in a single 2-ml microcentrifuge tube.

Note: As a general rule, one round of chromatography through an oligo(dT)-cellulose column is not sufficient to obtain a population of pure mRNAs. Typically, after one round of poly(A) RNA selection, residual amounts of rRNAs that are still present can be easily detected by electrophoresis on a formaldehyde–agarose gel (see pp. 80–82). Therefore, further purification of the mRNA population on a second oligo(dT)-cellulose column as described in the steps below is highly recommended.

Elution buffer for oligo(dT)-cellulose columns

Component and final concentration	Amount to add per 50 ml
2 mM EDTA	200 µl of 0.5 M (pH 8.0)
0.1% SDS	0.5 ml of 10%
DEPC-treated H$_2$O	49.3 ml

Mix the components in a 50-ml polypropylene tube. Store at room temperature for up to 1 month.

10. Reequilibrate the column with 5 ml of loading buffer.

11. Add 106 µl of 10 M LiCl to the 2 ml of poly(A)+ mRNA eluate (from step 9) and reload on the column. Collect the eluate.

12. Wash the column with 1 ml of loading buffer. Collect the eluate.

13. Combine the eluates from steps 11 and 12 and reload on the column.

14. Wash the column with 2 ml of medium-salt washing buffer.

15. Elute the poly(A)+ RNA with 2 ml of elution buffer. Collect the eluate in four 2-ml microcentrifuge tubes (~0.5 ml/tube).

16. Precipitate the RNA in each tube with 0.1 volume of 3 M sodium acetate (pH 7.0) and 2.5 volumes of absolute ethanol and dry the pellet under vacuum (see p. 71, step 10).

    Notes: If the RNA is not to be used immediately, it can be stored in ethanol at −20°C indefinitely.
    Be sure to use sodium acetate at pH 7.0, not 5.2, in this precipitation.

17. Dissolve the poly(A)+ RNA in DEPC-treated H_2O at a final concentration of approximately 0.5 mg/ml.

    Notes: The yield of poly(A)+ RNA should correspond to approximately 1–5% of the starting amount of total cellular RNA or cytoplasmic RNA.
    To quantitate the poly(A)+ RNA, dilute an aliquot in 1 ml of DEPC-treated H_2O and measure the OD_{260} (see Appendix). A 40-µg/ml solution of single-stranded RNA corresponds to an OD_{260} of 1 in a quartz cuvette with a 1-cm path length. Alternatively, estimate the amount of poly(A)+ RNA by electrophoresis of a small aliquot on a 1% agarose minigel. Include a known amount of an RNA standard (e.g., 100 ng of globin RNA).
    To reuse the aliquot of poly(A)+ RNA used for quantitation, the OD_{260} must be read in cuvettes that have been soaked in **concentrated HCl:methanol** (1:1) for 1 hour and then washed several times in DEPC-treated H_2O.
    Since a great deal of time and energy has been invested in producing the mRNA sample, it is imperative that RNase-free conditions be strictly maintained at this time.

    concentrated HCl, methanol (see Appendix for Caution)

in a glass shop and used in the protocol on pp. 106–107. The column should be calibrated with radiolabeled DNA molecular-weight markers as described on pp. 104–105 before use in size fractionating cDNAs. Size-selected cDNAs are then quantitated (pp. 108–109).

The remaining steps in preparing the cDNA for cloning are: ligating *Eco*RI adapters to size-selected cDNAs (pp. 110–111), digesting the ligated cDNA/ adapter with *Not*I (p. 112), removing excess *Eco*RI adapters (p. 113), and finally phosphorylating the 5′ end of the ligated *Eco*RI adapter (pp. 114–115).

Cloning Ligated cDNA/adapter into the Phagemid Vector

The final stages in constructing the cDNA library are ligation of the cDNA and the prepared phagemid vector DNA, amplification of the cDNA library, and gel purification of the library cDNA to eliminate nonrecombinant clones.

In the ligation on pp. 116–117, the vector DNA is present in only a slight excess over the cDNA and it is not dephosphorylated. This approach is favored here since dephosphorylation usually reduces cloning efficiencies. A vector DNA:cDNA molar ratio of 2:1 seems to be a good compromise.

Existing protocols for the preparation of competent bacteria (Dower et al. 1988) have been modified so that they reproducibly yield cells with very high electroporation efficiencies. These modifications include: (1) optimization of the OD_{600} at which the bacterial culture is harvested, (2) optimization of the conditions and materials used to centrifuge the bacterial culture and to perform the subsequent washes with glycerol, and (3) optimization of the cell density in the final aliquots. A detailed protocol for preparing *E. coli* DH10B (Life Technologies) with electroporation efficiencies of 1×10^{10} to 3×10^{10} cfu/μg of CsCl-purified supercoiled pUC19 control DNA is provided on pp. 118–119.

Three transformations by electroporation are typically performed with these competent bacteria. Each transformation uses 50 ng of cDNA ligated to 200 ng of vector DNA. The transformed cells are then propagated under selection with ampicillin and plasmid DNA is prepared from the culture. This procedure for amplification of the cDNA library is provided on pp. 119–120.

To eliminate completely from the library any background of nonrecombinant clones, the plasmid DNA preparation of the library can be linearized with *Not*I and loaded on an agarose gel. The linear recombinant molecules are purified from the gel by using β-agarase and are recircularized in a large-volume ligation under conditions that promote recircularization instead of intermolecular ligations. The exact reaction conditions to promote recircularization instead of intermolecular ligations can be determined by the following formula (Smith et al. 1987):

$$3.3/(\text{size in kb})^{1/2} \ \mu\text{g/ml}$$

Since the starting amount of plasmid DNA was 0.5 μg in the protocol on pp. 121–122, the average size of an insert is typically 1.5 kb, and the vector is 2.9 kb (2.9-kb vector + 1.5-kb insert = 4.4 kb), the reaction volume should be approximately 320 μl (3.3/[4.4 kb]$^{1/2}$ = 1.57 μg/ml, or 0.5 μg in 318 μl) for the recircularization reaction on pp. 122–123. After recircularization, the DNA is precipitated, competent DH10B bacteria are transformed by electroporation and propagated under selection with an appropriate antibiotic, and plasmid DNA is prepared. This procedure is provided on pp. 121–124.

PROTOCOL

Preparing Phagemid Vector DNA Digested with *Not*I and *Eco*RI

PURIFICATION OF SUPERCOILED pT7T3-Pac DNA BY CENTRIFUGATION ON CsCl–ETHIDIUM BROMIDE DENSITY GRADIENTS

1. Streak a YT agar plate containing ampicillin (75 μg/ml) with a pT7T3-Pac clone (in DH10B bacteria) from a frozen glycerol stock and incubate at 37°C for 16–24 hours.

   Note: The pT7T3-Pac vector can be prepared from the pT7T318D vector (Pharmacia) as described in Bonaldo et al. (1996).

2. Place 1 liter of 2x YT medium containing ampicillin (75 μg/ml) in a 2-liter flask and inoculate with a single colony from the plate in step 1. Incubate at 37°C with agitation at 250 rpm until the culture reaches saturation (usually overnight).

   Note: The OD_{600} of the culture should reach approximately 2 (i.e., 1.2×10^9 cells).

3. Divide the culture into 250- or 500-ml centrifuge bottles. Centrifuge in a Sorvall centrifuge 6000 rpm ($4664g$) at 4°C for 20 minutes to recover the cells. Discard the supernatant.

4. Isolate plasmid DNA (pT7T3-Pac DNA) from the cell pellet by using Qiagen-tip500 columns according to the manufacturer's instructions.

5. Dissolve the DNA in 1 ml of TE (pH 8.0).

   Note: The yield should be 1.5–2 mg of plasmid.

6. Add 1 g of solid ultrapure CsCl for each milliliter of DNA solution and mix by vortexing until all of the CsCl has dissolved.

7. Add 0.8 ml of **ethidium bromide** solution (10 mg/ml in H_2O) for each 10 ml of DNA/CsCl solution and mix thoroughly by vortexing.

   Note: The final density of the solution should be 1.55 g/ml (refractive index of 1.3860), and the concentration of ethidium bromide should be approximately 0.74 mg/ml.

   ethidium bromide (see Appendix for Caution)

8. Transfer the solution into a 35-ml ultracentrifuge tube (DuPont 03989). Fill the tube with either 78% (w/v) CsCl in TE (pH 8.0) or light paraffin oil and then seal the tube (Crimper; DuPont).

9. Centrifuge in either a Sorvall TFT50.38 or a Beckman 50Ti rotor at 40,000 rpm ($144,000g$) at 20°C for 56 hours or at 45,000 rpm ($183,000g$) at 20°C for 48 hours. Alternatively, centrifuge in a Beckman VTi65 rotor at 45,000 rpm ($194,000g$) at 20°C for 16 hours.

Note: When the tubes are removed from the centrifuge, two bands of DNA should be visible in the middle of the gradient in ordinary light. The upper band, which usually contains much less material, consists of nicked circular plasmid DNA and linear bacterial chromosomal DNA. (Note, however, that there should be practically no contaminating bacterial chromosomal DNA since the plasmid DNA was first purified through a Qiagen column in this protocol.) The lower band consists of supercoiled (closed-circular) plasmid DNA. The deep-red pellet on the bottom of the tube consists of ethidium bromide/RNA complexes.

10. Collect the band containing supercoiled plasmid DNA as follows: Insert a 21-gauge needle into the top of the tube to allow air to enter. Insert an 18-gauge needle (bevel side up) attached to a 10-ml syringe into the tube so that the open, beveled side of the needle is positioned just below the band of supercoiled plasmid DNA and parallel to it. Withdraw all of the clear, orange DNA band through the needle into the syringe and then transfer the DNA band into a sterile 10-ml polypropylene screwcap tube.

11. Extract as follows to remove the ethidium bromide from the DNA:

    a. Add 1 volume of **n-butanol** saturated with H_2O or 1 volume of isopropanol saturated with either H_2O or isoamyl alcohol and mix thoroughly.

       H_2O-Saturated n-*butanol*
       In a glass bottle, mix 100 ml of *n*-butanol with 100 ml of H_2O and allow it to stand at room temperature for 1 hour. The hydrated (saturated) *n*-butanol will be the upper phase, whereas the H_2O will be the lower phase. Store at room temperature for up to 6 weeks.

       Isopropanol saturated with H_2O or isoamyl alcohol
       Prepare saturated isopropanol as described above for saturating *n*-butanol.

       n-butanol (see Appendix for Caution)

    b. Centrifuge in a tabletop clinical centrifuge at 1500 rpm (300*g*) at room temperature for 3 minutes to separate the phases.

    c. Transfer the aqueous (lower) phase into a new tube. Decontaminate and discard the organic (upper) phase.

    d. Repeat steps a–c three to six times (or until all of the pink color disappears from both the aqueous and the organic phases).

12. Dialyze against 1 liter of TE (pH 8.0) at 4°C for 24–48 hours with three or four changes of buffer to remove the CsCl from the DNA solution.

    Notes: It is crucial to remove the CsCl from the solution. Otherwise, the CsCl will precipitate with the DNA and obscure the DNA pellet.
    For preparation of dialysis tubing, see Appendix.

13. Transfer the DNA solution into a 13-ml Sarstedt tube (55.518 and 65.816) and precipitate as follows:

    a. Add 0.1 volume of 3 M sodium acetate (pH 7.0) and 2 volumes of ice-cold absolute ethanol and mix thoroughly.

    b. Place at –20°C overnight.

c. Centrifuge in a Sorvall SS34 rotor at 10,000 rpm (12,000*g*) at 4°C for 15 minutes. Carefully remove the supernatant and discard.

Note: It is not necessary to wash the pellet with 70% ethanol.

d. Dry the DNA pellet under vacuum in a SpeedVac Concentrator.

14. Dissolve the purified pT7T3-Pac DNA in 1 ml of TE (pH 8.0). Measure the OD_{260} of the solution to determine the DNA concentration (see Appendix).

Note: The purified pT7T3-Pac DNA can be stored at –20°C for several months.

15. Digest the purified pT7T3-Pac DNA with *Eco*RI as described below.

DIGESTION OF THE PURIFIED pT7T3-Pac DNA WITH *Eco*RI

1. If necessary, adjust the concentration of the pT7T3-Pac DNA (from the previous protocol) to 0.5 mg/ml with TE (pH 8.0). Combine the components below (in the order listed) in a 1.5-ml microcentrifuge tube and mix thoroughly but gently. Centrifuge briefly to collect the liquid at the bottom of the tube.

pT7T3-Pac DNA	20 μl (10 μg)
10x *Eco*RI buffer (supplied with the enzyme)	20 μl
H_2O	158 μl
*Eco*RI (20,000 units/ml; New England Biolabs 101)	2 μl

2. Incubate at 37°C for 2 hours to linearize the DNA.

3. Add 8 μl of 0.5 M EDTA (pH 8.0) to stop the reaction. Mix thoroughly by vortexing. Heat at 65°C for 10 minutes.

4. Extract with 200 μl of **phenol:chloroform**:isoamyl alcohol (25:24:1) (see Appendix). Use phenol saturated with TE (pH 8.0).

phenol, chloroform (see Appendix for Caution)

5. Precipitate the DNA with 22 μl of 3 M sodium acetate (pH 7.0) and 450 μl of absolute ethanol and dry the pellet under vacuum as described for RNA on p. 71, step 10.

6. Dissolve the DNA in 20 μl of TE (pH 8.0) and proceed immediately with the purification of the linearized pT7T3-Pac DNA as described below.

Note: The DNA may religate if it is stored.

GEL PURIFICATION OF THE *Eco*RI-DIGESTED pT7T3-Pac DNA

1. Add 2 μl of 10x SDS/glycerol gel-loading solution to the linearized pT7T3-Pac DNA from the previous protocol. Heat at 65°C for 5 minutes.

2. Load on a regular 1% agarose minigel containing **ethidium bromide** (at a final concentration of 0.5 µg/ml). Also load DNA molecular-weight markers (0.5–10 kb). Perform electrophoresis under standard conditions until the bromophenol blue dye runs off the gel.

   Notes: This much DNA (10 µg) can be loaded in a single well of a minigel with wells that are 13 mm wide.

   Using regular agarose instead of low-melting-temperature agarose in this step provides greater resolution as well as a lower background in the final DNA preparation.

   ethidium bromide (see Appendix for Caution)

3. Locate the linearized DNA by using a long-wavelength **UV** transilluminator, and use a sterile scalpel or razor blade to excise it.

   Note: To avoid unnecessary nicking, be sure not to overexpose the DNA to the UV light in any of the steps in this protocol. As a precaution, place a protective shield between the gel and the UV transilluminator and set the long-wavelength light to the lowest intensity that still allows detection of the DNA.

   UV radiation (see Appendix for Caution)

4. Place the gel slice containing the linearized DNA near the top of an empty minigel mold and pour a 1% solution of low-melting-temperature agarose over it. Allow the low-melting-temperature agarose to solidify at 4°C for 20 minutes. Perform electrophoresis under standard conditions until the DNA has run into the low-melting-temperature gel.

5. Locate and excise the gel slice as described in step 3. Place the gel slice in a 1.5-ml microcentrifuge tube.

6. Heat at 68°C for 10 minutes and measure the volume of the gel slice. Add 1 volume of TE (pH 8.0) and heat at 68°C for an additional 10 minutes to melt the agarose.

7. Add 0.1 volume of 10x β-agarase buffer (supplied with the enzyme) and mix by vortexing briefly. Centrifuge briefly to collect the liquid at the bottom of the tube.

8. Immediately place the tube in a water bath set at 40°C.

9. Add 1 unit of β-agarase (New England Biolabs 392) for each 100 µl of agarose/β-agarase buffer and mix gently. Centrifuge briefly to collect the liquid at the bottom of the tube. Incubate at 40°C for 1–2 hours.

   Note: This step will digest the low-melting-temperature agarose, liberating the linearized DNA into the buffer.

10. Precipitate the DNA as follows:

    a. Add 0.1 volume of 3 M sodium acetate (pH 7.0) and mix. Place on ice for 15 minutes.

    b. Centrifuge in a microcentrifuge at 12,000*g* at room temperature for 15 minutes.

c. Transfer the DNA-containing supernatant into a new tube and discard the pellet.

d. Add 1 volume of isopropanol and mix. Place at –20°C for 30 minutes.

e. Recentrifuge for 20 minutes. Carefully remove the supernatant and discard.

   Note: It is not necessary to wash the pellet with 70% ethanol.

f. Dry the DNA pellet under vacuum in a SpeedVac Concentrator.

11. Dissolve the DNA in 20 μl of TE (pH 8.0) and proceed immediately with the digestion of the purified, linearized pT7T3-Pac with *Not*I as described below.

DIGESTION OF THE GEL-PURIFIED *Eco*RI-DIGESTED pT7T3-Pac DNA WITH *Not*I

1. Combine the components below in the order listed in a 1.5-ml microcentrifuge tube and mix thoroughly but gently. Centrifuge briefly to collect the liquid at the bottom of the tube.

pT7T3-Pac DNA (from the previous protocol)	20 μl
10x NEBuffer 3 (New England Biolabs)	20 μl
BSA (10 mg/ml; New England Biolabs)	2 μl
H$_2$O	156 μl
*Not*I (10,000 units/ml; New England Biolabs 189)	2 μl

2. Incubate at 37°C for 2 hours.

3. Follow steps 3–6 on p. 92.

   Note: If the DNA is not to be used immediately, it can be stored in ethanol at –20°C indefinitely.

4. Digest the *Eco*RI- and *Not*I-digested pT7T3-Pac DNA with *Bam*HI as described below.

DIGESTION OF THE *Eco*RI- AND *Not*I-DIGESTED pT7T3-Pac DNA WITH *Bam*HI

1. Combine the components below in the order listed in a 1.5-ml microcentrifuge tube and mix thoroughly but gently. Centrifuge briefly to collect the liquid at the bottom of the tube.

pT7T3-Pac DNA (from the previous protocol)	20 µl
10x NEBuffer *Bam*HI (supplied with the enzyme)	20 µl
BSA (10 mg/ml; New England Biolabs)	2 µl
H₂O	156 µl
*Bam*HI (20,000 units/ml; New England Biolabs 136)	2 µl

Note: Digestion with *Bam*HI will prevent recircularization (which could occur during the subsequent ligation of the vector DNA to the cDNA) of any *Eco*RI-digested molecule that failed to be digested with *Not*I.

2. Incubate at 37°C for 2 hours.

3. Follow steps 3–5 on p. 92.

   Note: If the DNA is not to be used immediately, it can be stored in ethanol at −20°C indefinitely.

4. Dissolve the DNA in 16.4 µl of TE (pH 8.0).

5. Purify the digested pT7T3-Pac DNA on a Sepharose CL-4B column as described below.

CHROMATOGRAPHY ON SEPHAROSE CL-4B COLUMNS FOR PURIFICATION OF THE 2.9-KB DIGESTED VECTOR DNA

1. Prepare a Sepharose CL-4B column as follows:

   a. Combine 10 ml of Sepharose CL-4B (Pharmacia 17-0150-01) and 20 ml of 400 mM NaCl in TE (pH 8.0).

      Note: It is advisable to combine the beads and buffer and then incubate at 37°C for approximately 1 hour. This helps degas the suspension and prevent bubble formation once the column is packed for running at room temperature.

      400 mM NaCl in TE (pH 8.0)

Component and final concentration	Amount to add per 500 ml
400 mM NaCl	40 ml of 5 M
10 mM Tris-Cl	5 ml of 1 M (pH 8.0 at 25°C)
1 mM EDTA	1 ml of 0.5 M (pH 8.0)
H₂O	454 ml

      Store at room temperature for up to 6 months.

   b. Pack the suspension to a 30-cm height in a sterile 5-ml disposable plastic pipette with glass wool at the bottom.

   c. Equilibrate the column with at least two column volumes of 400 mM NaCl in TE (pH 8.0).

Note: Load 10 μl of 10x SDS/glycerol gel-loading solution for the equilibration. The presence of the dye allows detection of the movement of the buffer through the column.

2. Add 1.6 μl of 5 M NaCl and 2 μl of 10x SDS/glycerol gel-loading solution to the digested pT7T3-Pac DNA (from the previous protocol) and mix thoroughly. Heat at 65°C for 5 minutes. Centrifuge briefly to collect the liquid at the bottom of the tube.

3. Load the sample on the column.

4. Immediately begin collecting 0.5-ml aliquots in 12 1.5-ml microcentrifuge tubes.

5. Analyze a 10-μl sample of each fraction along with the appropriate molecular-weight markers (e.g., bacteriophage λ DNA digested with *Hin*dIII) on a 1% agarose minigel to identify the fractions containing the 2.9-kb pT7T3-Pac DNA. Keep the remainder of each fraction on ice while the gel is running.

   Note: The vector DNA typically appears in fractions 6–8.

6. Precipitate the DNA in each fraction containing the vector with 1 ml of absolute ethanol and dry the pellet under vacuum as described for RNA on p. 71, step 10.

   Note: In this step, no additional salt is needed for the precipitation.

7. Dissolve each DNA pellet in 10 μl of TE (pH 8.0). Heat at 65°C for 5 minutes. Centrifuge briefly to collect the liquid at the bottom of the tube. Combine the dissolved pellets.

8. Estimate the final DNA concentration from the fluorescence intensity of the bound **ethidium bromide** by analyzing 0.5 μl of the DNA on a 1% agarose minigel along with a known amount of a DNA standard (see Appendix).

   ethidium bromide (see Appendix for Caution)

9. Check the quality of the vector preparation in control ligations and then store at –20°C until needed for ligation to the cDNA/adapter as described on pp. 116–117.

   Note: Two control ligations should be performed using the protocol on pp. 116–117. The first should contain vector DNA but no insert DNA to assess background levels. The second should contain vector DNA and a test insert to assess the cloning efficiency of a particular batch of vector. Typically, a total of 10^6 recombinant clones can be obtained from 25 ng of a 1.5-kb test insert (generated by digestion with *Not*I plus *Eco*RI) in a ligation containing 80 ng of 2.9-kb vector DNA, with a background level that is as low as 10^4 nonrecombinants per microgram of vector. Any cDNA clone from a library that is directionally cloned using *Not*I plus *Eco*RI can be used as source of a test insert for the control ligation.

PROTOCOL

First-strand Synthesis

"One-tube" first- and second-strand cDNA syntheses are performed essentially as described by D'Alessio et al. (1987). Be sure to observe the special considerations for working with RNA throughout this procedure (see pp. 67–68). Carry out reactions in silanized microcentrifuge tubes.

1. Carefully pipette the components below (in the order listed) into a 1.5-ml microcentrifuge tube (labeled A) and mix. Incubate at 37°C for 5–10 minutes (until needed for step 4). Centrifuge briefly to collect the liquid at the bottom of the tube.

5x first-strand cDNA buffer	4 µl
100 mM DTT	2 µl
10 mM dNTP mixture (prepared in **DEPC**-treated H_2O [see Appendix])	1 µl
*Not*I-(dT)$_{18}$ oligonucleotide primer (1 µg/µl solution in H_2O)	1 µl

    Notes: The *Not*I-(dT)$_{18}$ oligonucleotide primer is

    5'TGTTACCAATCTGAAGTGGGA*GCGGCCGC*ACAA(T)$_{18}$3'.

    The 21 nucleotides before the *Not*I site (in italics) increase the distance between the *Not*I site and the end of the molecule and thereby improve the efficiency of the digestion of the *Not*I at a later stage. The excess sequence also protects the *Not*I site from excessive trimming during the treatment with bacteriophage T4 DNA polymerase. The nucleotides (ACAA) present between the *Not*I site and the oligo(dT) track serve as a library tag.

    The final reaction volume will be 20 µl once tubes A, B, and C are combined.

    5x First-strand cDNA buffer

Component and final concentration	Amount to add per 1 ml
250 mM Tris-Cl	250 µl of 1 M (pH 8.3 at room temperature)
375 mM KCl	375 µl of 1 M
15 mM $MgCl_2$	15 µl of 1 M
DEPC-treated H_2O	360 µl

    Mix the components in a 1.5-ml microcentrifuge tube. Divide into aliquots and store at –20°C indefinitely. (The 5x buffer supplied with the reverse transcriptase can be used instead of this buffer.)

    DEPC (see Appendix for Caution)

2. In a second 1.5-ml microcentrifuge tube (labeled B), combine 1 µl (0.5 µg) of poly(A)$^+$ RNA (from pp. 83–85) and 9 µl of DEPC-treated H_2O. Incubate at 65°C for 5 minutes. Centrifuge briefly to collect the liquid at the bottom of the tube. Transfer the tube into a water bath set at 37°C, and add 1 µl

(40 units) of RNase inhibitor (Boehringer Mannheim 799017). Keep the tube at 37°C until needed in step 4.

3. To a third tube (labeled C), add 1 μl (200 units) of RNase H⁻ reverse transcriptase (Superscript II; Life Technologies 18064-014). Warm at 37°C for 2–5 minutes.

4. Pipette the mixture from tube A to tube B and mix gently by pipetting up and down. Centrifuge briefly to collect the liquid at the bottom of the tube. Incubate at 37°C for 1–2 minutes.

5. Transfer the prewarmed reverse transcriptase from tube C to tube B and mix gently by pipetting up and down. Immediately incubate at 37°C for 2 hours.

6. Place the first-strand cDNA (tube B) on ice and immediately proceed with second-strand synthesis on pp. 101–102.

   Notes: When it is necessary to assess the efficiency of first-strand synthesis (see below), proceed with the synthesis of the second strand of cDNA during the assessment.

   The second-strand synthesis is set up to use 19.5 μl of first-strand cDNA. It is not necessary to assess the efficiency of first-strand synthesis. All 20 μl of first-strand cDNA can be used but the volumes of the components used in the second-strand synthesis will have to be adjusted accordingly.

Comments

■ This protocol reproducibly yields first-strand cDNA of high quality. It should therefore *not* be necessary to assess the quality of the first-strand cDNA after every synthesis. However, assessment is recommended when this protocol is first implemented and when the quality of the final double-stranded cDNA will seem suboptimal (i.e., <50% conversion to cDNA). To assess the quality and to estimate the average length of the first-strand cDNA, the synthesis should be performed in the presence of trace amounts of [α-³²P]dCTP (i.e., include 1 μl of [α-³²P]dCTP [10 μCi/μl, 800 Ci/mmole; DuPont NEN Research Products] in the reaction in step 1 and decrease the volume of DEPC-treated H₂O to 8 μl in step 2 above). A 0.5-μl aliquot of the synthesized first-strand cDNA can then be analyzed on a denaturing alkaline agarose gel (McDonell et al. 1977) as described below. Note, however, that the calculations on pp. 108–109 for determining the final nanogram amounts of double-stranded cDNA are based on the assumption that [α-³²P]dCTP is present only during second-strand synthesis.

radioactive substances (see Appendix for Caution)

1. Prepare and, if necessary, seal the ends of the minigel mold according to the manufacturer's instructions.

2. Place 0.5 g of electrophoresis-grade agarose in 50 ml of H₂O in a 250-ml flask. Place in a microwave oven and boil to dissolve the agarose.

3. Allow to cool at room temperature to 60–65°C.

4. Add 250 µl of 10 N **NaOH** and 100 µl of 0.5 M EDTA (pH 8.0) (to make final concentrations of 0.05 N NaOH and 1 mM EDTA) and mix thoroughly by swirling the contents of the flask. Pour the mixture into the minigel mold.

   NaOH (see Appendix for Caution)

5. Position the desired comb to cast the wells. Allow the gel to solidify at room temperature for 30 minutes.

6. Gently transfer the solidified gel into an electrophoresis chamber and submerge it in 1x alkaline agarose electrophoresis running buffer.

   Note: Ethidium bromide is omitted from alkaline agarose gels because it does not bind to DNA at high pH.

   1x *Alkaline agarose electrophoresis running buffer*

Component and final concentration	Amount to add per 250 ml
0.05 N NaOH	1.25 ml of 10 N
1 mM EDTA	0.5 ml of 0.5 M (pH 8.0)
H₂O	248.25 ml

   Mix the components in a 0.5-liter bottle. Prepare this buffer just before use.

7. Precipitate the first-strand cDNA as follows:

   a. Add 2 µl of 3 M sodium acetate (pH 7.0) and 17.5 µl of H₂O to 0.5 µl of first-strand cDNA and mix thoroughly.

   b. Add 50 µl of ice-cold absolute ethanol and mix.

   c. Place in a dry-ice/ethanol bath for 30 minutes.

   d. Centrifuge in a microcentrifuge at 12,000g at 4°C for 20 minutes. Carefully remove the supernatant and discard.

      Note: It is not necessary to wash the pellet with 70% ethanol.

   e. Dry the first-strand cDNA pellet under vacuum in a SpeedVac Concentrator.

8. Dissolve the first-strand cDNA in 10 µl of 1x alkaline agarose electrophoresis running buffer. Heat at 65°C for 10 minutes, add 2 µl of 6x alkaline gel-loading solution, and load the first-strand cDNA sample in a well of the gel. Also combine 2 µl of 6x alkaline gel-loading solution and 12 µl of [α-³²P]dCTP-end-labeled *Hin*dIII-digested bacteriophage λ DNA and load in a separate well.

   Note: *Hin*dIII-digested bacteriophage λ DNA is available from New England Biolabs. It can be end-labeled as described in Ausubel et al. (1995, p. 3.5.7).

9. Perform electrophoresis at less than 0.25 V/cm until the bromocresol green dye has migrated approximately two thirds of the length of the gel.

   Note: Alkaline gels draw more current than neutral gels at comparable voltages. Standard minigels are run at 40 V for approximately 4 hours.

10. Remove the gel from the electrophoresis chamber and soak it in 7% **TCA** at room temperature for 30 minutes.

    TCA (see Appendix for Caution)

11. Place the gel on a glass plate, cover it with layers of paper towels and another glass plate, and dry for several hours. Alternatively, place the gel on a gel dryer, cover with a stack of blotting paper, and dry under vacuum for a few hours.

12. Autoradiograph the gel at room temperature or at −80°C overnight with an intensifying screen.

PROTOCOL

Second-strand Synthesis

"One-tube" first- and second-strand cDNA syntheses are performed essentially as described by D'Alessio et al. (1987). Carry out reactions in silanized microcentrifuge tubes.

1. Carefully pipette the components below (in the order listed) into a 1.5-ml microcentrifuge tube (labeled D) on ice. Mix each by pipetting up and down.

H_2O	87.5 µl
5x second-strand cDNA buffer	32 µl
7.5 mM β-NAD	3.2 µl
10 mM dNTP mixture (see Appendix)	3 µl
100 mM DTT (supplied with Superscript II)	6 µl
[α-^{32}P]dCTP (10 µCi/µl, 800 Ci/mmole; DuPont NEN Research Products)	1 µl
E. coli DNA ligase (6 units/µl; New England Biolabs 205)	2.5 µl
E. coli DNA polymerase I (10 units/µl; New England Biolabs 209)	4 µl
DNase-free RNase H (1.1 unit/µl; Pharmacia 27-0894)	1.3 µl

Note: The final reaction volume will be 160 µl once tubes B and D are combined.

5x Second-strand cDNA buffer

Component and final concentration	Amount to add per 1 ml
94 mM Tris-Cl	94 µl of 1 M (pH 6.9 at room temperature)
453 mM KCl	453 µl of 1 M
23 mM $MgCl_2$	23 µl of 1 M
50 mM ammonium sulfate	50 µl of 1 M
H_2O	380 µl

Mix the components in a 1.5-ml microcentrifuge tube. Divide into aliquots and store at −20°C indefinitely.

radioactive substances (see Appendix for Caution)

2. Transfer the ice-cold mixture from tube D to tube B (from p. 98, step 6) and mix gently by pipetting up and down. Centrifuge briefly to collect the liquid at the bottom of the tube.

3. Incubate at 16°C for 2 hours and then at room temperature for 30 minutes.

   Note: A constant temperature of 16°C is best achieved in a refrigerating water bath or in a water bath placed in the coldroom.

4. Immediately treat the cDNA in tube B to create blunt ends as described on p. 103.

PROTOCOL

Creating Blunt-ended Double-stranded cDNA

1. Pipette 1 μl of bacteriophage T4 DNA polymerase (3 units/μl; New England Biolabs 203) into tube B (from p. 102, step 4) and mix by gently tapping the tube. Incubate at room temperature for 10 minutes.

2. Add 6.5 μl of 0.5 M EDTA (pH 8.0) and mix thoroughly by vortexing. Centrifuge briefly to collect the liquid at the bottom of the tube. Incubate at 65ºC for 10 minutes.

3. Add 282.5 μl of TE (pH 8.0) and mix thoroughly by vortexing. Centrifuge briefly to collect the liquid at the bottom of the tube. Set aside 0.5 μl for determining the total cpm of [α-^{32}P]dCTP in the cDNA synthesis reaction (see p. 108, step 2).

   radioactive substances (see Appendix for Caution)

4. Extract the remaining cDNA with 450 μl of **phenol:chloroform**:isoamyl alcohol (25:24:1) (see Appendix). Use phenol saturated with TE (pH 8.0).

   phenol, chloroform (see Appendix for Caution)

5. Precipitate the cDNA in the aqueous phase as follows:

   a. Add 50 μl of 3 M sodium acetate (pH 7.0) and mix thoroughly.

   b. Add 1 ml of ice-cold absolute ethanol and mix by vigorous vortexing. Centrifuge briefly to collect the liquid at the bottom of the tube.

   c. Place at –20ºC overnight or in a dry-ice/ethanol bath for 30 minutes.

      Note: If the cDNA is not to be used immediately, it can be stored in ethanol at –20ºC indefinitely.

6. Size fractionate the cDNA as described on pp. 106–107 on a column calibrated as described on pp. 104–105.

PROTOCOL

Calibrating the Bio-Gel A-50m Column Used for Size Selection of cDNAs

This protocol is essentially that of Huynh et al. (1985). The column should be run at room temperature.

1. Prepare the column as follows:

   a. For maximum resolution, pack Bio-Gel A-50m agarose beads (100–200 mesh; Bio-Rad) in a piece of glass tubing 64 cm long with an internal diameter of 0.2 cm.

   Note: Bio-Gel A-50m beads can be packed in a standard 1-ml disposable pipette and equilibrated with 2 column volumes of 400 mM NaCl in TE (pH 8.0). This alternative provides a lower yet acceptable resolution.

   b. Equilibrate the column with 10 ml of 400 mM NaCl in TE (pH 8.0) (for preparation, see p. 95).

   Note: For convenience, connect the column directly to an upper reservoir containing the equilibration buffer and allow the column to run continuously at room temperature.

2. Prepare radiolabeled DNA molecular-weight markers for use in calibration as follows:

   a. Prepare 0.5–1 μg of a mixture containing equal parts of end-labeled pBR322 DNA digested with *Msp*I (9–622 bp; New England Biolabs 303-2) and end-labeled bacteriophage λ DNA digested with *Bst*EII (0.12–8.45 kb; New England Biolabs 301-4). Precipitate the mixture with ethanol and dry the pellet under vacuum (see Appendix).

   Note: [α-32P]dCTP and the Klenow fragment of *E. coli* DNA polymerase I can be used to end-label the pBR322 DNA and bacteriophage λ DNA (see Ausubel et al. 1995, p. 3.5.7).

   radioactive substances (see Appendix for Caution)

   b. Dissolve the DNA in 8.2 μl of TE (pH 8.0). Heat at 65°C for 5 minutes. Centrifuge briefly to collect the liquid at the bottom of the tube.

   c. Add 0.8 μl of 5 M NaCl and 1 μl of 10x SDS/glycerol gel-loading solution to the DNA and mix thoroughly by vortexing. Centrifuge briefly to collect the liquid at the bottom of the tube.

3. Load the 10-μl marker sample on the equilibrated column.

4. Immediately begin collecting 5-drop fractions (1 drop ≈ 16 μl; 1 drop can be collected in ~6 minutes) in 1.5-ml microcentrifuge tubes for a total of 20 fractions.

5. Precipitate the DNA in each fraction with 2 volumes of absolute ethanol and dry the pellet under vacuum as described for RNA on p. 71, step 10.

   Note: In this step, no additional salt is needed for the precipitation.

6. Dissolve each DNA pellet in 10 μl of TE (pH 8.0). Heat at 65°C for 5 minutes. Centrifuge briefly to collect the liquid at the bottom of each tube.

7. Add 1 μl of 10x SDS/glycerol gel-loading solution to each tube.

8. Load the samples along with 0.25 μg of a mixture of the same end-labeled DNA molecular-weight markers used in step 2a and analyze on a 1% agarose minigel in 1x TAE buffer. Perform electrophoresis at 80 V for approximately 2 hours or until the bromophenol blue dye has migrated approximately two thirds of the length of the gel.

9. Remove the gel from the electrophoresis chamber and soak it in 7% **TCA** at room temperature for 30 minutes.

   TCA (see Appendix for Caution)

10. Place the gel on a glass plate, cover it with layers of paper towels and another glass plate, and dry for several hours. Alternatively, place the gel on a gel dryer, cover with a stack of blotting paper, and dry under vacuum for a few hours.

11. Autoradiograph the gel at room temperature or at −80°C overnight with an intensifying screen.

    Note: This autoradiograph will provide information on which fraction each marker will appear. This in turn will provide an indication of where cDNAs larger than 350 bp can be expected to appear during analysis of experimental samples.

Comments

▪ If the column is only going to be used for a couple of libraries, it may be simpler not to calibrate the column. In such cases, analyze an aliquot (5%) of each of the cDNA fractions on a 1% agarose gel as described above.

PROTOCOL

Size Selection of Double-stranded cDNAs

1. Centrifuge the precipitated cDNA (from p. 103) in a microcentrifuge at 12,000*g* at 4°C for 20 minutes. Carefully remove all of the supernatant and discard.

2. Dry the cDNA pellet under vacuum in a SpeedVac Concentrator.

3. Dissolve the cDNA in 8.2 μl of TE (pH 8.0). Heat at 65°C for 5 minutes. Centrifuge briefly to collect the liquid at the bottom of the tube.

4. Add 0.8 μl of 5 M NaCl and 1 μl of 10x SDS/glycerol gel-loading solution and mix thoroughly. Centrifuge briefly to collect the liquid at the bottom of the tube.

5. Load on a Bio-Gel A-50m column equilibrated in 400 mM NaCl in TE (pH 8.0) at room temperature and precalibrated with DNA molecular-weight markers (see pp. 104–105).

   Note: A simpler but somewhat satisfactory alternative to size selecting the cDNAs on the 64-cm column is to run them through a Bio-Gel A-50m column prepared in a standard 1-ml pipette equilibrated in 400 mM NaCl in TE (pH 8.0). Even though the resolution of such small columns is not comparable to that of either the 32-cm or the 64-cm column, it is nonetheless still superior to the resolution of some of the spin columns that are used in commercially available kits for construction of cDNA libraries.

6. Immediately begin collecting 100-μl fractions in 1.5-ml microcentrifuge tubes for a total of 20 fractions.

7. Using the gel from the column calibration (pp. 104–105) as a guide, determine which fractions are expected to contain cDNAs at least 350 bp in length. Heat a 5-μl aliquot of each of these fractions at 65°C for 10 minutes, add 4 μl of TE (pH 8.0) and 1 μl of 10x SDS/glycerol gel-loading solution to each, and load on a 1% agarose gel. Analyze along with **radiolabeled DNA molecular-weight markers** as described on p. 105, steps 8–11, to verify their length and to estimate their overall average size. Keep the remainder of each fraction on ice while the gel is running.

   radioactive substances (see Appendix for Caution)

8. Combine the fractions with cDNAs at least 350 bp in length in a 2-ml microcentrifuge tube and mix thoroughly by vortexing. Set aside an aliquot (1–5 μl [use ≤1%]) in a silanized 1.5-ml microcentrifuge tube to determine the total cpm from the size-selected cDNA mixture (see pp. 108–109, step 4).

Notes: The combined volume is usually approximately 0.5 ml.

Typically, cDNAs with the size range of interest (i.e., >350 bp) will appear in fractions 9 through 14 (total volume ~600 μl).

9. Precipitate the remaining combined cDNAs with 2 volumes of absolute ethanol and dry the pellet under vacuum as described for RNA on p. 71, step 10.

   Notes: In this step, no additional salt is needed for the precipitation.

   If the cDNA is not to be used immediately, it can be stored in ethanol at –20°C indefinitely.

10. After calculating the amount of size-selected cDNA (pp. 108–109), ligate the cDNA to *Eco*RI adapters (pp. 110–111).

Comments

■ A different approach for size selection of cDNAs is to fractionate them in and recover them from a low-melting-temperature agarose gel (Sealey and Southern 1983). Although this method is more convenient than using a column, it is far less effective and there is a considerable risk of the cDNA fractions being contaminated with small molecules. These small molecules consist primarily of small fragments of double-stranded cDNA representing exclusively poly(A) tails of mRNAs or adapter molecules (if the size fractionation is being performed to remove excess adapters after the adapters are ligated to the blunt-ended cDNAs). These adapter molecules are present in vast molar excess and may smear backward in the gel into the higher-molecular-weight fractions (see Huynh et al. 1985).

PROTOCOL

Calculating the Amount of Size-selected cDNAs

1. Determine the total amount of dNTPs present in the synthesis reaction as follows: Each dNTP is present at a final concentration of 250 µM in the second-strand synthesis reaction mixture (4 µl of the 10 mM dNTP mixture in a final reaction volume of 160 µl). Accordingly, the final concentration of all four dNTPs is 1 mM (250 µM x 4), and the total nanomoles of dNTPs (*A*) present in the reaction mixture is 1.6 x 10² nmoles (since 1 mM is 10⁶ nmoles/liter, 1.6 x 10² nmoles is in 160 µl of reaction mixture).

   Note: Because the molar contribution of the radioactive dNTP is so negligible, it does not need to be accounted for in this calculation (1 µl of [α-³²P]dCTP [10 µCi/µl; 800 Ci/mmole] in 160 µl of reaction mixture results in a final concentration of 0.08 x 10⁻³ mM).

2. Determine the total cpm of **[α-³²P]dCTP** (incorporated plus unincorporated) present in the cDNA synthesis reaction as follows:

   a. Spot the 0.5-µl aliquot set aside on p. 103, step 3 on the center of a Whatman GF/C glass-fiber filter (2.4 cm in diameter). Allow to dry at room temperature or under a heat lamp.

   radioactive substances (see Appendix for Caution)

   b. Place the dried filter in a vial containing 3 ml of a **toluene**-based scintillation fluid, and measure the radioactivity in a liquid scintillation counter.

   toluene (see Appendix for Caution)

   c. Calculate the total cpm of [α-³²P]dCTP as follows: total cpm of [α-³²P] dCTP (incorporated plus unincorporated) in the final reaction volume of 450 µl (*B*) = (cpm from the 0.5-µl aliquot x 450 µl)/0.5 µl.

3. Calculate the specific activity of the synthesized cDNAs as follows: specific activity (in cpm/nmole of dNTPs) (*C*) = total cpm of [α-³²P]dCTP (*B*)/total nmole of dNTPs present in the reaction mixture (*A*).

4. Determine the total cpm from the size-selected cDNAs.

   Note: To measure accurately the proportion of the radioactive precursor that has been incorporated into cDNAs, precipitation with TCA should be performed. Under the conditions specified here, cDNA molecules longer than 50 bp will be precipitated by TCA, whereas unincorporated dNTPs will remain in solution.

   a. Add 1–10 µg of a nucleic acid carrier (e.g., sheared salmon sperm DNA) to the 1–5-µl aliquot set aside in a silanized 1.5-ml microcentrifuge tube on p. 106, step 8. Adjust the volume to 50 µl with TE (pH 8.0).

   b. Add 50 µl of 20% **TCA** and mix thoroughly. Place on ice for 5–10 minutes.

   TCA (see Appendix for Caution)

c. Spot the total volume (100 µl) on the center of a Whatman GF/C glass-fiber filter (2.4 cm in diameter) and place it on a commercially available vacuum-driven filtration apparatus. Wash the filter with 5 ml of 5% TCA and then with 1 ml of absolute ethanol.

d. Allow the filter to dry at room temperature or under a heat lamp.

e. Place the dried filter in a vial containing 3 ml of a toluene-based scintillation fluid. Measure the radioactivity (the total cpm of $[\alpha\text{-}^{32}P]dCTP$ that was incorporated during the cDNA synthesis reaction) in a liquid scintillation counter.

f. Calculate the total cpm from the size-selected cDNAs as follows: total cpm from the size-selected cDNAs (D) = (cpm from the aliquot x total volume of the combined fractions)/volume of the aliquot.

5. Determine the total mass of size-selected double-stranded cDNA as follows: The total nanomoles of dNTPs incorporated during cDNA synthesis (E) = total cpm from the size-selected cDNAs (D)/specific activity (C). Given that the molecular mass of a dNTP is 330 daltons, the total mass of size-selected double-stranded cDNAs (in ng) (F) = total nmoles of dTNPs (E) x 330 ng/nmole x 2 (because it is double stranded).

PROTOCOL

Ligating *Eco*RI Adapters to Size-selected cDNAs

1. Dissolve the size-selected cDNA (from pp. 106–107) in TE (pH 8.0) at a final concentration of 100 ng/μl. Heat at 65°C for 5 minutes. Centrifuge briefly to collect the liquid at the bottom of the tube.

   Note: There should be 0.5–0.7 μg of cDNA at this point (for calculation, see pp. 108–109).

2. Combine the components below in the order listed in a 1.5-ml microcentrifuge tube and mix gently to prepare a 10-μl ligation mixture. Centrifuge briefly to collect the liquid at the bottom of the tube.

cDNA (100 ng/μl)	2.5 μl (0.5 pmoles of ends)
*Eco*RI adapter (370 ng/μl; Pharmacia 27-7805-01)	5 μl (~250 pmoles)
10x bacteriophage T4 DNA ligase buffer (supplied with the enzyme)	1 μl
H₂O	0.5 μl
10x bacteriophage T4 DNA ligase (400 units/μl; New England Biolabs 202)	1 μl

   Notes: The *Eco*RI adapter is:

   5′-OH AATTCGGCACGAGG 3′-OH
   3′-OH GCCGTGCTCC 5′-PO₄

   Note that only one of the two strands of the adapter molecule is phosphorylated at the 5′ end to avoid formation of long concatemers during ligation (only dimers can be generated).

   As a general rule, adapter molecules should be present at a vast molar excess (~500-fold) over the cDNAs ends in this ligation. The total number of picomoles of cDNA ends can be determined on the basis of the estimated average length of the cDNAs (see p. 106, step 7) and the calculated total mass (see p. 109, step 5). Typically, the average length of the size-fractionated cDNAs is 1.5 kb and only 250 ng of it is used for ligation. Since 1 pmole of a 1.5-kb double-stranded cDNA ≈ 1500 x 660 pg or 990 ng, the number of picomoles represented in 250 ng of a 1.5-kb cDNA = 250 ng x 1 pmole/990 ng = 0.25 pmole of cDNA or 0.25 x 2 = 0.5 pmole of cDNA ends. For the *Eco*RI adapter, 1 pmole ≈ 7920 pg. Therefore, this ligation includes approximately 250 pmoles of adapter molecules and 0.25 pmole of cDNA (or 0.5 pmole of cDNA ends).

   Using bacteriophage T4 DNA ligase from different suppliers can dramatically affect the results.

3. Incubate at 16°C for 12–18 hours.

4. Add 1 μl of 0.5 M EDTA (pH 8.0) to stop the ligation. Mix thoroughly by vortexing. Centrifuge briefly to collect the liquid at the bottom of the tube. Heat at 65°C for 10 minutes.

5. Add 80 µl of TE (pH 8.0) and mix thoroughly.

6. Extract with 1 volume of **phenol:chloroform**:isoamyl alcohol (25:24:1) (see Appendix). Use phenol saturated with TE (pH 8.0).

   phenol, chloroform (see Appendix for Caution)

7. Precipitate the cDNA/adapter in the aqueous phase with 10 µl of 3 M sodium acetate (pH 7.0) and 200 µl of absolute ethanol and dry the pellet under vacuum as described for RNA on p. 71, step 10.

   Note: The cDNA/adapter can be stored in ethanol at –20°C indefinitely, but it is preferable to use it immediately.

8. Dissolve the cDNA/adapter in 10 µl of TE (pH 8.0).

9. Digest with *Not*I as described on p. 112.

PROTOCOL

Digesting Ligated cDNA/Adapter with *NotI*

1. Add the components below (in the order listed) to the 10 μl of cDNA/ adapter (from pp. 110–111) and mix thoroughly but gently. Centrifuge briefly to collect the liquid at the bottom of the tube.

10x NEBuffer 3 (New England Biolabs)	10 μl
H$_2$O	77 μl
BSA (10 mg/ml; New England Biolabs)	1 μl
NotI (10 units/μl; New England Biolabs 189)	2 μl

2. Incubate at 37°C for 4 hours.

   Note: Because the *NotI* site is so close to the 3′ end of the cDNAs, a prolonged incubation with *NotI* is absolutely necessary to ensure complete digestion. When this digestion is performed for only 1 hour, a significant fraction (10–35%) of the clones in the final library will not have the expected poly(A) tail at the 3′ end. Such clones, which are derived from mRNAs with an internal *NotI* site (and therefore should occur at very low frequencies in any library), become overrepresented. This overrepresentation results from the fact that most cDNA molecules (which only have a *NotI* site at their 3′ end) are rendered unclonable because they failed to be digested with *NotI* (or the *NotI* site was destroyed during the reactions, e.g., by the 5′→3′ exonucleolytic activity of *E. coli* DNA polymerase I or alternatively, the polymerase failed to make a full-length second-strand and as a result the *NotI* site may be lost) and therefore do not have a sticky end. However, if digestion with *NotI* is performed for 4 hours as recommended here, the frequency of clones without a tail in the library is lower (~10%), although still higher than expected.

3. Add 4 μl of 0.5 M EDTA (pH 8.0) to stop the digestion. Mix thoroughly by vortexing. Centrifuge briefly to collect the liquid at the bottom of the tube. Heat at 65°C for 10 minutes.

4. Extract with 1 volume of **phenol:chloroform**:isoamyl alcohol (25:24:1) (see Appendix). Use phenol saturated with TE (pH 8.0).

   phenol, chloroform (see Appendix for Caution)

5. Precipitate the digested cDNA/adapter in the aqueous phase with 10 μl of 3 M sodium acetate (pH 7.0) and 200 μl of absolute ethanol and dry the pellet under vacuum as described for RNA on p. 71, step 10.

6. Immediately proceed with the removal of the excess *Eco*I adapters as described on p. 113.

PROTOCOL

Removing Excess *Eco*RI Adapters

1. Prepare a Sepharose CL-4B column as described on pp. 95–96, step 1.

2. Add 0.8 μl of 5 M NaCl and 1 μl of 10x SDS/glycerol gel-loading solution and mix thoroughly. Centrifuge briefly to collect the liquid at the bottom of the tube.

3. Dissolve the digested cDNA/adapter (from p. 112) in 8.2 μl of TE (pH 8.0). Heat at 65°C for 5 minutes. Centrifuge briefly to collect the liquid at the bottom of the tube.

4. Load the sample on the column.

5. Immediately begin collecting 0.5-ml aliquots in 12 1.5-ml microcentrifuge tubes. For each fraction, combine a 10-μl aliquot of the fraction with 3 ml of a **toluene**-based scintillation fluid and measure the radioactivity (the total cpm of ^{32}P) in a liquid scintillation counter.

   Notes: There is no need to precipitate the DNA and count unincorporated radioactivity at this point.
   Determination of the cpm of ^{32}P allows identification of the cDNA-containing fractions and accurate quantitation of the amount of cDNA. Typically, the cDNA fragments appear in fractions 7 and 8.
   The excess adapters are not retained on the column. They should run straight through.

   toluene (see Appendix for Caution)

6. Precipitate the ligated cDNA/adapter in each cDNA-containing fraction with 1 ml of absolute ethanol and dry the pellet under vacuum as described for RNA on p. 71, step 10.

   Notes: In this step, no additional salt is needed for the precipitation.
   If the cDNA/adapter is not to be used immediately, it can be stored in ethanol at –20°C indefinitely.

7. Dissolve each cDNA/adapter pellet and combine. Use a total volume of 7.5 μl of TE (pH 8.0) for dissolving all of the pellets.

8. Phosphorylate the 5′ end of the ligated *Eco*I adapter as described on pp. 114–115.

PROTOCOL

Phosphorylating the 5′End of the Ligated *Eco*RI Adapter

1. Combine the components below (in the order listed) and mix gently. Centrifuge briefly to collect the liquid at the bottom of the tube.

purified cDNA/adapter (from p. 113)	7.5 μl
10x bacteriophage T4 polynucleotide kinase buffer (supplied with the enzyme)	1 μl
10 mM ATP (Boehringer Mannheim; Pharmacia)	1 μl
bacteriophage T4 polynucleotide kinase (10 units/μl; New England Biolabs 201)	0.5 μl

2. Incubate at 37°C for 30 minutes.

3. Digest with proteinase K as follows:

   a. Add the components below (in the order listed) and mix gently. Centrifuge briefly to collect the liquid at the bottom of the tube.

10x proteinase K buffer	2 μl
H_2O	7 μl
proteinase K (1 mg/ml; Boehringer Mannheim)	1 μl

      Note: The final concentration of proteinase K will be 50 μg/ml.

      10x Proteinase K buffer

Component and final concentration	Amount to add per 1 ml
100 mM Tris-Cl	100 μl of 1 M (pH 7.8 at 22°C)
50 mM EDTA	100 μl of 0.5 M (pH 8.0)
5% **SDS**	0.5 ml of 10%
H_2O	300 μl

      Mix the components in a 1-ml microcentrifuge tube. Store at room temperature for up to 6 months.

      SDS (see Appendix for Caution)

   b. Incubate at 50°C for 15 minutes.

4. Extract as follows:

   a. Add 80 μl of H$_2$O and 100 μl of **phenol:chloroform**:isoamyl alcohol (25:24:1) and mix thoroughly by vortexing. Use phenol saturated with TE (pH 8.0).

      phenol, chloroform (see Appendix for Caution)

   b. Centrifuge in a microcentrifuge at 12,000*g* at room temperature for 3 minutes to separate the phases.

   c. Transfer the aqueous (upper) phase into a new microcentrifuge tube.

   d. To "back extract" the organic (lower) phase, add 100 μl of TE (pH 8.0), mix by vortexing, and repeat steps b and c.

   e. Combine the aqueous phases from the two extractions (total volume of 200 μl). Combine an aliquot (~1%) with 3 ml of a **toluene**-based scintillation fluid and measure the radioactivity (the total cpm of ^{32}P) in a liquid scintillation counter. Calculate the amount of cDNA as described on pp. 108–109.

      Notes: There is no need to precipitate the DNA and count unincorporated radioactivity to calculate the amount of cDNA.
      At this point, there should be approximately 150 ng of cDNA.

      toluene (see Appendix for Caution)

5. Precipitate the phosphorylated cDNA/adapter in the combined aqueous phases with 22 μl of 3 M sodium acetate (pH 7.0) and 450 μl of absolute ethanol and dry the pellet under vacuum as described for RNA on p. 71, step 10.

   Note: The phosphorylated cDNA/adapter can be stored in ethanol at –20°C indefinitely, but it is preferable to dissolve the precipitated cDNA/adapter and then ligate immediately.

6. Ligate the phosphorylated cDNA/adapter to the prepared vector as described on pp. 116–117.

PROTOCOL

Ligating cDNA/Adapter to Phagemid Vector DNA Digested with *Eco*RI and *Not*I

1. Dissolve the phosphorylated cDNA/adapter from pp. 114–115 (~150 ng with an average length of 1.5 kb) in 4 µl of H_2O.

2. If necessary, adjust the concentration of the *Eco*RI- and *Not*I-digested pT7T3-Pac DNA (from pp. 90–96) to 150 ng/µl with TE (pH 8.0). Add 4 µl of *Eco*RI- and *Not*I-digested pT7T3-Pac vector DNA to the phosphorylated cDNA/adapter and mix thoroughly by vortexing.

   Note: Using 600 ng of 2.9-kb vector DNA and approximately 150 ng of insert cDNA with an average length of 1.5 kb in the 10-µl ligation provides a vector DNA:cDNA molar ratio of approximately 2:1 (0.3 pmole of vector to 0.15 pmole of cDNA).

3. Heat at 37°C for 5 minutes. Centrifuge briefly to collect the liquid at the bottom of the tube.

4. Add 1 µl of 10x bacteriophage T4 DNA ligase buffer (supplied with the enzyme) and then add 1 µl of bacteriophage T4 DNA ligase (400 units/µl; New England Biolabs 202). Mix gently. Centrifuge briefly to collect the liquid at the bottom of the tube.

   Note: Using bacteriophage T4 DNA ligase from different suppliers can dramatically affect the results.

5. Incubate at 16°C for 12–18 hours.

6. Add 0.5 µl of 0.5 M EDTA (pH 8.0) to stop the reaction. Mix thoroughly by vortexing. Centrifuge briefly to collect the liquid at the bottom of the tube. Heat at 65°C for 20 minutes. Centrifuge briefly to collect the liquid at the bottom of the tube.

7. Add 80 µl of TE (pH 8.0) and mix thoroughly.

8. Extract with 1 volume of **phenol:chloroform**:isoamyl alcohol (25:24:1) (see Appendix). Use phenol saturated with TE (pH 8.0).

   phenol, chloroform (see Appendix for Caution)

9. Precipitate the DNA in the aqueous phase with 10 µl of 3 M sodium acetate (pH 7.0) and 200 µl of absolute ethanol and dry the pellet under vacuum as described for RNA on p. 71, step 10.

   Note: The DNA can be stored in ethanol at –20°C indefinitely, but it is preferable to amplify the cDNA library immediately.

10. Dissolve the ligated DNAs (which is the unamplified cDNA library) in 3 μl of TE (pH 8.0). Heat at 37°C for 5 minutes. Centrifuge briefly to collect the liquid at the bottom of the tube.

11. Amplify the cDNA library as described on pp. 119–120.

PROTOCOL

Amplifying the cDNA Library by Electroporation

PREPARING BACTERIA COMPETENT FOR ELECTROPORATION

1. Place 100 ml of 2x YT medium containing streptomycin (25 µg/ml) in a 500-ml flask and inoculate with a single colony of DH10B (Life Technologies). Incubate at 37°C with constant agitation at 225 rpm overnight.

2. Place 1 liter of 2x YT medium in a 2-liter flask and inoculate with 10 ml of the overnight culture from step 1. Incubate at 37°C with constant agitation at 225 rpm until the culture reaches an OD_{600} of 0.2–0.25 and then immediately place the culture on ice to prevent any further growth.

   Notes: Harvesting the culture at a higher OD_{600} may result in an impairment of electroporation efficiencies.

   It usually takes approximately 4 hours for the culture to reach an OD_{600} of 0.2–0.25 (i.e., 1.2–1.5 x 10^8 cells/ml).

   Keep the cells on ice as much as possible throughout the rest of the procedure.

3. Divide the 1-liter culture into six sterile 200-ml polypropylene conical bottles (Nalge Nunc 376813) and centrifuge in a Sorvall GSA rotor at 10,000 rpm (16,274g) at 4°C for 10 minutes. Decant the supernatants.

   Note: Since the cells do not adhere very tightly to the sides of these polypropylene conical bottles, special care should be taken to avoid cell losses when the supernatants are decanted. However, the fact that the cells do not adhere tightly and can therefore be readily resuspended makes these bottles ideal for this step.

4. Resuspend each cell pellet in 50 ml of sterile ice-cold 10% glycerol (in H_2O) by gently pipetting up and down.

5. Combine three of the resuspended cell pellets in each of two bottles (150 ml/bottle).

6. Centrifuge in a Sorvall GSA rotor at 10,000 rpm (16,274g) at 4°C for 15 minutes. Immediately, but gently, decant the supernatants.

7. Resuspend each cell pellet in 100 ml of sterile ice-cold 10% glycerol by gently pipetting up and down.

8. Repeat steps 6–7.

9. Repeat steps 6–7 again but resuspend each cell pellet in 50 ml of sterile ice-cold 10% glycerol.

10. Combine the resuspended cell pellets in one bottle.

11. Repeat step 6 but centrifuge for 20 minutes.

12. Resuspend the cell pellet and adjust the final cell density as follows:

    a. Resuspend the cell pellet in 1 ml of sterile ice-cold 10% glycerol.

    b. Dilute 25 μl of the resuspended cell pellet with 10 ml of 10% glycerol and determine the OD_{600}.

    c. If the OD_{600} of the dilution in step b is greater than 0.15, add more sterile ice-cold 10% glycerol to the resuspended cell pellet in step a and repeat step b. Repeat this process until the OD_{600} of the dilution in step b is 0.15.

       Note: Loss of cells occurs during this protocol. To obtain the desired cell density of 3.6 x 10^{10} cells/ml, it is advisable to resuspend the cell pellet with a small volume of glycerol and then to increase the volume in small increments. Typically, a total of 1–2 ml of 10% glycerol must be added to the cell pellet and the final volume of the resuspended cells is approximately 2.5 ml.

13. Divide the final cell suspension into 25-μl aliquots. Quick-freeze the cells in a dry-ice/ethanol bath (do not use liquid nitrogen since it may lyse some of the cells).

    Note: Cells can be stored at –80°C indefinitely.

14. Transform the cells by electroporation as described below.

ELECTROPORATION

This protocol is essentially that described by Dower et al. (1988).

1. Place the library cDNA (from pp. 116–117), three 25-μl aliquots of DH10B bacteria competent for electroporation (prepared as described in the previous protocol), and three cuvettes (0.2-cm electrode gap; Bio-Rad 165-2086) on ice. Keep them on ice as much as possible during step 2.

2. Perform three transformations with the library cDNA as described below. Transform the bacteria in one aliquot before starting the next transformation.

   a. Transfer 1 μl of library cDNA into an aliquot of competent bacteria, mix gently, and then transfer into a cuvette. Make sure the mixture is all at the bottom of the cuvette.

      Note: Each transformation will contain 50 ng of cDNA ligated to 200 ng of vector.

   b. Place the cuvette in the chamber of an electroporation device (Gene Pulser with a pulse controller; Bio-Rad). Set the following conditions on the electroporation device: a resistance of 200 ohms, a capacitance of 25 microfarads, a voltage of 2.5 kV, and a time constant of 4.7–4.8 milliseconds. Apply the pulse.

   c. Immediately add 1 ml of 2x YT medium to the transformation mixture.

d. Inoculate 250 ml of 2x YT medium in a 1-liter flask with the diluted transformation mixture.

e. Follow steps a–d for each of the other two transformations. In step d, inoculate the same flask of 2x YT medium with all three transformation mixtures.

3. Incubate the culture at 37°C with agitation at 250 rpm for 1 hour.

4. Determine the total number of recombinant clones in the library as follows:

a. Spread 1 μl and 10 μl of the culture on separate 2x YT agar plates containing ampicillin (75 μg/ml).

b. Incubate at 37°C overnight.

c. Count the number of ampicillin-resistant colonies on each plate. The number of recombinant clones obtained from the transformation is: (number of ampicillin-resistant colonies/plate) x (culture volume/volume of aliquot spread on plate).

5. To the remainder of the culture, add 187.5 μl of ampicillin solution (100 mg/ml) and continue the incubation at 37°C with agitation at 250 rpm until the culture reaches saturation.

Note: It takes 8–12 hours for the culture to reach saturation (i.e., an OD_{600} of 2 or 6 x 10^8 cells/ml).

6. Prepare a frozen glycerol stock of the bacterial culture as follows:

a. Transfer 1 ml of the culture into a sterile microcentrifuge tube and centrifuge in a microcentrifuge at 12,000*g* at 4°C for 15 seconds. Discard the medium.

b. Resuspend the cells in 0.5 ml of sterile 10% glycerol (in H_2O). Store at –80°C indefinitely.

7. Prepare high-quality supercoiled plasmid DNA by purifying the remainder of the culture through a Qiagen-tip100 column according to the manufacturer's instructions. Store the library cDNA in TE (pH 8.0) at –20°C indefinitely.

8. To eliminate from the library all of the nonrecombinant ("empty vector") molecules, gel purify the library cDNA as described on pp. 121–123.

Note: Gel purification is recommended at this point, but some investigators choose not to perform these steps before normalization of the library.

PROTOCOL

Gel Purifying Library cDNA to Eliminate Nonrecombinant Clones

This protocol is essentially that described by Soares (1994).

DIGESTION OF A PLASMID DNA PREPARATION OF THE LIBRARY WITH *NotI* AND GEL PURIFICATION OF THE DNA

1. If necessary, adjust the concentration of the plasmid DNA (from pp. 119–120) to 0.25 µg/µl with TE (pH 8.0). Combine the components below (in the order listed) in a 1.5-ml microcentrifuge tube and mix thoroughly but gently. Centrifuge briefly to collect the liquid at the bottom of the tube.

plasmid DNA	2 µl (0.5 µg)
10x NEBuffer 3 (New England Biolabs)	2 µl
BSA (1 mg/ml)	2 µl
H₂O	13 µl
NotI (10,000 units/ml; New England Biolabs 189)	1 µl

2. Incubate at 37°C for 2 hours.

3. Add 2 µl of 10x SDS/glycerol gel-loading solution and heat at 65°C for 10 minutes.

4. Load the digested plasmid DNA on a regular 1% agarose minigel containing **ethidium bromide** (at a final concentration of 0.5 µg/ml). Also load DNA molecular-weight markers (1–15 kb). Perform electrophoresis under standard conditions until the bromophenol blue dye runs off the gel.

   Note: Using regular agarose instead of low-melting-temperature agarose in this step provides greater resolution as well as a lower background in the final DNA preparation.

   ethidium bromide (see Appendix for Caution)

5. Detect the DNA by using a long-wavelength **UV** transilluminator. Use a sterile scalpel or razor blade to excise a gel slice (~1–2 cm) containing the smear of library cDNA in which all of the cDNA clones have inserts larger than 350 bp. Leave behind any residual nonrecombinant vector molecules.

   Notes: To avoid unnecessary nicking, be sure not to overexpose the DNA to the UV light in any of the steps in this protocol. As a precaution, place a protective shield between the gel and the UV transilluminator and set the long-wavelength light to the lowest intensity that still allows detection of the DNA.
   Since the vector is 2.9 kb, all clones larger than 3.3 kb (2.9-kb vector + 0.4-kb insert) should be excised in the gel slice.

   UV radiation (see Appendix for Caution)

6. Place the gel slice backward (i.e., place the lower-molecular-weight DNAs at the top) on the uppermost edge of an empty minigel mold and pour a 1% solution of low-melting-temperature agarose over it. Allow the low-melting-temperature agarose to solidify at 4°C for 20 minutes. Perform electrophoresis under standard conditions until the DNA smear has entered the low-melting-temperature gel (as determined by visual inspection under long-wavelength UV light).

   Note: Since the gel slice is cast in the low-melting-temperature agarose with the lower-molecular-weight DNAs at the top, the DNA smear will become progressively sharper as the gel runs. At the point at which the DNA starts to enter the low-melting-temperature gel, it should be reduced to a single band.

7. Detect and excise the gel slice containing the linearized plasmid DNA as described in step 5. Place the gel slice in a 1.5-ml microcentrifuge tube.

8. Follow steps 6–11 on pp. 93–94 and proceed immediately with the recircularization of the purified, linearized DNA as described below.

RECIRCULARIZATION OF THE PURIFIED, LINEARIZED LIBRARY cDNA

1. Combine the components below in the order listed in a 1.5-ml microcentrifuge tube and mix gently. Centrifuge briefly to collect the liquid at the bottom of the tube.

linearized plasmid DNA (from the previous protocol)	20 µl
10x bacteriophage T4 DNA ligase buffer (supplied with the enzyme)	32 µl
H$_2$O	265 µl
bacteriophage T4 DNA ligase (2,000,000 units/ml; New England Biolabs 202CS)	3 µl

   Notes: The exact reaction conditions to promote recircularization instead of intermolecular ligations can be determined as described on pp. 88–89.

   Using bacteriophage T4 DNA ligase from different suppliers can dramatically affect the results.

2. Incubate at 16°C overnight.

3. Add 13 µl of 0.5 M EDTA (pH 8.0) to stop the reaction. Mix thoroughly by vortexing. Centrifuge briefly to collect the liquid at the bottom of the tube. Heat at 65°C for 20 minutes.

4. Extract with 1 volume (333 µl) of **phenol:chloroform**:isoamyl alcohol (25:24:1) (see Appendix). Use phenol saturated with TE (pH 8.0).

   phenol, chloroform (see Appendix for Caution)

5. Precipitate the library cDNA in the aqueous phase with 0.1 volume of 3 M sodium acetate (pH 7.0) and 2 volumes of absolute ethanol and dry the pellet under vacuum as described for RNA on p. 71, step 10.

Note: The library cDNA can be stored in ethanol at –20°C indefinitely, but it is preferable to amplify the library cDNA immediately.

6. Dissolve the purified, recircularized library cDNA in TE (pH 8.0) to make a final concentration of ≤0.1 µg/µl.

7. Amplify the purified, recircularized library cDNA as described below.

AMPLIFICATION OF THE PURIFIED, RECIRCULARIZED LIBRARY cDNA

1. Place the purified, recircularized library cDNA (from the previous protocol), five 25-µl aliquots of DH10B bacteria competent for electroporation (prepared as described on pp. 118–119), and five cuvettes (0.2-cm electrode gap; Bio-Rad 165-2086) on ice. Keep them on ice as much as possible during step 2.

2. Perform five transformations with the purified, recircularized library cDNA as described on pp. 119–120, step 2. Transform the bacteria in one aliquot before starting the next transformation.

   Note: Each transformation will contain ≤0.1 µg of purified, recircularized library cDNA.

3. Follow steps 3–7 on p. 120.

   Note: The cDNA library should be stored in TE (pH 8.0) at approximately 0.5 µg/µl. It should be stable at –20°C indefinitely.

4. Normalize the cDNA library as described on pp. 124–148.

NORMALIZATION OF DIRECTIONALLY CLONED cDNA LIBRARIES

In this procedure for normalizing cDNA libraries (see Figure 4) (Soares et al. 1994), single-stranded DNA is first prepared from a directionally cloned cDNA library by using helper bacteriophage M13KO7 (pp. 127–128). For efficient normalization, the single-stranded circular DNA (library cDNA) should be absolutely free of any contaminating double-stranded replicative form DNA and preferably free of single-stranded helper bacteriophage DNA. Single-stranded DNA of suitable quality for the normalization procedure provided here can be obtained by a series of purification steps involving chromatography on HAP columns and agarose gel electrophoresis. The purpose of purification on the HAP column (pp. 129–132) is to eliminate any contaminating double-stranded replicative form DNA from the preparation of single-stranded DNA. The gel-purification step (pp. 133–134) is intended to purify the library from the helper bacteriophage M13KO7 single-stranded DNA.

HAP consists of calcium phosphate crystals produced in a controlled precipitation. Nucleic acids bind to HAP by virtue of interactions between the phosphate groups of the polynucleotide backbone and calcium residues in the resin. The useful capacity of most HAP batches (i.e., the maximum amount of DNA that will be totally adsorbed by the particular batch of HAP) is approximately 200 µg of DNA per milliliter of HAP (1 ml of HAP weighs 400 mg) (Britten et al. 1974). The capacity of HAP for partially paired or poorly matched DNA is somewhat lower than that for DNA with perfectly matched base pairs. The definition of capacity is thus somewhat arbitrary, and it should be measured for the system being examined.

Single- and double-stranded DNAs have different affinities for HAP. At low phosphate ion concentrations (10–30 mM sodium phosphate buffer), both single-stranded and double-stranded DNAs bind to HAP. At intermediate concentrations (120–140 mM sodium phosphate buffer), double-stranded DNA binds to HAP, whereas single-stranded DNA flows through. At high concentrations of sodium phosphate buffer (0.4–0.5 M), double-stranded DNAs elute from the column. For most purposes, the best procedure is to pass the sample over the column at the temperature and the sodium phosphate buffer concentration that allows single-stranded DNA to flow through (i.e., 120 mM sodium phosphate buffer at 60ºC or 140 mM sodium phosphate buffer at 50ºC). Note that these phosphate buffer concentrations apply only for sodium phosphate buffer.

Because linearized double-stranded DNA binds more quantitatively to HAP than supercoiled plasmid DNA does under standard conditions (120 mM sodium phosphate buffer, 10 mM EDTA [pH 8.0], 1% SDS at 60ºC), the preparation of single-stranded DNA is routinely digested with *Pvu*II to linearize any contaminating double-stranded DNA before it is passed through the HAP column. *Pvu*II is chosen for two reasons: There are two *Pvu*II sites in the pT7T3-Pac vector (one is 216 bp 5′ of the *Eco*RI site and the other is 123 bp downstream from the *Not*I site in the multiple cloning site shown in Figure 5), and *Pvu*II does not cleave single-stranded DNA.

Once a suitable single-stranded DNA preparation has been purified, the purified DNA is annealed to an oligo(dT)$_{12-18}$ primer and thus converted into partially double-stranded circular DNA in a controlled primer-extension reaction with the large (Klenow) fragment of *E. coli* DNA polymerase I in the presence of dNTPs and ddNTPs (pp. 135–136). The reaction conditions are such that the duplex region does not exceed 200 ± 20 bp in length and thus corresponds primarily to 3'-noncoding sequences. The purpose of generating partial duplexes that are limited to 3'-noncoding sequences of mRNAs is to address the problem first recognized by Ko (1990) that cross-hybridization of coding regions belonging to unequally represented members of oligo- or multigene families could result in the elimination of rarer members from the population during the normalization process. In contrast, the 3'-noncoding region is almost always unique to the transcript that it represents and is therefore expected to anneal only to its complement.

At this point, the partially double-stranded circular DNA is purified by chromatography on HAP columns to separate it from any remaining single-stranded DNA (see pp. 137–139). This is necessary since clones derived from mRNAs with internal *Not*I sites do not have an oligo(dA) tract at the 3' end and consequently are not converted into partial duplexes during the primer-extension reaction.

In the protocol on pp. 140–141, the flow-through single-stranded circular DNA from the HAP column (from pp. 137–138) is partially converted into double strands to improve the electroporation efficiency. Bacteria are transformed by electroporation and propagated under selection with an appropriate anti-biotic. Finally, plasmid DNA is prepared. This DNA represents a sublibrary of clones without an oligo(dA) tract at the 3' end. These clones therefore correspond to mRNAs with an internal *Not*I site. This DNA is not combined with the DNA from the normalized library prepared on pp. 144–146, but it is kept to preserve the representation of mRNAs with internal *Not*I sites in the final collection of clones.

The HAP-bound double-stranded DNA from pp. 137–139 is melted and reassociated to a relatively low C_0t (~5–20 seconds-moles/liter) on pp. 144–146. This HAP-bound fraction comprises the partially double-stranded circular DNA generated by the controlled primer-extension reaction on pp. 135–136. The fraction that remains single-stranded after reassociation represents the unamplified normalized library. (Calculations of the mass of DNA synthesized in the controlled primer-extension reaction and the length of hybridization required to achieve the desired C_0t value are found on pp. 142–143. For a discussion of the theoretical considerations in normalizing cDNA populations and various parameters that affect the rate of reassociation in a reaction, see pp. 61–63.)

In the final steps (pp. 147–148), this single-stranded fraction is purified from the reassociated molecules by chromatography on HAP columns and then partially converted into double strands by primer extension as described for the sublibrary. Bacteria are transformed by electroporation and propagated under selection with an appropriate antibiotic. Finally, plasmid DNA is prepared. This plasmid preparation represents the amplified normalized cDNA library. Although only 3'-noncoding sequences participate in the reassociation reaction, the resulting normalized library consists of clones with large inserts encompassing both coding and noncoding sequences.

This method (Soares et al. 1994) can be applied to any library constructed in a phagemid vector or in any other vector with an f1 origin of replication such that library cDNA can be obtained in the form of single-stranded circular DNA after superinfection of a growing culture with a helper bacteriophage.

Preparation of Single-stranded DNA

1. Place the supercoiled plasmid DNA (from pp. 121–123), a 25-µl aliquot of *E. coli* DH5αF′ competent for electroporation (Life Technologies), and a cuvette (0.2-cm electrode gap; Bio-Rad 165-2086) on ice.

2. Use 50 ng (1 µl) of the DNA to transform the 25-µl aliquot of DH5αF′ bacteria as described on pp. 119–120, step 2a–d, but use 100 ml of medium in a 500-ml flask in step 2d.

   Note: The supercoiled plasmid DNA used for the transformation should represent the entire cDNA library.

3. Incubate the culture at 37°C with constant agitation at 250 rpm for 1 hour.

4. Add 75 µl of ampicillin solution (100 mg/ml) to make a final concentration of 75 µg/ml and continue the incubation overnight.

   Note: For libraries that are not constructed in the pT7T3-Pac vector, the concentration and type of antibiotic will vary.

5. Dilute 1 ml of the overnight culture from step 4 in 98 ml of 2x YT medium. (Discard the remainder of the overnight culture.) Add 1 ml of 20% glucose and 75 µl of ampicillin solution (100 mg/ml). Incubate at 37°C with constant agitation at 250 rpm for approximately 2–4 hours (i.e., until the culture reaches an OD_{600} of 0.2).

6. Superinfect the culture with helper bacteriophage M13KO7 (Pharmacia) at an m.o.i. of 10 (i.e., superinfect 1.2×10^{10} cells with 1.2×10^{11} bacteriophage pfu). Incubate the culture at 37°C with constant agitation at 250 rpm for 2 hours.

   Note: Since 6×10^8 cells/ml corresponds to an OD_{600} of 1, 1.2×10^8 cells/ml corresponds to an OD_{600} of 0.2. Therefore, the total number of cells in 100 ml of a culture with an OD_{600} of 0.2 is 1.2×10^{10} cells.

7. Transfer the culture into a sterile 200-ml polypropylene conical bottle (Nalge Nunc 376813) and centrifuge in a Sorvall GSA rotor at 10,000 rpm ($16,274g$) at 4°C for 20 minutes.

8. Transfer the supernatant into a new bottle and discard the pellet. Recentrifuge for 10 minutes.

   Notes: The bacteriophage will infect the *E. coli* and be released from them. Thus, the single-stranded bacteriophage DNA will be found in the supernatant at this point.

   Unlike other bacteriophages, M13KO7 does not lyse the cells. Its replication occurs in harmony with that of the host bacterium. The infected cells therefore continue to grow but at one half to three fourths of their normal rate.

9. Transfer the supernatant into a new bottle and discard the pellet. Add 4 g of PEG 8000 (Sigma P 2139) and 2.92 g of NaCl per 100 ml of supernatant. Mix by vigorous vortexing until until all of the PEG and NaCl have dissolved.

10. Incubate at 4°C overnight.

    Note: The PEG and NaCl will enable the bacteriophage DNA to precipitate from the supernatant during this incubation.

11. Centrifuge as described in step 7. Carefully remove the supernatant and discard.

12. Dissolve the bacteriophage pellet in 0.5 ml of TE (pH 8.0) by pipetting up and down, and then transfer it into a 1.5-ml microcentrifuge tube.

13. Extract sequentially with 0.5 ml of **phenol**, with 0.5 ml of phenol:**chloroform**: isoamyl alcohol (25:24:1), and then with 0.5 ml of chloroform:isoamyl alcohol (24:1) (see Appendix). Use phenol saturated with TE (pH 8.0).

    phenol, chloroform (see Appendix for Caution)

14. Precipitate the DNA in the aqueous phase as follows:

    a. Add 0.1 volume of 3 M sodium acetate (pH 7.0) and 1 volume of isopropanol and mix. Place at −20°C for 1 hour.

    b. Centrifuge in a microcentrifuge at 12,000g at 4°C for 20 minutes. Carefully remove the supernatant and discard.

       Note: It is not necessary to wash the pellet with 70% ethanol.

    c. Dry the DNA pellet under vacuum in a SpeedVac Concentrator.

15. Dissolve the single-stranded DNA in 100 μl of TE (pH 8.0). Analyze 1 μl of the DNA and appropriate concentration markers on a 1% agarose minigel to check the yield of the DNA.

    Notes: The single-stranded DNA adheres fairly tightly to the sides of the tube. To obtain maximum recovery, it is absolutely necessary to wash all of the inner side walls of the tube very thoroughly with TE while dissolving the DNA. Expect a yield of approximately 150 μg of single-stranded DNA (1.5 μg per ml of culture).

    The DNA can be stored at −20°C indefinitely.

16. Purify the single-stranded DNA as described on pp. 129–132.

PROTOCOL

Purification of Single-stranded DNA on HAP Columns

1. Prepare the HAP column as follows:

   a. Suspend 400 mg of HAP (Bio-Rad) in 5 ml of 120 mM sodium phosphate/EDTA/SDS. Pour the suspension into a 60°C jacketed column.

      Note: This protocol requires a jacketed column, which can be manufactured in a local glass shop. The column should be jacketed over its entire length as shown in Figure 6. A piece of tubing with a width similar to that of the column is connected to the top of the column to apply pressure from an air pump. For further details of preparing and running the column, see Britten et al. (1974); see also pp. 124–125.

      120 mM Sodium phosphate/EDTA/SDS

Component and final concentration	Amount to add per 50 ml
120 mM sodium phosphate	6 ml of 1 M sodium phosphate buffer (pH 6.8) (see Appendix)
10 mM EDTA	1 ml of 0.5 M (pH 8.0)
1% **SDS**	5 ml of 10%
H₂O	38 ml

      Mix the components in a 50-ml polypropylene tube. Store at room temperature for up to 1 month.

      SDS (see Appendix for Caution)

   b. Wash the column three times with 2 ml of 120 mM sodium phosphate/EDTA/SDS.

      Notes: Positive pressure should be applied from the top of the column to obtain a flow rate of approximately 2–5 ml/minute in all steps.
      The column should not be left dry for any length of time.

2. Combine the components below (in the order listed) in a silanized 1.5-ml microcentrifuge tube and mix gently. Centrifuge briefly to collect the liquid at the bottom of the tube.

single-stranded DNA (~1.5 μg/μl; from pp. 127–128)	7 μl (~10 μg)
10x NEBuffer 2 (New England Biolabs)	5 μl
H₂O	36 μl
*Pvu*II (10,000 units/ml; New England Biolabs 151)	2 μl

   Note: Silanized tubes should be used throughout the protocol. Bacteriophage DNA sticks to plastic tubes unless they are silanized. Tubes can be silanized as described on p. 70, step 6, Note.

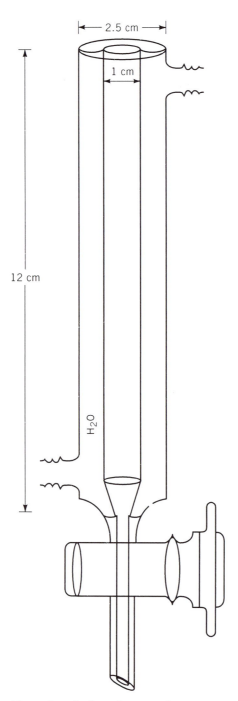

Figure 6 A jacketed HAP column (Britten et al. 1974). This column can be manufactured in a local glass shop. It should be jacketed over its entire length. A piece of tubing of similar width to that of the column can be connected to the top of the column to apply pressure from an air pump.

3. Incubate at 37°C for 2 hours.

4. Extract with 50 μl of **phenol:chloroform**:isoamyl alcohol (25:24:1) (see Appendix). Use phenol:chloroform:isoamyl alcohol saturated with TE (pH 8.0).

   phenol, chloroform (see Appendix for Caution)

5. Transfer the aqueous (upper) phase into a silanized 5-ml (or larger) plastic tube. Add 2 ml of 120 mM sodium phosphate/EDTA/SDS and mix gently. Heat at 60°C for 2–5 minutes to equilibrate the sample to the temperature of the column.

6. Load the 2-ml sample on the column and pass it through the column, collecting the flow-through.

   Note: In 120 mM sodium phosphate/EDTA/SDS at 60°C, double-stranded replicative form DNA that is at least 100 bp in length binds quantitatively to HAP, whereas single-stranded DNA does not bind and can be found in the flow-through.

7. Heat the flow-through at 60°C for 3 minutes and then reload it on the column. Pass it through the column, collecting the flow-through. Set aside the flow-through in a silanized 50-ml polypropylene tube (Falcon 2098).

8. Wash the column three times with 2 ml of 120 mM sodium phosphate/EDTA/SDS, collecting the flow-through from each wash. Pool the flow-through from the washes with the flow-through in the tube set aside in step 7.

9. Extract the 8 ml of pooled flow-through with butanol as follows:

   a. Add 30 ml of H$_2$O-saturated **sec-butanol** (for preparation of H$_2$O-saturated butanol, see p. 91) and mix by vigorous vortexing.

      sec-butanol (see Appendix for Caution)

   b. Centrifuge in a tabletop clinical centrifuge at 3500 rpm (1900g) at room temperature for 5 minutes to separate the phases. Discard the organic (upper) phase.

   c. Repeat steps a–b to reextract the aqueous (lower) phase twice with 30 ml of H$_2$O-saturated *sec*-butanol and once with 30 ml of *sec*-butanol that has not been saturated with H$_2$O.

10. Extract with ether as follows:

   a. Add 20 ml of H$_2$O-saturated ether to the aqueous phase and mix.

      H$_2$O-saturated ether
      Add 10 ml of H$_2$O to 50 ml of anhydrous **diethyl ether** in a 15-ml plastic centrifuge tube. Cap the tube and mix by vortexing. Prepare the mixture in a chemical fume hood just before use. Be sure to vent the tube before opening it fully.

      diethyl ether (see Appendix for Caution)

   b. Centrifuge in a clinical centrifuge at 3500 rpm (1900g) at room temperature for 5 minutes to separate the phases.

   c. Carefully remove the organic (upper) phase with a pipette and discard. Either allow the tube to stand open at 65°C for 10 minutes or apply a vacuum so that all of the remaining ether evaporates.

   d. Repeat steps a–c two more times.

11. Desalt the sample through a Nensorb-20 purification cartridge (binding capacity of 20 μg; DuPont NEN Research Products) according to the manufacturer's instructions.

12. Dry the eluted single-stranded DNA sample under vacuum in a SpeedVac Concentrator until the volume is approximately 350 μl.

    Note: Drying should take approximately 30 minutes.

13. Precipitate the single-stranded DNA with 0.1 volume of 3 M sodium acetate (pH 7.0) and 2 volumes of absolute ethanol and dry the pellet under vacuum as described for RNA on p. 71, step 10.

    Note: If the DNA is not to be used immediately, it can be stored in ethanol at −20°C indefinitely.

14. Dissolve the single-stranded DNA in 20 μl of TE (pH 8.0).

    Note: Assume that approximately 80% of the DNA will be recovered (~8 μg).

15. Gel purify the single-stranded DNA as described on pp. 133–134.

PROTOCOL

Gel Purification of Single-stranded DNA

1. Add 2 μl of 10x SDS/glycerol gel-loading solution to the 20 μl (~8 μg) of single-stranded DNA from pp. 129–132. Heat at 65°C for 10 minutes.

2. Load on a regular 1% agarose minigel containing **ethidium bromide** (at a final concentration of 0.5 μg/ml). Also load double-stranded bacteriophage λ DNA digested with *Hin*dIII as a marker (125 bp–23.1 kb). Perform electrophoresis under standard conditions until the bromophenol blue dye runs off the gel.

   Notes: This much DNA can be loaded in a single well of a minigel with wells that are 13 mm wide.

   The single-stranded DNA will run faster than the 2-kb marker band but slower than the 565-bp marker band.

   Using regular agarose instead of low-melting-temperature agarose in this step provides greater resolution as well as a lower background in the final DNA preparation.

   The digested bacteriophage λ DNA is not a true molecular-weight marker here because it is double-stranded. However, it serves as a reference.

   ethidium bromide (see Appendix for Caution)

3. Excise a narrow strip along the entire length of the gel lane (~5% of the lane or ~0.5 μg). Detect the single-stranded DNA in this gel strip by using a long-wavelength **UV** transilluminator, and use a clean needle to mark the location of the DNA on the strip.

   Note: Do not expose the entire gel to UV light! This will nick the single-stranded DNA, rendering it inappropriate for the normalization procedure.

   UV radiation (see Appendix for Caution)

4. Place the exposed gel strip next to the unexposed gel. Use a sterile scalpel or razor blade to excise the region of the unexposed gel that corresponds to the single-stranded DNA located in the exposed strip.

5. Place both the exposed and the unexposed gel slices side by side and backward (i.e., place the lower-molecular-weight DNAs at the top) on the uppermost edge of an empty minigel mold and pour a 1% solution of low-melting-temperature agarose over them. Allow the low-melting-temperature agarose to solidify at 4°C for 20 minutes. Perform electrophoresis under standard conditions until the DNA has entered the low-melting-temperature gel.

   Note: Since the gel slice is cast in the low-melting-temperature agarose with the lower-molecular-weight DNAs at the top, the DNA smear will become progressively sharper as the gel runs. At the point that the DNA starts to enter the low-melting-temperature gel, it should be reduced to a single band.

6. Excise the previously exposed gel strip. Detect the single-stranded DNA in this gel strip by using a long-wavelength UV transilluminator, mark the location of the DNA, and excise the region of the unexposed low-melting-temperature gel that corresponds to the single-stranded DNA located in the exposed strip as described above. Place the purified, unexposed single-stranded DNA in a 1.5-ml microcentrifuge tube.

7. Follow steps 6–10 on pp. 93–94.

8. Dissolve the purified single-stranded DNA in 10–20 µl of TE (pH 8.0) to make a final concentration of approximately 0.5 µg/µl.

   Notes: Assume that approximately 75% of the DNA will be recovered.
   The purified single-stranded DNA can be stored at –20°C until needed.

9. Perform the controlled primer-extension reaction with the purified single-stranded DNA as described on pp. 135–136.

PROTOCOL

Controlled Primer-extension Reaction

1. Combine the components below in a silanized 0.5-ml microcentrifuge tube
   and mix. Centrifuge briefly to collect the liquid at the bottom of the tube.

purified single-stranded DNA (~0.5 µg/µl; from pp. 133–134)	5 µl
oligo(dT)$_{12-18}$ primer (10 ng/µl; Pharmacia 27-7610-01)	7 µl
10x primer-extension buffer	10 µl
100 mM DTT	10 µl
10 mM dNTP mixture (see Appendix)	10 µl
25 mM ddNTP mixture	25 µl
[α-^{32}P]dCTP (10 µCi/µl, 800 Ci/mmole; DuPont NEN Research Products)	5 µl
H$_2$O	20.5 µl

Note: This 100-µl reaction mixture will contain 1.7–1.9 pmoles of single-stranded library
cDNA and approximately 14 pmoles of primer. The final concentration of each dNTP will be
1 mM. The final concentration of each ddNTP will be 6.25 mM.

10x Primer-extension buffer

Component and final concentration	Amount to add per 1 ml
300 mM Tris-Cl	300 µl of 1 M (pH 7.5 at 22°C)
0.5 M NaCl	100 µl of 5 M
150 mM MgCl$_2$	150 µl of 1 M
H$_2$O	450 µl

Mix the components in a 1.5-ml microcentrifuge tube. Store at
−20°C indefinitely.

25 mM (ddATP, ddCTP, ddGTP) ddNTP mixture

Component and final concentration	Amount to add per 40 µl
25 mM ddATP	10 µl of 100 mM
25 mM ddCTP	10 µl of 100 mM
25 mM ddGTP	10 µl of 100 mM
H$_2$O	10 µl

Mix the components in a 1.5-ml microcentrifuge tube. Store at
−20°C for up to 6 months.

radioactive substances (see Appendix for Caution)

2. Overlay with 100 µl of light mineral oil (Sigma M 5904).

Notes: This step is not necessary if a thermal cycler with a top heating block will be used in step 3.

If a mineral oil overlay is used, insert a pipette through the oil to add reagents or to remove the sample (but not the oil) in the steps below.

3. Incubate sequentially as follows:

   60°C for 5 minutes
   50°C for 15 minutes
   37°C for 2 minutes

   Note: Although these incubations can be performed in a water bath, use of a thermal cycler may be more convenient.

4. Add 7.5 µl of the Klenow fragment of *E. coli* DNA polymerase I (5000 units/ml; Amersham) and mix gently. Centrifuge briefly to collect the liquid at the bottom of the tube.

5. Immediately incubate at 37°C for 30 minutes.

6. Spot a 0.5-µl aliquot of the reaction mixture on the center of a Whatman GF/C glass-fiber filter (2.4 cm in diameter). Store the filter at room temperature until needed for determination of the total cpm of $[\alpha\text{-}^{32}\text{P}]\text{dCTP}$ on p. 142.

7. To the remainder of the reaction mixture, add 4 µl of 0.5 M EDTA (pH 8.0) and mix thoroughly by vortexing. Centrifuge briefly to collect the liquid at the bottom of the tube.

8. Extract with 100 µl of **phenol:chloroform**:isoamyl alcohol (25:24:1) (see Appendix). (Use phenol saturated with TE [pH 8.0].) Transfer the aqueous (upper) phase (~100 µl) into a silanized plastic tube.

   Note: The sample can be stored at –20°C until needed.

   phenol, chloroform (see Appendix for Caution)

9. Purify partially double-stranded circular DNA on a HAP column as described on pp. 137–139.

PROTOCOL

Purification of Partially Double-stranded Circular DNA on HAP Columns

1. Prepare the HAP column as described on p. 129, step 1.

   Note: For further details on running the HAP column, see pp. 124–125 and 129–132.

2. Add the following components to the sample from pp. 135–136 and mix gently:

sonicated, denatured salmon sperm DNA (10 μg/μl)	5 μl
1 M sodium phosphate buffer (for preparation, see Appendix)	14.3 μl
120 mM sodium phosphate/EDTA/ SDS (for preparation, see p. 129)	2 ml

   Note: The 1 M sodium phosphate buffer is added to provide a final concentration of 120 mM sodium phosphate.

3. Heat at 60°C for 2–5 minutes to equilibrate the sample to the temperature of the column.

4. Load the approximately 2-ml sample on the column and pass it through the column, collecting the flow-through. Set aside the flow-through in a silanized 50-ml polypropylene tube (Falcon 2098) (tube A).

   Note: Silanized tubes should be used throughout the protocol.

5. Wash the column three times with 2 ml of 120 mM sodium phosphate/ EDTA/SDS, collecting the flow-through from each wash. Pool the flow-through from the washes with the flow-through in tube A.

   Note: The partially double-stranded circular DNA binds to HAP, whereas any single-stranded circular DNA flows through in steps 4–5. The flow-through in tube A will contain all of the clones without an oligo(dA) tract at the 3′ end.

6. Wash the column three times with 2 ml of 400 mM sodium phosphate/ EDTA/SDS, collecting the eluate. Pool all three eluates in a second silanized 50-ml polypropylene tube (tube B).

   Note: The partially double-stranded circular DNA (the HAP-bound fraction) should elute in these washes.

400 mM Sodium phosphate/EDTA/SDS

Component and final concentration	Amount to add per 50 ml
400 mM sodium phosphate	20 ml of 1 M sodium phosphate buffer (pH 6.8) (see Appendix)
10 mM EDTA	1 ml of 0.5 M (pH 8.0)
1% **SDS**	5 ml of 10%
H_2O	24 ml

Mix the components in a 50-ml polypropylene tube. Store at room temperature for up to 1 month.

SDS (see Appendix for Caution)

7. Repurify the 6 ml of pooled eluate in tube B.

   Note: This repurification step minimizes the background of single-stranded circular DNA (caused by nonspecific binding to the column) in the HAP-bound fraction.

   a. Add 14 ml of H_2O to tube B (to adjust the sodium phosphate buffer concentration to 120 mM), and then add 50 µg of sonicated, denatured salmon sperm DNA.

   b. Reequilibrate the HAP column by washing it three times with 2 ml of 120 mM sodium phosphate/EDTA/SDS.

   c. Heat the sample from step a at 60°C for 2 minutes.

   d. Load the sample on the column and pass it through the column. Discard the flow-through.

   e. Wash the column three times with 2 ml of 120 mM sodium phosphate/EDTA/SDS. Discard the wash.

   f. Wash the column three times with 2 ml of 400 mM sodium phosphate/EDTA/SDS, collecting the eluate. Pool all three eluates in a third silanized 50-ml polypropylene tube (tube C).

      Note: The partially double-stranded circular DNA should elute in these washes.

8. Process the samples in tube A and tube C separately.

   For the flow-through from the first HAP column (tube A):

   a. Follow steps 9–13 on pp. 131–132.

      Note: If the DNA is not to be used immediately, it can be stored in ethanol at –20°C indefinitely.

   b. Dissolve the DNA in 5 µl of TE (pH 8.0).

   c. Proceed as described on pp. 140–141.

For the final eluate from the column in step 7 (i.e., the HAP-bound DNA; tube C):

a. Follow steps 9–12 on pp. 131–132.

b. Measure the exact final volume of the desalted HAP-bound sample. Set aside 1 µl of the desalted HAP-bound sample until needed on p. 142 for determination of the cpm of α-^{32}P and the mass of DNA available for reassociation.

   Note: The sample can be stored at −20°C until needed.

c. Use the remainder of the sample to proceed as described on pp. 144–146.

PROTOCOL

Primer-extension Reaction for Converting Single-stranded Circular DNA into Double-stranded Circular DNA for a Sublibrary

1. Combine the components below in a silanized 0.5-ml microcentrifuge tube and mix. Centrifuge briefly to collect the liquid at the bottom of the tube.

single-stranded circular DNA (tube A from pp. 137–138)	5 µl
10x primer-extension buffer (for preparation, see p. 135)	1 µl
10 mM DTT	1 µl
10 mM dNTP mixture (see Appendix)	1 µl
bacteriophage M13 sequencing primer (0.01 µg/µl)	1 µl

   Notes: The bacteriophage M13 sequencing primer is 5'GTAAAACGACGGCCAGT3'.
       Silanized tubes should be used throughout the protocol.

2. Overlay with 100 µl of light mineral oil (Sigma M 5904).

   Notes: This step is not necessary if a thermal cycler with a top heating block will be used in step 3.
       If a mineral oil overlay is used, insert a pipette through the oil to add reagents or to remove the sample (but not the oil) in the steps below.

3. Incubate sequentially as follows:

   65°C for 5 minutes
   50°C for 15 minutes
   37°C for 2 minutes

   Note: Although these incubations can be performed in a water bath, use of a thermal cycler may be more convenient.

4. Add 1 µl of the Klenow fragment of *E. coli* DNA polymerase I (5000 units/ml; Amersham) and mix gently. Centrifuge briefly to collect the liquid at the bottom of the tube.

5. Immediately incubate at 37°C for 15 minutes.

6. Add 1 µl of 0.5 M EDTA (pH 8.0) and mix thoroughly by vortexing. Centrifuge briefly to collect the liquid at the bottom of the tube.

7. Add 90 µl of TE (pH 8.0) and 100 µl of **phenol:chloroform**:isoamyl alcohol (25:24:1) and extract (see Appendix). Use phenol saturated with TE (pH

8.0). Transfer the aqueous (upper) phase into a new silanized micro-centrifuge tube.

phenol, chloroform (see Appendix for Caution)

8. Precipitate the DNA with 0.1 volume of 3 M sodium acetate (pH 7.0) and 2 volumes of absolute ethanol and dry the pellet under vacuum as described for RNA on p. 71, step 10.

   Note: If the DNA is not to be used immediately, it can be stored in ethanol at −20°C indefinitely.

9. Dissolve the DNA in TE (pH 8.0) at a final concentration of ≤50 ng/ml.

10. Use 1 µl of the DNA to transform each of three 25-µl aliquots of DH10B bacteria competent for electroporation, and then prepare plasmid DNA. Follow the protocol on pp. 119–120, steps 1–7, but use 100 ml of medium in a 500-ml flask in step 2d and 75 µl of ampicillin solution in step 5.

    Notes: Each transformation will contain ≤50 ng/ml of DNA.
        This DNA represents a sublibrary of clones without an oligo(dA) tract at the 3′ end. These clones therefore correspond to mRNAs with an internal *Not*I site. This DNA is not combined with the DNA from the normalized library prepared on pp. 144–146, but it is kept to preserve the representation of mRNAs with internal *Not*I sites in the final collection of clones.

PROTOCOL

Calculation of the Mass of DNA Synthesized in the Controlled Primer-extension Reaction and the Length of Hybridization Required for the Desired C_0t

1. Calculate the mass of DNA synthesized in the controlled primer-extension reaction.

   Note: It is necessary to calculate the amount of DNA synthesized before the length of hybridization required to achieve the desired C_0t can be determined. The reassociation reaction described on pp. 144–146 should be performed to a C_0t of 5–20 seconds-moles/liter. Given the amounts specified in the protocol, this C_0t can typically be achieved within 12–24 hours. However, to calculate how long the reaction should be allowed to proceed to achieve a C_0t of 5–20 seconds-moles/liter, calculate the exact amount of hybridizable DNA.

   a. Determine the total amount of dNTPs present in the 100-µl reaction mixture in step 1, p. 135 as follows: Each dNTP is present at a final concentration of 1 mM in the 100-µl reaction mixture. Accordingly, there are 100 nmoles of each dNTP or 400 nmoles of total dNTPs (*A*) present in the reaction mixture.

   b. Determine the cpm of [α-³²P]dCTP (incorporated plus unincorporated) in the 0.5-µl aliquot spotted on the filter in step 6, p. 136, by placing the dried filter in a vial containing 3 ml of a **toluene**-based scintillation fluid and measuring the radioactivity in a liquid scintillation counter. Calculate the total cpm of [α-³²P]dCTP as follows: total cpm of [α-³²P]dCTP (incorporated plus unincorporated) in the final reaction volume of 100 µl (*B*) = (cpm from the 0.5-µl aliquot x 100 µl)/0.5 µl.

   radioactive substances, toluene (see Appendix for Caution)

   c. Calculate the specific activity of the synthesized cDNAs as follows: specific activity (in cpm/nmole of dNTPs) (*C*) = total cpm of [α-³²P]dCTP (*B*)/total nmole of dNTPs present in the reaction mixture (*A*).

   d. Determine the cpm of α-³²P in the 1-µl aliquot of the desalted HAP-bound sample set aside in step 8b, p. 139, as described on pp. 108–109, step 4. Calculate the total cpm in the HAP-bound fraction as follows: total cpm (*D*) = (cpm from the 1-µl aliquot x total measured volume of the desalted HAP-bound sample)/1 µl volume of the aliquot.

   e. Determine the total mass of double-stranded DNA as follows: The total nanomoles of dNTPs incorporated during the primer-extension reaction (*E*) = total cpm in the HAP-bound fraction (*D*)/specific activity (*C*). Given that the molecular mass of a dNTP is 330 daltons, the total mass (in ng) of single-stranded DNA synthesized in the reaction (*F*) = total nmoles of dNTPs incorporated (*E*) x 330 ng/nmole. To obtain the total mass (in ng) of double-stranded DNA (*G*), multiply this result by 2.

2. Determine the length of hybridization required to achieve a C_0t of 5–20 seconds-moles/liter from the following formula:

time in hours = C_0t ÷ ([OD$_{260}$/2]×[correction factor for 50% formamide]×[correction factor for salt])

Notes: An OD reading of 0.99 is extrapolated from the calculated amount of synthesized DNA (the reassociation reaction mixture typically contains 150 ng of complementary single-stranded DNA in a volume of 5 μl and an OD of 1 for single-stranded DNA corresponds to a 36 μg/ml solution). The correction factor for 50% formamide is 0.45. The correction factor for the salt concentration is 0.5 M/0.12 M, or 4.2. If the length of the hybridization is 16 hours, the calculated C_0t = 0.99/2 × 0.45 × 4.2 × 16 = 15 seconds-moles/liter. For further discussion, see pp. 61–63.

Reassociation

1. Verify that the extension products in the desalted HAP-bound sample (tube C from pp. 137–139) are of the desired length (200 ± 20 bp).

   Note: The desalted HAP-bound sample contains the partially double-stranded circular DNA generated by the controlled primer-extension reaction.

   a. Prepare a 5% acrylamide/8.3 M urea gel in 1x TBE buffer.

      i. Combine the components below in a 200-ml beaker. Stir until the urea is completely dissolved. (The solution will be slightly cool.) Filter through Whatman No. 1 paper.

urea (USB 23040)	50 g
5x TBE buffer	20 ml
40% acrylamide:bisacrylamide (19:1) solution	12.5 ml
H_2O	to make 100 ml

      40% Acrylamide:bisacrylamide (19:1) solution

Component and final concentration	Amount to add per 0.5 liter
19 parts **acrylamide** (Boehringer Mannheim 100 137)	190 g
1 part **bisacrylamide** (m.w. = 154.17; Boehringer Mannheim 100 140)	10 g
H_2O	to make 0.5 liter

      Mix the components in a 1-liter beaker. Filter through Whatman No. 1 paper. Store in dark bottles at room temperature for up to 6 months.

      acrylamide, bisacrylamide (see Appendix for Caution)

      ii. Just before the gel is cast, add the following:

10% ammonium persulfate	0.625 ml
TEMED (Bio-Rad 161-0800)	70 µl

      10% Ammonium persulfate
      In a 15-ml polypropylene tube, dissolve 1 g of ammonium persulfate (IBI 70080) in 10 ml of H_2O. Store at 4ºC for up to 1 week. (Ammonium persulfate breaks down rapidly in solution to produce forms that are inactive in polymerizing acrylamide.)

      iii. Pour the gel and allow the solution to polymerize. Use the gel immediately.

b. Transfer 10 μl of the desalted HAP-bound sample into a silanized 1.5-ml microcentrifuge tube and dry under vacuum in a SpeedVac Concentrator.

Note: Silanized tubes should be used throughout the protocol.

c. Dissolve the DNA in 10 μl of deionized formamide (for preparation, see p. 82). Heat at 80°C for 3 minutes, and then add 1 μl of BP/XC/formamide gel-loading solution.

BP/XC/Formamide gel-loading solution

Component and final concentration	Amount to add per 10 ml
0.2% bromophenol blue	20 mg
0.2% xylene cyanole FF	20 mg
deionized **formamide**	10 ml

Mix the components in a 15-ml polypropylene tube to dissolve the bromophenol blue and the xylene cyanole FF. Divide into 1-ml aliquots in 1.5-ml microcentrifuge tubes and store at –20°C for up to 6 months.

formamide (see Appendix for Caution)

d. Load the sample on the 5% acrylamide/8.3 M urea gel along with an end-labeled DNA molecular-weight marker (e.g., *Msp*I-digested pBR322 DNA; 9–622 bp). Perform electrophoresis until the bromophenol blue dye reaches the bottom of the gel.

Note: Electrophoresis can be performed on a Hoefer SE400 apparatus (or equivalent) with 18-cm x 16-cm glass plates (model SE6119SM-50), 1-mm thick spacers, and 8-mm wide wells.

e. Remove one of the glass plates from the gel, cover the gel with layers of paper towels and another glass plate, and dry for several hours. Alternatively, place the gel on a gel dryer, cover with a stack of blotting paper, and dry under vacuum for a few hours. Cover the dried gel with plastic wrap and X-ray film.

f. Determine the size of the extension product by autoradiography at –80°C overnight with an intensifying screen.

2. Precipitate the DNA in the remaining desalted HAP-bound sample (from pp. 137–139) with 0.1 volume of 3 M sodium acetate (pH 7.0) and 2 volumes of absolute ethanol and dry the pellet under vacuum as described for RNA on p. 71, step 10.

Note: If the DNA is not to be used immediately, it can be stored in ethanol at –20°C indefinitely.

3. Dissolve the DNA in 2.5 μl of deionized formamide.

4. Set up and perform the reassociation as follows:

a. Overlay the 2.5 μl of DNA with 10 μl of light mineral oil.

Note: There are typically 100–200 ng of double-stranded DNA (or 0.75–1.5 pmoles). (The amount of double-stranded DNA can be calculated as described on p. 142.) If more DNA is present at the start of the reaction, it will take less time to reassociate to the desired C_0t of approximately 5–20 seconds-moles/liter in step f.

b. Heat at 80°C for 3 minutes.

c. Add 1 μl (~550 pmoles) of oligo(dT)$_{25-30}$ (5 μg/μl; Pharmacia 27-7839-01) and mix. Centrifuge briefly to collect the liquid at the bottom of the tube.

Notes: This oligonucleotide is used to block the oligo(dA)$_{18-36}$ tracts present at the 3′ end of the single-stranded circular DNA and thus prevent the reassociation between a circular DNA and an extension product through these tails.
  To add each reagent to the DNA in these steps, pipette the reagent under the oil, mix by tapping the tube, and centrifuge briefly to collect the liquid.

d. Heat at 80°C for 1 minute.

e. Add the components below in the order listed and mix. Centrifuge briefly to collect the liquid at the bottom of the tube.

5 M NaCl	0.5 μl
100 mM Tris-Cl (pH 8.0)/	
100 mM EDTA (pH 8.0)	0.5 μl
H$_2$O	0.5 μl

Note: Varying the formamide (added in step 3) or NaCl concentration to make the hybridization more permissive is not easy. If hybridization is not performed under very stringent conditions, some hybrids (cross-hybrids) may melt while they are in the HAP column buffer before they have a chance to bind to the column.

100 mM Tris-Cl (pH 8.0)/100 mM EDTA (pH 8.0)

Component and final concentration	Amount to add per 10 ml
100 mM Tris-Cl	1 ml of 1 M (pH 8.0 at 25°C)
100 mM EDTA	2 ml of 0.5 M (pH 8.0)
H$_2$O	7 ml

Store at –20°C for 6 months.

f. Incubate at 42°C for 12–24 hours.

Notes: The sample can be stored at –20°C until needed.
  The length of hybridization required to achieved a C_0t of approximately 5–20 seconds-moles/liter can be calculated as described on p. 143.

5. Purify the remaining single-stranded circular DNA and prepare the amplified normalized cDNA library as described on pp. 147–148.

Note: The fraction that remains single-stranded after reassociation to a C_0t of approximately 5–20 seconds-moles/liter represents the unamplified normalized library.

PROTOCOL

Purification of the Remaining Single-stranded Circular DNA and Conversion into Double-stranded Circular DNA to Create an Amplified Normalized cDNA Library

1. Prepare two HAP columns as described on p. 129, step 1.

   Note: For further details on running the HAP column, see pp. 124–125 and 129–132.

2. Transfer the 5-μl reassociation reaction mixture (from pp. 144–146) into a new silanized plastic tube to remove the oil overlay.

   Note: To transfer the reaction mixture, insert a pipette through the mineral oil overlay and transfer the sample but not the oil. Alternatively, transfer the reaction mixture and the oil onto a piece of Parafilm M and then transfer the reaction mixture into the tube.

3. Add the following components to the sample and mix gently by inverting the tube:

sonicated, denatured salmon sperm DNA (10 μg/μl)	5 μl
120 mM sodium phosphate/EDTA/SDS (for preparation, see p. 129)	2 ml

4. Heat at 60°C for 2–5 minutes to equilibrate the sample to the temperature of the column.

5. Load the approximately 2-ml sample (from step 4) on the first column and pass it through the column, collecting the flow-through. Set aside the flow-through in a silanized 50-ml polypropylene tube (Falcon 2098).

   Note: The flow-through contains the unreassociated (single-stranded) DNA.

6. Wash the column three times with 2 ml of 120 mM sodium phosphate/EDTA/SDS, collecting the flow-through from each wash. Pool the flow-through from the washes with the flow-through in the tube set aside in step 5.

7. Repurify the 8 ml of pooled flow-through on the second HAP column.

   Note: This step is routinely performed to minimize the possibility of contamination of the single-stranded fraction with small amounts of reassociated double-stranded DNA may have flowed through some (invisible) small channels in the HAP column. Using a fresh column is recommended instead of reequilibrating the original column.

   a. Heat the sample at 60°C for 2 minutes.

   b. Load the sample on the column and pass it through the column, collecting the 8 ml of flow-through.

c. Wash the column three times with 2 ml of 120 mM sodium phosphate/ EDTA/SDS, collecting the flow-through from each wash. Pool the flow-through from the washes with the flow-through from step b.

8. Follow steps 9–13 on pp. 131–132 to extract, desalt, and precipitate the 14 ml of pooled, repurified flow-through. In step 9c, however, reextract with 30 ml of **sec-butanol** that has not been saturated with H_2O until the volume of the aqueous phase is 1–5 ml (instead of just one extraction).

Note: If the DNA is not to be used immediately, it can be stored in ethanol at –20°C indefinitely.

sec-butanol (see Appendix for Caution)

9. Dissolve the DNA in 5 μl of TE (pH 8.0).

10. Convert the purified single-stranded circular DNA (i.e., the normalized cDNA population) into double strands by primer extension as described for the sublibrary on pp. 140–141, steps 1–9.

Note: This procedure will improve the electroporation efficiency.

11. Use 1 μl of the DNA to transform each of three 25-μl aliquots of DH10B bacteria competent for electroporation, and then prepare plasmid DNA. Follow the protocol on pp. 119–120, steps 1–5 and step 7, but use 100 ml of medium in a 500-ml flask in step 2d and 75 μl of ampicillin solution in step 5. There is no need to prepare a frozen glycerol stock of the bacterial culture.

Note: The final plasmid preparation represents the amplified normalized cDNA library.

SCREENING NORMALIZED cDNA LIBRARIES CONSTRUCTED IN PHAGEMID VECTORS

cDNA libraries constructed in plasmid vectors can be screened by colony hybrization with specific probes according to standard protocols (Grunstein and Hogness 1975). This procedure can be facilitated by the use of normalized libraries, since fewer colonies need to be screened to isolate a clone that may otherwise be underrepresented in the starting (nonnormalized) library. As a general rule, screening a total of 10^5 recombinants should be sufficient for isolating the great majority of clones from a normalized library. Briefly, the procedure involves the transfer of bacterial colonies from agar plates onto a solid support (nitrocellulose filters or nylon membranes), where the cells are lysed and their DNA is denatured and immobilized for subsequent detection by hybridization to labeled probes.

Nylon membranes are more durable than nitrocellulose filters and will withstand several rounds of hybridization and washing at elevated temperatures. They are therefore preferred when colonies are to be screened sequentially with a number of different probes. As a rule of thumb, all colony-hybridization experiments should be performed with duplicate membranes/filters to avoid the pursuit of false positives.

PROTOCOL

Screening Normalized cDNA Libraries by Colony Hybridization

1. Determine the titer of the bacterial culture of the normalized cDNA library (in cfu/μl).

   Note: The colonies should be clearly visible and should be approximately the size of a pin head.

2. Spread 20,000 bacteria on each of five dry prewarmed (37°C) 2x YT agar plates (15-cm diameter) containing ampicillin (75 μg/ml). Number the plates from 1–5. Incubate at 37°C for approximately 12 hours. Do not overgrow the colonies.

   Note: A turntable or equivalent should be used to facilitate spreading the colonies evenly.

3. Chill the plates at 4°C for 1 hour.

4. Label five pairs of nylon membrane disks (DuPont NEN Research Products NEF-978A) 1A–5A and 1B–5B. Prepare replica membranes for each of the five numbered plates using a pair of labeled disks (A and B) as follows:

   a. Lay the first nylon disk (the master membrane A) on top of the colonies. Allow the disk to become evenly wetted, and then allow it to sit on the plate for 1 minute.

      Notes: Wear gloves at all times when handling the membranes.
         The center of the disk should be allowed to contact the center of the plate first. The rest of the disk should then be carefully spread over the rest of the plate so that air bubbles do not become trapped.
         The disk should not be prewetted but it should immediately become wet as it contacts the agar.

   b. Moisten the tip of a needle with India ink. Mark the disk position by stabbing the needle through the disk and into the agar at three asymmetric positions.

      Note: The ink will facilitate detection of the holes.

   c. Use forceps to lift the membrane and transfer it onto a stack of paper towels with the bacterial colonies facing up.

      Note: There should be no visible colonies left on the original plate at this point. All of the colonies should have been transferred onto the membrane.

   d. Incubate the plate at 37°C for 6–8 hours to regrow the colonies.

   e. Wrap the plate with plastic wrap and store at 4°C for up to 2 weeks.

      Note: This is the master plate for step 13.

f. Prewet the second nylon disk (B) by laying it on a prewarmed (37ºC) dry 2x YT agar plate containing ampicillin (75 µg/ml). Immediately lay the wetted disk on top of the master membrane.

g. Place a 1/2-inch stack of paper towels on top of the pair of membrane disks and apply even pressure downward for 15–30 seconds.

    Note: Be sure to promote even contact over the entire surface of the two disks.

h. Remove the upper stack of paper towels. Flip over the pair of membrane disks and locate the three holes on the master membrane (disk A). Use the same needle moistened with India ink to mark disk B at exactly the same three positions.

i. Peel the membranes apart. Lay disk B, with the (almost invisible) colonies facing up, on the same prewarmed dry 2x YT agar plate containing ampicillin (75 µg/ml) that was used to prewet it in step f. Store the master membrane at room temperature until needed for step 5.

    Notes: The master membrane (disk A) can be used as a template to produce additional replica membranes as desired.
        There should be no need to grow the colonies on the master membrane any further.

j. Incubate the plate with disk B at 37ºC for 6–8 hours to grow the colonies.

5. Denature the DNA and fix it to the membranes as follows:

    a. Lay a sheet of plastic wrap on the bench top. Pipette 20 (two pools for each of the five pairs of disks) 0.75-ml pools of denaturing solution for colony hybridization onto the plastic wrap.

        Denaturing solution for colony hybridization

Component and final concentration	Amount to add per 1 liter
1.5 M NaCl	87.7 g
0.5 N **NaOH**	50 ml of 10 N
H_2O	to make 1 liter

        In a 1-liter beaker, dissolve the NaCl in 0.8 liter of H_2O. Add the NaOH and mix thoroughly. Adjust the volume to 1 liter with H_2O. Store at 4ºC for up to 1 month.

        NaOH (see Appendix for Caution)

    b. Place each disk, colony side up, onto a pool of denaturing solution. Stretch the plastic wrap to make sure that all of the disks wet evenly. Allow the disks to sit in the denaturing solution for 2 minutes.

    c. Repeat step a and transfer each disk, colony side up, onto a fresh pool of denaturing solution. Allow the disks to sit in the denaturing solution for 1 minute.

    d. Pipette 20 0.75-ml pools of neutralizing solution for colony hybridization onto a new sheet of plastic wrap.

Neutralizing solution for colony hybridization

Component and final concentration	Amount to add per 1 liter
1.5 M NaCl	87.7 g
0.5 M Tris-Cl (pH 8.0 at 22°C)	60.6 g of Tris base
H₂O	to make 1 liter

In a 1-liter beaker, dissolve the NaCl and the Tris base in 0.8 liter of H₂O. Adjust the pH to 8.0 with **concentrated HCl** and mix thoroughly. Adjust the volume to 1 liter with H₂O. Store at 4°C for up to 1 month.

concentrated acids (see Appendix for Caution)

e. Place each disk, colony side up, onto a pool of neutralizing solution. Allow the disks to sit in the neutralizing solution for 2 minutes.

f. Repeat step d and transfer each disk, colony side up, onto a fresh pool of neutralizing solution for 1 minute.

g. Pipette 20 0.75-ml pools of 2x SSC onto a new sheet of plastic wrap.

h. Place each disk, colony side up, onto a pool of 2x SSC. Allow the disks to sit in the SSC for 2 minutes.

i. Repeat step g and transfer each disk, colony side up, onto a fresh pool of 2x SSC for 1 minute.

j. Place the disks, colony side up, on a sheet of blotting paper (e.g., VWR Scientific 238) to remove excess 2x SSC. Transfer the disks onto a new sheet of blotting paper and allow to dry at room temperature.

Note: In a humid environment, air drying may not always be complete. Therefore, baking at 80°C for 1–2 hours is recommended to fix the DNA to the membrane and to minimize possible loss of target nucleic acids. Alternatively, cross-linking can be performed with **UV** light (e.g., by using a Stratalinker [Stratagene] as specified by the manufacturer).

UV radiation (see Appendix for Caution)

k. Store the dried membrane disks in a sealed plastic bag at room temperature until needed for prehybridization.

6. Immediately before the membranes are prehybridized, immerse them in prewashing solution at 55°C for 30 minutes.

Prewashing solution for colony hybridization

Component and final concentration	Amount to add per 1 liter
1.5 M NaCl	87.7 g
0.1% **SDS**	1 g
50 mM Tris-Cl	50 ml of 1 M (pH 8.0 at 22°C)
1 mM EDTA	2 ml of 0.5 M (pH 8.0)
H₂O	to make 1 liter

In a 1-liter beaker, dissolve the NaCl in 0.8 liter of H_2O. Add the SDS and stir slowly on a hot plate until dissolved. Add the Tris-Cl and the EDTA. Continue to stir slowly and adjust the volume to 1 liter with H_2O. Sterilize the solution by passing it through a 0.22-μm filter. Store at 22°C for up to 6 months.

SDS (see Appendix for Caution)

7. Prehybridize the membranes in prehybridization solution for colony hybridization at 42°C for 30 minutes. Use at least 2 ml of prehybridization solution per membrane (typically, five pairs of membranes in 20 ml of prehybridization solution).

Prehybridization solution for colony hybridization

Component and final concentration	Amount to add per 20 ml
6x SSC	6 ml of 20x
2% SDS	4 ml of 10%
50% deionized **formamide**	10 ml
5x Denhardt's reagent	1 ml of 100x
100 μg/ml sonicated, denatured salmon sperm DNA	2 mg
H_2O	to make 20 ml

Mix the components in a 50-ml plastic tube by inverting the tube. Make fresh each time.

formamide (see Appendix for Caution)

8. Transfer the membranes into hybridization solution for colony hybridization and incubate at 42°C overnight. Use at least 2 ml of hybridization solution per membrane (typically, five pairs of membranes in 20 ml of hybridization solution).

Notes: If an RNA probe is used instead of a DNA probe, the incubation should be performed at 50–55°C.

cDNA probes generated from gel-purified inserts at least 500 bp in length yield very strong hybridization signals that are typically detected after a 2–4-hour exposure.

Hybridization solution for colony hybridization
Add denatured, [α-^{32}P]dCTP-labeled DNA probe to prehybridization solution for colony hybridization to make a final probe concentration of 5 ng/ml. Probes should be labeled by random priming and denatured by heating at 95°C for 10 minutes (Feinberg and Vogelstein 1983, 1984; see also Wolff and Gemmill 1997, pp. 60–61).

radioactive substances (see Appendix for Caution)

9. Wash the membranes (with constant agitation) in 100–200 ml of 2x SSC/2% SDS once at room temperature for 20 minutes and twice at 65°C for 15 minutes each. Use a fresh batch of washing solution for each wash.

10. Wash the membranes in 0.5x SSC/0.5% SDS at 65°C for 15 minutes.

11. Place the membranes on blotting paper to remove excess washing solution (do not allow the membranes to dry out). Lay the membranes, colony side up, on top of a used piece of X-ray film covered with plastic wrap. Cover the membranes with a sheet of plastic wrap. Spot radioactive India ink on a few pieces of tape and allow to air dry. Place the pieces of tape asymmetrically around the membranes. In the darkroom, place an unexposed piece of X-ray film on top.

    Notes: To prepare radioactive ink, see Sambrook et al. (1989, p. 1.103).
    It may be necessary to arrange the membranes so that more than one piece of film is exposed.

12. Autoradiograph at −80°C for 1–4 hours with an intensifying screen.

13. Using the spots of radioactive ink as a guide, align the film(s) with the membranes to identify all positive (hybridizing) colonies. To select colonies for rescreening, prepare each positive colony as follows:

    a. For each region of the membrane that contains a positive colony, use a sterile toothpick or pipette tip or loop to transfer the corresponding region from the master plate (stored in step 4e) into 1 ml of 2x YT medium containing ampicillin (75 µg/ml) in a 5-ml tube. Incubate at 37°C with agitation at 225 rpm for 2 hours.

       Note: The region of hybridization is not likely to align with only one colony. Be sure to align the plate carefully and transfer a colony in the exact region of the hybridization signal.

    b. Mix 1 µl of each culture from step a with 250 µl of 2x YT medium. Also prepare several tenfold dilutions of each culture from step a and mix 1 µl of each with 250 µl of 2x YT medium. Spread each mixture on a separate large (15-cm diameter) 2x YT agar plate containing ampicillin (75 µg/ml). Incubate at 37°C overnight.

    c. Select plates containing approximately 250 colonies each for rescreening. Prepare replica membranes and hybridize them as described in steps 3–12 above.

       Note: This time each positive hybridization signal on the autoradiograph should be clearly identifiable as a single isolated colony. Several of these isolated colonies will contain the same insert since they are clones of the original colony with the positive hybridization signal in the original screening.

    d. Select the most isolated colony for propagation in 2x YT under selection with ampicillin and purification of the plasmid DNA as described on p. 90, steps 1–4. Use the DNA for sequence analysis of the isolated transcript.

REFERENCES

Adams, M.D., A.R. Kerlavage, C. Fields, and J.C. Venter. 1993a. 3,400 expressed sequence tags identify diversity of transcripts in human brain. *Nat. Genet.* **4:** 256–267.

Adams, M.D., M.B. Soares, A.R. Kerlavage, C. Fields, and J.C. Venter. 1993b. Rapid cDNA sequencing (expressed sequence tags) from a directionally cloned human infant brain cDNA library. *Nat. Genet.* **4:** 373–380.

Adams, M.D., M. Dubnick, A.R. Kerlavage, R. Moreno, J.M. Kelley, T.R. Utterback, J.W. Nagle, C. Fields, and J.C. Venter. 1992. Sequence identification of 2,375 human brain genes. *Nature* **355:** 632–634.

Adams, M.D., J.M. Kelley, J.D. Gocayne, M. Dubnick, M.H. Polymeropoulos, H. Xiao, C.R. Merril, A. Wu, B. Olde, R.F. Moreno, A.R. Kerlavage, W.R. McCombie, and J.C. Venter. 1991. Complementary DNA sequencing: Expressed sequence tags and Human Genome Project. *Science* **252:** 1651–1656.

Ausubel, F.M., R. Brent, R.E. Kingston, D.D. Moore, J.G. Seidman, J.A. Smith, and K. Struhl, eds. 1995. *Current protocols in molecular biology,* vol. 1. Greene/Wiley-Interscience, New York.

Bishop, J.O., J.G. Morton, M. Rosbash, and M. Richardson. 1974. Three abundance classes in HeLa cell messenger RNA. *Nature* **250:** 199–204.

Bonaldo, M.F., G. Lennon, and M.B. Soares. 1996. Normalization and subtraction: Two approaches to facilitate gene discovery. *Genome Res.* **6:** 791–806.

Britten, R.J., D.E. Graham, and B.R. Neufeld. 1974. Analysis of repeating DNA sequences by reassociation. *Methods Enzymol.* **29:** 363–441.

Chirgwin, J.M., A.E. Przybyla, R.J. MacDonald, and W.J. Rutter. 1979. Isolation of biologically active ribonucleic acid from sources enriched in ribonucleases. *Biochemistry* **18:** 5294–5299.

Chomczynski, P. and ??. Mackey. 1995. Modification of the TRI Reagent™ procedure for isolation of RNA from polysaccharide- and proteoglycan-rich sources. *Biotechniques* **19:** 942–945.

Chomczynski, P. and N. Sacchi. 1987. Single-step method of RNA isolation by acid guanidinium thiocyanate–phenol-chloroform extraction. *Anal. Biochem.* **162:** 156–159.

Compton, T. 1990. Degenerate primers for DNA amplification. In *PCR protocols: A guide to methods and applications* (ed. M.A. Innis et al.), pp. 39–45. Academic Press, New York.

D'Alessio, J.M. and G.F. Gerard. 1988. Second-strand cDNA synthesis with *E. coli* DNA polymerase I and RNase H: The fate of information at the mRNA 5′ terminus and the effect of *E. coli* DNA ligase. *Nucleic Acids Res.* **16:** 1999–2014.

D'Alessio, J.M., M.C. Noon, H.L. Ley III, and G.F. Gerard. 1987. One-tube double-stranded cDNA synthesis using cloned M-MLV reverse transcriptase. *Focus* **9:** 1–4.

Davidson, E.H. and R.J. Britten. 1979. Regulation of gene expression: Possible role of repetitive sequences. *Science* **204:** 1052–1059.

Dente, L. and R. Cortese. 1987. pEMBL: A new family of single-stranded plasmids for sequencing DNA. *Methods Enzymol.* **155:** 111–119.

Dotto, G.P., V. Enea, and N.D. Zinder. 1981. Functional analysis of bacteriophage f1 intergenic region. *Virology* **114:** 463–473.

Dower, W.J., J.F. Miller, and C.W. Ragsdale. 1988. High efficiency transformation of *E. coli* by high voltage electroporation. *Nucleic Acids Res.* **16:** 6127–6145.

Efstratiadis, A., F.C. Kafatos, A.M. Maxam, and T. Maniatis. 1976. Enzymatic in vitro synthesis of globin genes. *Cell* **7:** 279–288.

Enea, V. and N.D. Zinder. 1982. Interference resistant mutants of phage f1. *Virology* **122:** 222–226.

Fanning, S. and R.A. Gibbs. 1997. PCR in genome analysis. In *Genome analysis: A laboratory manual.* Vol. 1 *Analysing DNA* (ed. B. Birren et al.), pp. 249–299. Cold Spring Harbor Laboratory Press, Cold Spring Harbor, New York.

Favaloro, J., R. Treisman, and R. Kamen. 1980. Transcription maps of polyoma virus-specific RNA: Analysis by two-dimensional nuclease S1 gel mapping. *Methods Enzymol.* **65:** 718–749.

Feinberg, A.P. and B. Vogelstein. 1983. A technique for radiolabeling DNA restriction endonuclease fragments to high specific activity. *Anal. Biochem.* **132:** 6–13.

———. 1984. A technique for radiolabeling DNA restriction endonuclease fragments to high specific activity. Addendum. *Anal. Biochem.* **137:** 266–267.

Frohman, M.A. 1990. RACE: Rapid amplification of cDNA ends. In *PCR protocols: A guide to methods and applications* (ed. M.A. Innis et al.), pp. 28–38. Academic Press, New York.

Frohman, M.A., M.K. Dush, and G.R. Martin. 1988. Rapid production of full-length cDNAs from rare transcripts: Amplification using a single gene-specific oligonucleotide primer. *Proc. Natl. Acad. Sci.* **85:** 8998–9002.

Galau, G.A., W.H. Klein, R.J. Britten, and E.H. Davidson. 1977. Significance of rare mRNA sequences in liver. *Arch. Biochem. Biophys.* **179:** 584–599.

Glisin, V., R. Crkvenjakov, and C. Byus. 1974. Ribonucleic acid isolated by cesium chloride centrifugation. *Biochemistry* **13:** 2633–2637.

Goldberg, D.A. 1980. Isolation and partial characterization of the *Drosophila* alcohol dehydrogenase gene. *Proc. Natl. Acad. Sci.* **77:** 5794–5798.

Grunstein, M. and D.S. Hogness. 1975. Colony hybridization: A method for the isolation of cloned DNAs that contain a specific gene. *Proc. Natl. Acad. Sci.* **72:**

3961–3965.

Gubler, U. and B.J. Hoffman. 1983. A simple and very efficient method for generating cDNA libraries. *Gene* **25:** 263–269.

Heidecker, G. and J. Messing. 1987. A method for cloning full-length cDNA in plasmid vectors. *Methods Enzymol.* **154:** 28–41.

Huynh, T.V., R.A. Young, and R.W. Davis. 1985. Constructing and screening cDNA libraries in lambda gt10 and lambda gt11. In *DNA cloning* (ed. D.M. Glover), vol. I, pp. 49–78. IRL Press, Oxford.

Khan, A.S., A.S. Wilcox, M.H. Polymerapoulos, J.A. Hopkins, T.J. Stevens, M. Robinson, A.K. Orpana, and J.M. Sikela. 1992. Single pass sequencing and physical and genetic mapping of human brain cDNAs. *Nat. Genet.* **2:** 180–185.

Ko, M.S.H. 1990. An equalized cDNA library by the reassociation of short double-stranded cDNAs. *Nucleic Acids Res.* **18:** 5705–5711.

Kornberg, A. and T.A. Baker. 1992. RNA-directed DNA polymerases: Reverse transcriptases (RTs) and telomerase. In *DNA replication,* pp. 217–222. Freeman, New York.

Kroczek, R.A. 1989. Immediate visualization of blotted RNA in northern analysis. *Nucleic Acids Res.* **17:** 9497.

Land, H., M. Grez, H. Hauser, W. Lindenmaier, and G. Schutz. 1981. 5′-Terminal sequences of eukaryotic mRNA can be cloned with high efficiency. *Nucleic Acids Res.* **9:** 2251–2266.

Lehrach, H., D. Diamond, J.M. Wozney, and H. Boedtker. 1977. RNA molecular weight determinations by gel electrophoresis under denaturing conditions, a critical reexamination. *Biochemistry* **16:** 4743–4751.

Leis, J.P., I. Berkower, and J. Hurwitz. 1973. Mechanism of action of ribonuclease H from avian myeloblastosis virus and *Escherichia coli. Proc. Natl. Acad. Sci.* **70:** 466–470.

Lewin, B. 1994. Ribosomes provide a translation factory. In *Genes V,* pp. 233–252. Oxford University Press, New York.

Matoba, R., K. Okubo, N. Hori, A. Fukushima, and K. Matsubara. 1994. The addition of 5′-coding information to a 3′-directed cDNA library improves analysis of gene expression. *Gene* **146:** 199–207.

Matsubara, K. and K. Okubo. 1993. cDNA analyses in the human genome project. *Gene* **135:** 265–274.

McDonell, M.W., M.N. Simon, and F.W. Studier. 1977. Analysis of restriction fragments of T7 DNA and determination of molecular weights by electrophoresis in neutral and alkaline gels. *J. Mol. Biol.* **110:** 119–146.

Mead, D. and B. Kemper. 1986. In *Vectors: A survey of molecular cloning vectors and their uses* (ed. R.L. Rodriguez and D.T. Denhardt). Butterworth, Massachusetts.

————. 1988. Chimeric single-stranded DNA phage–plasmid cloning vectors. *Bio/Technology* **10:** 85–102.

Mead, D.A., E. Szczesna-Skorupa, and B. Kemper. 1986. Single-stranded DNA "blue" T7 promoter plasmids: A versatile tandem promoter system for cloning and protein engineering. *Protein Eng.* **1:** 67–74.

Murray, N.E., W.J. Brammar, and K. Murray. 1977. Lamboid phages that simplify the recovery of in vitro recombinants. *Mol. Gen. Genet.* **150:** 53–61.

Okayama, H. and P. Berg. 1982. High-efficiency cloning of full-length cDNA. *Mol. Cell. Biol.* **2:** 161–170.

Okubo, K. 1994. Body mapping of human genes. *Nippon Rinsho* **52:** 1095–1109.

Okubo, K., N. Hori, R. Matoba, T. Niiyama, A. Fukushima, Y. Kojima, and K. Matsubara. 1992. Large scale cDNA sequencing for analysis of quantitative and qualitative aspects of gene expression. *Nature Genet.* **2:** 173–179.

Omer, C.A. and A.J. Faras. 1982. Mechanism of release of the avian rotavirus tRNATrp primer molecule from viral DNA by ribonuclease H during reverse transcription. *Cell* **30:** 797–805.

Palazzolo, M.J., B.A. Hamilton, D. Ding, C.H. Martin, D.A. Mead, R.C. Mierendorf, K.V. Raghavan, E.M. Meyerowitz, and H.D. Lipshitz. 1990. Phage lambda cDNA cloning vectors for subtractive hybridization, fusion-protein synthesis and Cre–loxP automatic plasmid subcloning. *Gene* **88:** 25–36.

Patanjali, S.R., S. Parimoo, and S.M. Weissman. 1991. Construction of a uniform-abundance (normalized) cDNA library. *Proc. Natl. Acad. Sci.* **88:** 1943–1947.

Resnick, R., C.A. Omer, and A.J. Faras. 1984. Involvement of retrovirus reverse transcriptase-associated RNase H in the initiation of strong-stop (+) DNA synthesis and the generation of the long terminal repeat. *J. Virol.* **51:** 813–821.

Rosen, K.M. and L. Villa-Komaroff. 1990. An alternative method for the visualization of RNA in formaldehyde agarose gels. *Focus* **12:** 23–24.

Rougeon, F. and B. Mach. 1976. Stepwise biosynthesis in vitro of globin genes from globin mRNA by DNA polymerase of avian myeloblastosis virus. *Proc. Natl. Acad. Sci.* **73:** 3418–3422.

Saiki, R.K., D.H. Gelfand, S. Stoffel, S.J. Scharf, R. Higuchi, G.T. Horn, K.B. Mullis, and H.A. Erlich. 1988. Primer-directed enzymatic amplification of DNA with a thermostable DNA polymerase. *Science* **239:** 487–491.

Sambrook, J., E.F. Fritsch, and T. Maniatis. 1989. *Molecular cloning: A laboratory manual,* 2nd edition. Cold Spring Harbor Laboratory Press, Cold Spring Harbor, New York.

Sasaki, Y.F., D. Ayusawa, and M. Oishi. 1994. Construction of a normalized cDNA library by introduction of a semi-solid mRNA–cDNA hybridization system. *Nucleic Acids Res.* **22:** 987–992.

Sealey, P.G. and E.M. Southern. 1983. In *Gel electrophoresis of nucleic acids: A practical approach* (ed. D. Rickwood and B.D. Hames), p. 39. IRL Press, Oxford.

Short, J.M., J.M. Fernandez, J.A. Sorge, and W.D. Huse. 1988. Lambda ZAP: A bacteriophage lambda expression vector with in vivo excision properties. *Nucleic Acids Res.* **16:** 7583–7600.

Smith, C.L., S.K. Lawrance, G.A. Gillespie, C.R. Cantor, S.M. Weissman, and F.S. Collins. 1987. Strategies for mapping and cloning macroregions of mammalian genomes. *Methods Enzymol.* **151:** 461–489.

Soares, M.B. 1994. Construction of directionally cloned cDNA libraries in phagemid vectors. In *Automated DNA sequencing and analysis* (ed. M.D. Adams et al.), pp. 110–114. Academic Press, New York.

Soares, M.B., M.F. Bonaldo, P. Jelenc, L. Su, L. Lawton, and A. Efstratiadis. 1994. Construction and characterization of a normalized cDNA library. *Proc. Natl. Acad. Sci.* **91:** 9228–9232.

Thomas, P.S. 1983. Hybridization of denatured RNA transferred or dotted to nitrocellulose paper. *Methods Enzymol.* **100:** 255–266.

Ullrich, A., J. Shine, J. Chirgwin, R. Pictet, E. Tischer, W.J. Rutter, and H.M. Goodman. 1977. Rat insulin genes: Construction of plasmids containing the coding sequences. *Science* **196:** 1313–1319.

Vieira, J. and J. Messing. 1982. The pUC plasmids, an M13mp7-derived system for insertion mutagenesis with synthetic universal primers. *Gene* **19:** 259–268.

———. 1987. Production of single-stranded plasmid DNA. *Methods Enzymol.* **153:** 3–11.

———. 1991. New pUC-derived cloning vectors with different selectable markers and DNA replication origins. *Gene* **100:** 189–194.

Weissman, S.M. 1987. Molecular genetic techniques for mapping the human genome. *Mol. Biol. Med.* **4:** 133–143.

Young, R.A. and R.W. Davis. 1983a. Efficient isolation of genes by using antibody probes. *Proc. Natl. Acad. Sci.* **80:** 1194–1198.

———. 1983b. Yeast RNA polymerase II genes: Isolation with antibody probes. *Science* **222:** 778–782.

Wolff, R. and R. Gemmill. 1997. *Genome analysis: A laboratory manual.* Vol. 1 *Analyzing DNA* (ed. B. Birren et al.), pp. 1–81. Cold Spring Harbor Laboratory Press, Cold Spring Harbor, New York.

Zinder, N.D. and J.D. Boeke. 1982. The filamentous phage (Ff) as vectors for recombinant DNA: A review. *Gene* **19:** 1–10.

3

Direct cDNA Selection

ANDREW S. PETERSON

The ability of nucleic acids to form specific duplex hybrids is essential for DNA replication and for the synthesis of mRNA from DNA templates. Direct cDNA selection takes advantage of this specificity to purify transcribed sequences from complex mixtures of cDNA (Lovett et al. 1991; Parimoo et al. 1991; Lovett 1994). In this technique, cloned, double-stranded genomic DNAs and cDNAs derived from genes transcribed from these cloned genomic DNAs are separately denatured into their constituent strands and are then combined and allowed to anneal. Most of the cloned genomic DNA fragments (which are in excess) reanneal to each other, but some hybridize to the cDNA molecules. These DNA:cDNA hybrids are then isolated from the mixture and amplified by PCR.

In this chapter, the considerations involved in the choice and use of different sources of cloned genomic DNA and cDNA are discussed. The protocols for performing the technique and procedures for analyzing the recovered cDNAs are also presented.

Overview of Direct cDNA Selection

The steps in direct cDNA selection (Figure 1) are

- ligation of primer-adapters to the cDNA and preamplification of the cDNA
- prehybridization of uncloned genomic DNA to repetitive sequences present in the cDNA population
- hybridization of biotinylated, cloned genomic DNA to the preamplified, prehybridized cDNA
- selective purification of genomic DNA hybridized to cDNA (DNA:cDNA hybrids) using streptavidin-coated beads
- selective amplification of the cDNA
- a second round of prehybridization, hybridization, selective purification, and selective amplification of the cDNA

The cDNAs are then cloned and analyzed.

HYBRIDIZATION CONDITIONS USED IN DIRECT cDNA SELECTION

Because nucleic acid hybridization is the most critical part of the direct cDNA selection procedure, it is useful to understand this process first. The events that occur during annealing reactions between two complementary single-stranded DNAs in aqueous solution are generally well understood. These events are considered here. Less is known about the hybridization reactions that occur between an immobilized DNA strand (e.g., a DNA strand bound to a nylon membrane or nitrocellulose filter) and a complementary strand that is in solution. However, use of an immobilized strand of DNA in the hybridization steps has also allowed genes to be isolated successfully in variations of the basic cDNA selection procedure (Parimoo et al. 1991).

Two different types of hybridization reactions are used during direct cDNA selection. The first reaction is performed with cDNA and uncloned genomic DNA. The goal is to use the uncloned genomic DNA to block (prehybridize) repetitive sequences that are present in the cDNA population. If these repetitive sequences are not blocked, they hybridize to similar repeats in the cloned genomic DNA in the subsequent step of direct cDNA selection and contribute to nonspecific background clones in the selected products. The second hybridization reaction is performed with the prehybridized cDNA and cloned genomic DNA. Here, the goal is to allow the cloned genomic DNA to find its complementary segments in the cDNA population.

For optimal results, the blocking DNA should hybridize as completely as possible to any repetitive sequences in the cDNA but leave the unique sequences in the cDNA population single-stranded and available for hybridization to the cloned genomic DNA. If either the DNA concentration or the time needed for prehybridization is underestimated in the prehybridization step, more nonspecific background clones will be isolated; if either is overestimated, some of the unique-copy sequences in the cDNAs may be unavailable for hybridization to the cloned genomic DNA.

A standard set of reaction conditions can be used for each of the hybridization steps in direct cDNA selection. These standard reaction conditions are incubation at 65°C at a total DNA concentration of 1–2 mg/ml in 120 mM sodium phosphate buffer (pH 7.4) for DNA with a fragment length of approximately 500 bp.

It is worth discussing one of the hybridization steps in detail so that the conditions chosen are understandable. Under the standard reaction conditions, a second-order hybridization reaction for a single type of DNA sequence (i.e., DNA of uniform complexity—not a mixture of complexities) can be described by the following equation:

$$C_0 t_{1/2} = (2.4 \times 10^{-6})(X)$$

where C_0 is the concentration of single-stranded DNA at time 0, $t_{1/2}$ is the time (in seconds) required for half of the DNA to reanneal, 2.4×10^{-6} includes the second-order rate constant, and X is the complexity of the DNA sample in bases (i.e., the length in bases of the unique sequences). $C_0 t_{1/2}$ is in units of seconds-moles/liter. For historical reasons, moles refers to the number of moles of phosphate in the DNA, which is equivalent to the number of bases in the solution. If the DNA concentration is in the more conventional terms of micrograms per milliliter, the moles of phosphate can be calculated by using the average figure of 330 daltons (g/mole) for each dNMP. DNAs are 90% reannealed at approximately ten times the $C_0 t_{1/2}$.

Hybridization rates have a linear dependence on DNA concentration, which means that they can be controlled by altering the DNA concentrations. In

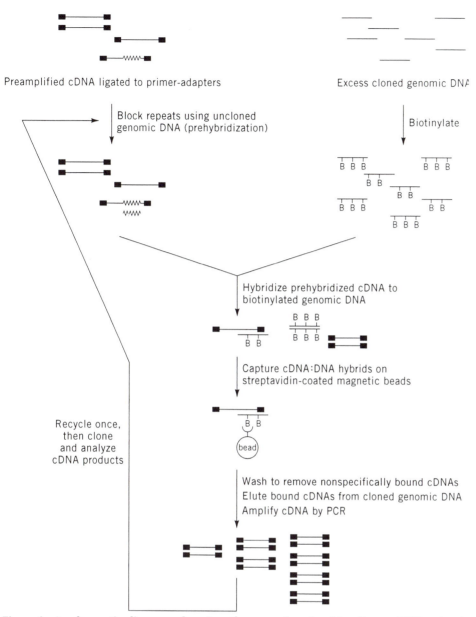

Preamplified cDNA ligated to primer-adapters

Excess cloned genomic DNA

Block repeats using uncloned genomic DNA (prehybridization)

Biotinylate

Hybridize prehybridized cDNA to biotinylated genomic DNA

Capture cDNA:DNA hybrids on streptavidin-coated magnetic beads

Recycle once, then clone and analyze cDNA products

Wash to remove nonspecifically bound cDNAs
Elute bound cDNAs from cloned genomic DNA
Amplify cDNA by PCR

Figure 1 A schematic diagram showing the steps involved in direct cDNA selection that are described in the text. A key part of the protocol is the ligation of primer-adapters to the cDNA so that it can be amplified by PCR (*top left*). Another important step in the preparation of the cDNA is blocking of repetitive sequences by prehybridization. Biotinylated genomic DNA is used so that it can be recovered following hybridization to the cDNA (*top right*). Any cDNA that is bound to the genomic DNA is then amplified and either recycled or cloned and analyzed (central part of the figure).

addition, hybridization rates of different types of sequences (unique versus repetitive sequences) are independent, making it easier to understand and control hybridization reactions.

In the first hybridization reaction in direct cDNA selection (which blocks all of the repetitive se-

quences in the cDNA), a relatively high concentration of cloned genomic DNA is used so that the rate of hybridization between the repetitive sequences is determined by the genomic DNA concentration and is independent of the cDNA concentration. The rate of hybridization between two unique cDNA sequences is determined by the concentration of the cDNA and is independent of the uncloned genomic DNA concentration. The independence of the two reaction rates from each other means that two separate rate calculations can be made (or a single equation can be used to describe the whole reaction).

The repetitive fraction of mammalian DNA, which comprises approximately 40% of the DNA, has been found empirically to reanneal at a C_0t of 10 seconds-moles/liter. If the repetitive DNA concentration is 0.6 mg/ml (i.e., 40% of 1.5 mg/ml), there are 1.8×10^{-3} moles/liter of phosphate. Therefore, the time required for the repeats to be 90% reannealed is:

$$(1.8 \times 10^{-3} \text{ moles/liter})(t) = 10 \text{ seconds-moles/liter}$$
$$t = 10/(1.8 \times 10^{-3}) = 5.5 \times 10^3 \text{ seconds} = {\sim}1.5 \text{ hours}$$

ADVANTAGES AND LIMITATIONS OF DIRECT cDNA SELECTION

One of the major advantages of direct cDNA selection is that its success is not highly dependent on the level of gene expression. When performed correctly, direct cDNA selection has a normalizing effect on the distribution of end products, such that high- and low-abundance mRNAs become represented in the selected end products (i.e., amplified DNA:cDNA) at levels that are much more similar to each other than they were originally. This normalizing feature makes it possible to use the procedure to isolate cDNAs corresponding to genes from complex tissues or from mixtures of mRNA.

Another advantage is that several variations on the procedure for direct cDNA selection have been reported, indicating that the technique is robust and can tolerate a wide range of experimental conditions. Lovett et al. (1991) and Parimoo et al. (1991) independently developed the technique by using very different conditions for the hybridization reaction that captures the cDNA products from complex mixtures (hybridization either to labeled DNA or to immobilized DNA).

The major technical limitation of direct cDNA selection and most other gene-isolation procedures is that genes must be expressed to be isolated. This means that multiple tissue sources that express the genes in the cloned genomic region must be available. Because the identities and expression patterns of all of the genes in a region are not known, it is not possible to predict what fraction of the genes will be isolated when a particular set of tissues is used. However, this problem is partly alleviated by the normalization that occurs in direct cDNA selection. Normalization allows the isolation of cDNAs from tissues in which the corresponding mRNA is present at levels as low as 1 in 10^6 mRNAs (Korn et al. 1992). In addition, because cDNA synthesized from mRNA instead of cDNA from cDNA libraries can be used as the source of expressed sequences, sources for selecting cDNAs can be prepared more easily from several tissues and thus the fraction of genes that can be isolated is increased.

Choosing Sources of Cloned Genomic DNA and cDNA for Direct cDNA Selection

SOURCE OF CLONED GENOMIC DNA

Positional cloning and other gene-isolation strategies that use methods like direct cDNA selection depend on the use of cloned fragments derived from genomic DNA for isolating cDNA fragments by specific hybridization. Most genomic regions that are available in cloned contigs are in the form of YAC, cosmid, BAC, or bacteriophage P1 clones or pools of clones. Although direct cDNA selection should be applicable with cloned genomic DNA from any source, there are advantages and disadvantages to each cloning vector. For the purposes in this chapter, the two most relevant considerations are the sizes of the genomic DNA inserts and the purity of the cloned genomic DNA that can be obtained with a particular cloning vector.

The size of the genomic DNA insert is an important determinant of the rate of hybridization in direct cDNA selection. At a given DNA concentration, the larger the DNA insert, the longer the time required for the hybridization reaction to proceed to completion. Because the time required for the hybridization reaction is a fraction of the total time needed to carry out a complete transcription analysis, hybridization time is not likely to be a serious consideration in choosing a cloning vector. Of greater significance is the effect of the size of the genomic DNA insert on the complexity of the cDNA that will be isolated with the procedure.

Larger cloned genomic DNA inserts are more likely to have multiple transcription units than are smaller ones. Although distinguishing between cDNAs derived from genes transcribed from different genomic DNA clones requires a significant amount of effort after cDNA selection is performed, this task is greatly simplified by using smaller genomic DNA clones. In addition, there appears to be competition for enrichment between different cDNAs. If the complexity of the cDNA is increased during enrichment by using larger cloned genomic DNA inserts, problems that result from this competition increase. The only disadvantage of using smaller genomic DNA inserts is that more clones must be used to cover the same genomic region. Thus, more hybridization and PCR assays must be performed. However, since these tests are relatively easy to perform, the problems with larger inserts (>100 kb) overshadow the additional effort required for smaller inserts (<100 bp).

The other issue to consider in choosing a source of cloned genomic DNA is its ease of purification and the level of purity that can be obtained. Although unpurified DNA from cells containing a genomic clone has been used to isolate cDNAs by direct cDNA selection (Parimoo et al. 1993), there are two advantages to using highly purified cloned genomic DNA. Host-cell DNA and RNAs that contaminate a genomic DNA clone can result in the isolation of cDNAs that do not correspond to the cloned insert. The most common example of this type of contamination occurs when YAC clones are used for direct cDNA selection. The presence of small amounts of yeast rRNA genes contaminating the purified YAC often leads to the enrichment of cDNAs derived from human, mouse, or other non-yeast rRNA genes during cDNA selection. In general, smaller DNA clones are more easily purified than larger ones, and DNA is more easily purified from *E. coli* than from *S. cerevisiae*.

SOURCE OF mRNA/cDNA

Although it may often be convenient to use a cDNA library as a source of cDNA for this technique, there are two major disadvantages to this type of cDNA source: (1) The cloning step in producing the library introduces a limitation on the complexity of the starting cDNA. Although direct cDNA selection can enrich sequences present at levels of less than 1 in 10^6 sequences, most cDNA libraries

are constructed with complexities of only 10^5 to 10^6 clones. When these libraries are amplified for distribution to multiple users, their complexities usually decrease. (2) Although most cDNA libraries are constructed with the goal of obtaining inserts larger than 1 kb, the PCR steps in direct cDNA selection are strongly biased toward amplifying smaller fragments, with fragments 200–500 bp in length being more efficiently amplified than larger fragments. The necessary use of PCR in direct cDNA selection thus further reduces the complexity of the cDNA.

To avoid the problems involved in using cDNA libraries in the direct cDNA selection procedure, use double-stranded cDNA that has not been ligated into a cloning vector. The best cDNA for direct cDNA selection is double-stranded cDNA that is made from RNA from one or more tissue sources and that has oligonucleotide adapters ligated to both ends to provide priming sites for amplification by PCR. The cDNA can be generated from picogram to microgram amounts of RNA, with synthesis primed by using random primers instead of oligo(dT) so that there is no bias for synthesizing sequences originating from the 3' ends of the mRNAs. Because the amounts of mRNA that are needed for performing direct cDNA selection are less than those normally needed for cDNA library construction, only a few circumstances (e.g., when tissue cannot be obtained) exist where the use of a cDNA library is necessary. Although investigators who have not generated cDNA libraries may be reluctant to synthesize their own cDNA, it should be noted that synthesis of suitable cDNA from multiple tissues for direct cDNA selection is much less demanding than constructing a library since smaller amounts of mRNA can be used, some degradation of the mRNA is tolerable, and the difficult steps of efficient ligation of the cDNA into a vector and transformation are avoided.

The choice of a tissue source for RNA purification depends on the availability of tissues and on the goals of the transcript-isolation project. If the goal is to construct the most complete transcript map possible for a genomic region, three alternatives are available: (1) a single tissue with a complex pattern of gene expression (e.g., fetal brain or whole embryo), (2) multiple tissues and/or cell lines examined as a mixture of RNAs, or (3) multiple tissues or cell lines examined as individual RNAs. The ability of direct cDNA selection to isolate cDNAs present at low abundance makes the simplicity of the first two alternatives attractive. However, the lower the complexity of the RNA population (i.e., the fewer distinct RNAs), the more effective the purification will be. Thus, although it is difficult to predict how effective any alternative will be in direct cDNA selection, the safest approach is to use the third alternative.

If the goal is to isolate a gene in which the pattern of expression can be predicted, it is advantageous to use RNA prepared from the tissue with the highest level of expression. Unfortunately, information on the level of expression is often not available. A reasonable alternative is to pool mRNAs from those tissues that are probable sites of expression. Another alternative is to use a tissue with a complex pattern of gene expression (e.g., fetal brain).

After a source for the RNA is chosen, care must be taken in preparing the RNA. The RNA must be free of unspliced, immature mRNAs and contaminating genomic DNA. The enrichment of intron or intergenic DNA sequences from these contaminants adds considerably to the labor involved in analyzing the cDNA products that result from direct cDNA selection. Both unspliced mRNA and genomic DNA are restricted to the nucleus, whereas mature mRNA is present predominantly in the cytoplasm. Therefore, mRNA should be prepared whenever possible from cytoplasm separated from the nucleus instead of from total cellular RNA that contains both (Sambrook et al. 1989). The RNA preparation should be treated with RNase-free DNase before the cDNA is synthesized. This will remove contaminating nuclear or mitochondrial DNA that copurifies with the RNA. In addition, chromatography should be performed on oligo(dT) columns to reduce the amount of rRNA present in the mRNA preparation. Although it is impossible to prepare mRNA that is completely free of rRNA, performing multiple rounds of chromatography will produce RNA pure enough that direct cDNA selection will isolate a relatively small fraction of clones that are derived from rRNA. Any contaminating rRNA-derived clones can be quickly eliminated from further analysis by a simple hybridization experiment (see Sambrook et al. 1989).

PREPARATION OF BIOTINYLATED, CLONED GENOMIC DNA

The cloned genomic DNA that is used for direct cDNA selection is currently a YAC, cosmid, bacteriophage P1, or BAC clone or pools of clones. It should be obtained in purified form, free of host DNA and other contaminants. Standard procedures (Sambrook et al. 1989) for purifying cloned DNA from bacterial sources provide sufficiently pure DNA for performing direct cDNA selection. PFGE should be used to enrich or purify YAC clones that are used for direct cDNA selection (see Green et al. 1998).

Biotinylated precursor dNTPs can easily be incorporated into cloned genomic DNA during polymerization (DNA synthesis) reactions. Two alternative biotinylation protocols are provided on pp. 166–168 and pp. 169–171. The first is a random-priming reaction that is useful for labeling DNA clones for which microgram amounts of template DNA are easily purified (e.g., cosmid, BAC, or bacteriophage P1 clones or pools of clones). In the case of YACs, the amount of DNA needed may be more than is easily obtained. Because of this difficulty, the alternative protocol included on pp. 169–171 uses PCR to amplify small amounts (nanograms) of purified YAC DNA and at the same time to incorporate biotin into the DNA. In this second protocol, YAC DNA is excised from a PFG. Each of several aliquots is treated with a restriction enzyme that cleaves at a 4-bp recognition site and PCR primer-adapters are ligated to the digested DNA. This is similar to the procedure for ligating the primer-adapter that is used on p. 175 for the preamplification of the cDNA, but different adapters are used so that no YAC DNA is amplified when the cDNA is amplified.

PROTOCOL

Biotinylation Using Random Primers

This protocol is adapted from Feinberg and Vogelstein (1983, 1984).

1. Add 5–50 ng of cosmid, bacteriophage P1, BAC, or YAC DNA to a 1.5-ml screwcap microcentrifuge tube and adjust the volume to 32 µl with H_2O. Heat at 95°C for 5 minutes to denature the DNA.

2. Immediately place the tube on ice to prevent reannealing of the DNA strands.

3. Keep the tube on ice while adding the following in the order listed:

5x oligonucleotide labeling buffer	10 µl
acetylated BSA (100 mg/ml)	2 µl
Klenow fragment of *E. coli* DNA polymerase I (5 units/µl)	1 µl

5x Oligonucleotide labeling buffer
Mix the solutions below in a ratio of 100:250:150 of solution A:B:C. Divide into 100-µl aliquots and store at –20°C for up to 5 years.

Solution A

2 M Tris-Cl (pH 8.0 at 25°C)	0.625 ml
1 M $MgCl_2$	125 µl
H_2O	250 µl
β-mercaptoethanol	18 µl
100 mM dCTP	5 µl
100 mM dTTP	5 µl
100 mM dGTP	5 µl

Store at –20°C for up to 5 years.

Solution B
Adjust the pH of 2 M HEPES to pH 6.6 with 10 N **NaOH**. Store at room temperature for up to 5 years.

Solution C
Dissolve 50 OD units of the random hexamers pd(N)$_6$ (Pharmacia) in 555 µl of H_2O. Store at –20°C indefinitely.

β-mercaptoethanol, NaOH (see Appendix for Caution)

4. Add 5 µl of 0.5 mM dATP–biotin-14-dATP mixture and mix.

0.5 mM dATP–biotin-14-dATP mixture

Component and final concentration	Amount to add per 1 ml
495 μM dATP	0.495 μl of 1 M
5 μM biotin-14-dATP (Life Technologies 19524-016)	12.5 μl of 400 μM
H$_2$O	492.5 μl

Store at –20°C for up to 5 years.

5. Centrifuge briefly to collect the liquid at the bottom of the tube.

6. Incubate at room temperature for 4–20 hours.

   Note: This incubation period can be varied for convenience of scheduling.

7. While the sample is incubating, prepare a Sephadex G-50 spin column as follows:

   a. Dip glass wool in **Sigmacote** (Sigma SL-2), allow it to dry, and then autoclave.

      Sigmacote (see Appendix for Caution)

   b. Discard the syringe plunger and cap from a 1-ml tuberculin syringe. Place a small amount of the treated glass wool at the bottom of the syringe.

   c. Break the surface tension in the glass wool by applying approximately 1 ml of 95% ethanol. Rinse several times with approximately 3-ml volumes of sterile distilled H$_2$O.

   d. Fill the column with a slurry of Sephadex G-50 (DNA grade; Pharmacia 17-0045-01) (swollen in TE [pH 7.5] containing 0.1% **SDS** and autoclaved). Avoid creating air bubbles.

      SDS (see Appendix for Caution)

   e. Centrifuge the column at 200*g* at room temperature for 1 minute.

   f. Add more Sephadex G-50 slurry to fill the column and centrifuge for an additional 4 minutes to pack the column.

8. Adjust the volume of the reaction mixture to 100 μl with TE (pH 7.5) and apply the mixture to a Sephadex G-50 spin column placed over a 1.5-ml screwcap microcentrifuge tube. Centrifuge at 200*g* at room temperature for 4 minutes and collect the eluate in the tube.

9. Adjust the eluate volume to 100 μl with TE (pH 7.5) and precipitate the DNA as follows:

   a. Add 10 μl of 7 M ammonium acetate, 1 μg of linear polyacrylamide, and 65 μl of isopropanol and mix well. Incubate at room temperature for 15 minutes.

5 mg/ml Linear polyacrylamide stock solution
Polymerize 1 ml of a 5% (50 mg/ml) solution of **acrylamide** monomer by adding 5 μl of 10% ammonium persulfate (in H_2O) and 1 μl of TEMED. The solution will become viscous but will not form a gel. After the acrylamide is completely polymerized, dilute the linear polyacrylamide with H_2O to make a final volume of 10 ml, divide into 100-μl aliquots, and store at −20°C indefinitely.

acrylamide   (see Appendix for Caution)

b. Centrifuge in a microcentrifuge at 12,000*g* at room temperature for 5 minutes.

c. Wash the pellet by adding 1 ml of 70% ethanol, recentrifuging for 5 minutes, and discarding the supernatant.

d. Air dry the DNA pellet.

   Note: Drying under vacuum may overdry the pellet and make it difficult to dissolve the DNA.

10. Dissolve the DNA in 10 μl of TE (pH 7.5). Determine the DNA concentration and adjust with TE to a final concentration of 200 ng/μl for YACs, 20 ng/μl for cosmids, and 40 ng/μl for bacteriophage P1 or BACs.

11. Store the biotinylated DNA at −20°C until needed for the hybridization reaction (see pp. 180–182).

PROTOCOL

Biotinylation Using PCR

1. For each YAC, perform PFGE with yeast DNA from a strain carrying the YAC in a 1% low-melting-temperature agarose gel. Include the appropriate molecular-weight markers (e.g., *S. cerevisiae* size standard; Bio-Rad 170-3605).

2. Detect the DNA by using a long-wavelength **UV** light source. Excise the YAC DNA band in as small a volume of agarose as possible (≤250 µl or 0.25 cm³) and place it in a 1.5-ml microcentrifuge tube.

   UV radiation (see Appendix for Caution)

3. Equilibrate the agarose with TE (pH 7.5) as follows:

   a. Fill the tube with TE (pH 7.5).

   b. Incubate at room temperature for 1 hour.

   c. Discard the TE and repeat steps a–b.

4. Carefully remove all excess liquid from the tube with a pipette. Place the tube in a water bath or a heating block set at 70ºC and incubate for 10 minutes to melt the agarose.

5. Adjust the volume to 250 µl with H_2O and place the tube in a water bath set at 37ºC to keep the agarose melted.

6. Set up six labeled tubes containing the components below and equilibrate to 37ºC before adding agarose in step 7. Include a different restriction enzyme in each tube. Use any six of the following enzymes with 4-bp recognition sites: *Alu*I, *Dpn*II, *Hae*III, *Rsa*I, *Sau*96I, *Dde*I, *Hha*I, and *Hin*fI.

restriction enzyme (2 units/µl)	5 µl
10x restriction enzyme buffer (supplied with the enzyme)	5 µl

   Note: Control experiments have shown that a particular single-copy sequence has a 50% probability of being present in the amplified material produced after digestion with a particular restriction enzyme. Those sequences that are not present presumably fall outside the 200–500-bp size range that is efficiently amplified under the conditions below. The use of four different enzymes here provides approximately a 93% representation of the YAC; use of six enzymes gives approximately a 98% representation.

7. Add 40 µl of the melted agarose to each of the six tubes and incubate at 37ºC for 2 hours.

8. Heat at 70°C for 15 minutes to inactivate the enzyme.

9. While the tubes are at 70°C, prepare the following end-repair mixture:

bacteriophage T4 DNA polymerase
  (30 units/6.5 µl; e.g., New
  England Biolabs, Life Technologies)
	6.5 µl

10x bacteriophage T4 DNA polymerase
  reaction buffer (supplied with
  the enzyme)
	6.5 µl

2.5 mM dNTP mixture (see Appendix)	6.5 µl
H_2O	15 µl

10. Add 5 µl of the end-repair mixture to each tube and incubate at 37°C for 15 minutes.

11. Heat at 70°C for an additional 15 minutes to stop the reaction. While the tubes are at 70°C, prepare the following ligation mixture:

10x ligase buffer (supplied with the enzyme)	50 µl
ORM 1564 (1 µg/µl)	33 µl
ORM 1565 (1 µg/µl)	33 µl

bacteriophage T4 DNA ligase
  (400 units/µl; New England
  Biolabs)
	5 µl

H_2O	9 µl

Notes: ORM 1564 is 5´CCATTGTGCTGGTCTAGATCGCACA3´.
      ORM 1565 is 5´CGATCTAGACCAGCACAATGG3´.

12. Add 20 µl of the ligation mixture to each tube and incubate at 15°C overnight.

13. Heat at 70°C for 15 minutes to stop the reaction.

14. While the tubes are at 70°C, set up six thin-walled PCR tubes so that each contains a different reaction mixture from step 13. For each tube, add 90 µl of the PCR mixture below to a PCR tube, place the tube on ice, and then add 10 µl of one of the mixtures from step 13.

10x PCR buffer	65 µl
2.5 mM dNTP mixture	32.5 µl
400 µM biotin-14-dATP (Life Technologies 19524-016)	6.5 µl
Taq DNA polymerase (5 units/µl)	3.25 µl
ORM 1565 (1 µg/µl)	3.25 µl
H_2O	474.5 µl

Note: For thermal cyclers that do not have a top heating block, overlay each reaction mixture with several drops of light mineral oil and then cover the tube.

10x PCR buffer

Component and final concentration	Amount to add per 100 ml
0.5 M KCl	50 ml of 1 M
100 mM Tris-Cl	10 ml of 1 M (pH 8.3 at room temperature)
15 mM MgCl$_2$	15 ml of 100 mM
0.1% (w/v) gelatin	100 mg
H$_2$O	to make 100 ml

Store at −20°C for up to 6 months.

15. Preheat the thermal cycler to 94°C. Perform thermal cycling as follows for 5 cycles:

   94°C for 30 seconds
   58°C for 30 seconds
   72°C for 45 seconds

   At the end of the last cycle, hold the samples at 4°C.

   Note: These amplification conditions are optimal for a GeneAmp PCR System 9600 (Perkin-Elmer) thermal cycler.

16. For each sample of amplified DNA, prepare a Sephadex G-50 spin column as described on p. 167, step 7. Apply each sample to a spin column placed over a 1.5-ml screwcap microcentrifuge tube. Centrifuge at 200g at room temperature for 4 minutes and collect the eluate in the tube.

   Note: If a mineral oil overlay was used, avoid transferring the oil with the sample.

17. Adjust the eluate volume to 100 µl with TE (pH 7.5) and precipitate the DNA with 10 µl of 7 M ammonium acetate, 1 µg of linear polyacrylamide, and 65 µl of isopropanol as described on pp. 167–168, step 9.

18. Dissolve the DNA in 5 µl of TE (pH 7.5). Determine the DNA concentration and adjust with TE to a final concentration of 200 ng/µl for YACs.

19. Store the biotinylated DNA at −20°C until needed for the hybridization reaction (see pp. 180–182).

PREPARATION OF cDNA

cDNA synthesis can be performed according to the manufacturer's instructions accompanying reverse transcriptase (e.g., the MMLV enzyme; Life Technologies), although random hexamers should be substituted for the oligo(dT) normally recommended. Alternatively, the protocol on pp. 173–174 can be used with AMV reverse transcriptase.

Before the cDNA and genomic DNA are prehybridized or hybridized (see p. 179 or p. 180), the cDNA must be ligated to primer-adapters and then preamplified as described on p. 175 and pp. 176–177. This will ensure efficient amplification of the cDNA after hybridization. Without this preamplification step, cDNA that lacks adapters and cannot be amplified may compete with amplifiable cDNA for hybridization to the genomic DNA.

In general, cDNA from a library is not recommended in place of the synthesized cDNA (see pp. 163–164).

PROTOCOL

cDNA Synthesis

1. Place 4 μg of mRNA (in a volume of 70 μl or less) in a 1.5-ml micro-centrifuge tube. Heat at approximately 100°C for 30 seconds to denature the mRNA fully (see Sambrook et al. 1989).

   Note: Working with RNA requires the use of appropriate precautions to maintain RNase-free conditions. These precautions include using **DEPC**-treated H_2O for preparing solutions (except those containing Tris), wearing gloves, and using sterile plasticware instead of glassware.

   DEPC (see Appendix for Caution)

2. Immediately place the tube on ice.

3. Adjust the volume to 70 μl with RNase-free H_2O.

4. Add the following in the order listed:

first strand reverse transcription buffer	20 μl
1 M DTT	1 μl
RNase inhibitor (25 units/μl; e.g., Boehringer Mannheim 799017)	2 μl
random primer (5 μg/μl; Pharmacia 27-2166-01)	1 μl
20 mM dNTP mixture (see Appendix)	2.5 μl
RT-XL (Life Sciences)	4 μl

   First strand reverse transcription buffer

Component and final concentration	Amount to add per 5 ml
250 mM Tris-Cl	1.25 ml of 1 M (pH 8.8 at 25°C, pH 8.2 at 42°C)
250 mM KCl	1.25 ml of 1 M
30 mM $MgCl_2$	150 μl of 1 M
H_2O	2.35 ml

   Divide into 0.5-ml aliquots and store at –20°C indefinitely.

5. Incubate at 42°C for 40 minutes.

6. Heat at 70°C for 10 minutes to inactivate the reverse transcriptase.

7. Add the following:

RNase-free H$_2$O	320 µl
second strand reverse transcription buffer	80 µl
E. coli DNA polymerase I (endonuclease-free, 5 units/µl; Boehringer Mannheim 642711)	5 µl
RNase H (DNase-free, 2 units/µl; e.g., Life Technologies)	2 µl

Second strand reverse transcription buffer

Component and final concentration	Amount to add per 5 ml
100 mM Tris-Cl	0.5 ml of 1 M (pH 7.5 at 25°C)
250 mM MgCl$_2$	1.25 ml of 1 M
0.5 M KCl	2.5 ml of 1 M
250 µg/ml acetylated BSA	12.5 µl of 100 mg/ml
50 mM DTT	250 µl of 1 M
H$_2$O	487.5 µl

Divide into 0.5-ml aliquots and store at −20°C indefinitely.

8. Incubate at 15°C for 1 hour and then at room temperature for 1 hour.

9. Add 20 µl of 0.5 M EDTA (pH 8.0) and extract with 1 volume of **phenol** (see Appendix). Use phenol saturated with TE (pH 8.0).

phenol (see Appendix for Caution)

10. Precipitate the cDNA in the aqueous phase with 0.1 volume of 3 M sodium acetate (pH 5.2) and 2–2.5 volumes of absolute ethanol and dry the pellet under vacuum (see Appendix).

11. Store the dried cDNA at −20°C until needed for ligation to the primer-adapter (see p. 175).

PROTOCOL

Ligating Primer-adapters to Double-stranded cDNA

1. Dissolve the dried cDNA (e.g., from pp. 173–174) in 270 μl of TE (pH 7.5).

2. Add the following:

10x ligase buffer (supplied with the enzyme)	30 μl
ORM 29 (0.75 μg/μl)	4.25 μg
ORM 28 (0.75 μg/μl)	5 μg
bacteriophage T4 DNA ligase (400 units/μl; New England Biolabs)	1 μl

   Notes: ORM 29 is 5′TAGTCCGAATTCAAGCAAGAGCACA3′.
   ORM 28 is 5′CTCTTGCTTGAATTCGGACTA3′.
   The pair of oligonucleotides consists of a 21-mer and a 25-mer. The additional bases on the 25-mer prevent the formation of concatemers during the ligation by preventing blunt-end ligation.

3. Incubate at 15°C overnight.

4. Store the ligated cDNA/primer-adapter at –20°C until needed for preamplification (see pp. 176–177).

PROTOCOL
_____

Preamplification of the cDNA

1. Add 5 µl of the ligated cDNA/primer-adapter (from p. 175) to each of two thin-walled PCR tubes and 5 µl of H$_2$O to a third tube as a control. Place the tubes on ice.

   Note: The control tube should show whether or not there is any nonspecific contamination.

2. Prepare the following PCR mixture and place on ice:

10x PCR buffer (for preparation, see p. 171)	30 µl
20 mM dNTP mixture (see Appendix)	1.875 µl
ORM 28 (0.75 µg/µl)	1.5 µl
Taq DNA polymerase (5 units/µl)	1.5 µl
H$_2$O	252 µl

   Note: ORM 28 is 5´CTCTTGCTTGAATTCGGACTA3´.

3. Add 95 µl of the PCR mixture to each of the three tubes on ice.

   Note: For thermal cyclers that do not have a top heating block, overlay each reaction mixture with several drops of light mineral oil and then cover the tube.

4. Preheat the thermal cycler to 94°C. Perform thermal cycling as follows for 20 cycles:

   > 94°C for 30 seconds
   > 58.5°C for 30 seconds
   > 72°C for 45 seconds

   At the end of the last cycle, hold the samples at 4°C.

   Note: These amplification conditions are optimal for a GeneAmp PCR System 9600 (Perkin-Elmer) thermal cycler.

5. Analyze 5 µl of each PCR product and the appropriate molecular-weight markers (e.g., 1-kb ladder) on a 1.5% agarose gel.

   Notes: There should be a smear of amplified cDNA, extending from approximately 200 bp to approximately 1 kb in size, with the majority of the products in the 200–500-bp range.
   If a mineral oil overlay was used, avoid transferring the oil with the sample.

6. Precipitate the cDNA in each sample with 0.1 volume of 3 M sodium acetate (pH 5.2) and 2–2.5 volumes of absolute ethanol and dry the pellet under vacuum (see Appendix).

7. Dissolve each cDNA pellet in 10 μl of TE (pH 7.5) and determine the concentration.

   Note: Yield should be 1–10 μg.

8. Store the cDNA at –20°C until needed for prehybridization (see p. 179).

HYBRIDIZATION REACTIONS AND ENRICHMENT OF THE cDNA

Two hybridization reactions are used in direct cDNA selection (for further discussion of hybridization, see pp. 160–162). In the first reaction, cDNA is hybridized to nonbiotinylated, uncloned genomic DNA to block repetitive sequences in the cDNA. The hybridization conditions in the protocol on p. 179 allow both components of the reaction mixture to be hybridized to a $C_0t_{1/2}$ of 20 seconds-moles/liter (which is ten times the $C_0t_{1/2}$ for repetitive sequences). This degree of hybridization is sufficient to hybridize the highly repetitive sequences in the blocking genomic DNA to those in the cDNA but leave the nonrepetitive cDNA primarily as single strands.

The second hybridization reaction (p. 180) is between the prehybridized cDNA and biotinylated, cloned genomic DNA. This is the first step in the enrichment of the cDNA. This reaction indirectly labels the cDNAs with biotin. In this reaction, the cloned genomic DNA is hybridized to a $C_0t_{1/2}$ of 120 seconds-moles/liter (which is ten times the $C_0t_{1/2}$ for single-copy sequences). Hybridization to this $C_0t_{1/2}$ is sufficient to hybridize the single strands of the biotinylated, cloned DNA to each other or to any cDNA strands that are complementary to the cloned DNA strands.

To complete the first enrichment protocol (pp. 180–182), the cloned genomic DNA and any associated cDNA from the second hybridization reaction is recovered with streptavidin-coated magnetic beads. The beads are then washed to remove cDNAs that are nonspecifically bound and to ensure that only DNA:cDNA hybrids are recovered.

The cDNA recovered from the magnetic beads is then amplified by PCR and subjected to a second round of prehybridization, hybridization, washing, and amplification as described on p. 183. After this second cycle of enrichment, the cDNAs are cloned and analyzed.

PROTOCOL

Prehybridization to Block Repeated Sequences in the cDNA

1. Mix 5 µg of preamplified cDNA (from pp. 176–177) and 5 µg of sheared human or mouse genomic DNA (~500 bp) in a 1.5-ml screwcap microcentrifuge tube.

2. Precipitate the DNAs with 0.1 volume of 3 M sodium acetate (pH 5.2) and 2–2.5 volumes of absolute ethanol and dry the pellet under vacuum (see Appendix).

3. Completely redissolve the DNAs in 3.5 µl of H_2O and then overlay with 50–100 µl of mineral oil to prevent evaporation during prehybridization.

   Note: The DNAs can be stored at –20°C indefinitely.

4. Heat at 95°C for 5 minutes to separate the DNA strands.

5. Immediately place the tube on ice to stop the reaction.

6. Add 1.25 µl of 480 mM sodium phosphate buffer (pH 7.4) and incubate at 65°C for 90 minutes to prehybridize the repetitive sequences in the cDNA.

   480 mM Sodium phosphate buffer
   Combine 18.25 ml of 1 M NaH_2PO_4 (monobasic), 77.75 ml of 1 M Na_2HPO_4 (dibasic), and 4 ml of H_2O. Store at room temperature indefinitely.

7. Place the tube on ice until needed for hybridization (see p. 180).

   Note: The prehybridized cDNA can be stored at –20°C indefinitely.

PROTOCOL

Primary Enrichment of the cDNA by Hybridization to Cloned Genomic DNA and Recovery on Beads

1. Heat 1 μl of the biotinylated DNA from pp. 166–168 or pp. 169–171 at 95°C for 5 minutes to separate the DNA strands.

2. Immediately place the tube on ice to stop the reaction.

3. Combine the following in a 1.5-ml screwcap microcentrifuge tube and over lay with minieral oil:

denatured, biotinylated DNA	1 μl
240 mM sodium phosphate buffer (pH 7.4) (for preparation, see p. 179)	2 μl
prehybridized cDNA (from p. 179)	1 μl

   Note: If necessary, adjust the concentration of the biotinylated DNA so that 1 μl contains 0.5–1 ng of DNA for each kilobase of genomic clone length (e.g., 20–40 ng of cosmid DNA).

4. Incubate at 65°C for 40 hours to hybridize the DNAs.

5. Wash streptavidin-coated magnetic beads (Dynabeads M-280 streptavidin beads; Dynal 112.05) as follows:

   a. Thoroughly resuspend the beads by vortexing in the vial provided by the manufacturer.

   b. Transfer 5 μl of the streptavidin-coated magnetic beads into a 0.5-ml microcentrifuge tube.

   c. Add 100 μl of 10 mM Tris-Cl/1 mM EDTA/1 M NaCl buffer and mix by vortexing.

   10 mM Tris-Cl/1 mM EDTA/1 M NaCl buffer

Component and final concentration	Amount to add per 100 ml
10 mM Tris-Cl	1 ml of 1 M (pH 7.5 at room temperature)
1 mM EDTA	200 μl of 0.5 M (pH 8.0)
1 M NaCl	20 ml of 5 M
H_2O	78.8 ml

   Store at room temperature indefinitely.

  d. Place the tube on a magnet (Dynal MPC; Promega) for at least 30 seconds to allow the beads to settle as they are attracted to the magnet. The supernatant should become clear.

  e. Keep the tube on the magnet and use a pipette to remove and discard the supernatant. Avoid removing any of the magnetic beads.

6. Add 100 μl of 10 mM Tris-Cl/1 mM EDTA/1 M NaCl buffer to the hybridized DNAs from step 4.

7. Transfer the diluted hybridized DNAs into the 0.5-ml microcentrifuge tube containing the beads and incubate at room temperature for 15 minutes. Keep the contents mixed by continuously inverting the tube throughout the incubation, and then repeat steps 5d–e.

8. Add 100 μl of 0.1x SSC/0.1% SDS and incubate at room temperature for 15 minutes. Keep the contents mixed by continuously inverting the tube throughout the incubation, and then repeat steps 5d–e.

0.1x SSC/0.1% SDS

Component and final concentration	Amount to add per 100 ml
0.1x SSC	0.5 ml of 20x SSC
0.1% **SDS**	1 ml of 10%
H$_2$O	98.5 ml

Store at room temperature indefinitely.

SDS (see Appendix for Caution)

9. Repeat step 8.

10. Add 100 μl of 0.1x SSC/0.1% SDS and incubate at 65°C for 15 minutes. Keep the contents mixed by continuously inverting the tube throughout the incubation and then repeat steps 5d–e.

11. Repeat step 10 two more times.

12. Add 50 μl of 0.05 N **NaOH** to the beads and incubate at room temperature for 5 minutes. Keep the contents mixed by continuously inverting the tube throughout the incubation and then repeat steps 5d–e.

   NaOH (see Appendix for Caution)

13. Add 50 μl of 1 M Tris-Cl (pH 7.0) to neutralize the supernatant. Keep the contents mixed by continuously inverting the tube throughout the incubation.

14. Prepare a Sephadex G-50 spin column as described on p. 167, step 7. Apply the sample to the spin column placed over a 1.5-ml screwcap micro-

centrifuge tube. Centrifuge at 200*g* at room temperature for 4 minutes and collect the eluate in the tube.

15. Adjust the eluate volume to 100 µl with TE (pH 7.5) and precipitate the cDNA with 10 µl of 7 M ammonium acetate, 1 µg of linear polyacrylamide, and 65 µl of isopropanol as described on pp. 167–168, step 9.

16. Dissolve the cDNA in 50 µl of TE (pH 7.5).

17. Store the cDNA at −20°C until needed for secondary enrichment (see p. 183).

PROTOCOL

Secondary Enrichment of the cDNA by Amplification of Selected cDNA

1. Add 5 μl of the recovered cDNA (from pp. 180–182) to a thin-walled PCR tube and 5 μl of H_2O to a second tube as a control. Place the tubes on ice.

   Note: The control tube should show whether or not there is any nonspecific contamination.

2. Prepare the following PCR mixture and place on ice:

10x PCR buffer (for preparation, see p. 171)	20 μl
20 mM dNTP mixture (see Appendix)	1.25 μl
ORM 28 (0.75 μg/μl)	1 μl
Taq DNA polymerase (5 units/μl)	1 μl
H_2O	168 μl

3. Follow steps 3–7 on pp. 176–177 for each of the two tubes on ice.

   Note: The cDNA can be stored at –20°C indefinitely. The yield should be 1–10 μg.

4. Repeat the prehybridization protocol on p. 179, the primary enrichment protocol on pp. 180–182, and steps 1–3 of this secondary enrichment protocol.

5. Clone and analyze the cDNA products as described on pp. 184–189.

CLONING PCR PRODUCTS OBTAINED BY DIRECT cDNA SELECTION AND PREPARING THE PRODUCTS FOR ANALYSIS

Cloning PCR Products

After the cDNA products are enriched and amplified, the PCR products should be cloned (see Fanning and Gibbs 1997). Conventional methods for cloning blunt-ended fragments can be used for this purpose (see Sambrook et al. 1989), but they are often unsuccessful, apparently because *Taq* DNA polymerase adds 3′ extensions. A number of kits are available commercially that are designed specifically for cloning PCR products (e.g., TA Cloning Kit [Invitrogen K2000-01] and pCR-Script SK[+] Cloning Kit [Stratagene 211190 or 211192]). Since the ability to detect plasmid clones that contain cDNA inserts greatly simplifies the analysis of the clones, it is convenient to use a vector that provides insertional inactivation of the *lacZ* gene such as the vectors in the kit.

Preparing cDNA Products for Analysis

During the analysis of the products of direct cDNA selection, it is often advantageous to have the clones arrayed in a 96-well format as described on p. 185 (384-well formats may also be convenient). This multiwell format simplifies the processing of large numbers of clones and the identification of a desired subset. The use of a multichannel pipettor and a thermal cycler with a 96-well format (e.g., a GeneAmp PCR System 9600 thermal cycler; Perkin-Elmer) is necessary to take full advantage of the arrayed cDNA clones.

A good starting point is to array one cloned PCR product for each kilobase pair of genomic insert. The cDNA inserts can then be amplified from the cloned and arrayed PCR products and analyzed by agarose gel electrophoresis as described on p. 186. Blots can also be prepared for further analysis.

PROTOCOL

Arraying Colonies from Cloned PCR Products

1. Transfer 100 µl of TB medium containing the appropriate antibiotic into each well of a round-bottom 96-well plate (Falcon 3918).

2. Prepare a humidified chamber by soaking several paper towels in H_2O and placing them in a plastic box with a tight-fitting lid.

   Note: The paper towels should be saturated but all of the H_2O should be absorbed into the towels.

3. Inoculate the TB medium in each well with a single freshly grown bacterial colony from a cloned PCR product (for a discussion of cloning, see p. 184).

4. Place a cover (Falcon 3071) on the 96-well plate, transfer it into the humidified chamber, and incubate at 37ºC overnight.

   Note: There should be obvious growth after the incubation.

5. To store the plates, add 50 µl of a mixture of 50% glycerol and 50% TB medium to each well, seal the wells with Seal Plate (Costar [USA/Scientific]), and place the cover (Falcon 3071) on the plate.

   Note: The plates can be stored at −80ºC for 1 year.

PROTOCOL

Amplifying cDNA Inserts from Cloned PCR Products

1. Prepare the following PCR mixture, pipette 50 μl into each well of a 96-well plate, and place on ice:

10x PCR buffer (for preparation, see p. 171)	550 μl
10% NP-40	550 μl
2.5 μM dNTP mixture (see Appendix)	275 μl
ORM 28 (0.75 μg/μl)	55 μl
Taq DNA polymerase (5 units/μl)	27.5 μl
H_2O	4.04 ml

   Notes: ORM 28 is 5′CTCTTGCTTGAATTCGGACTA3′.

   These steps are most easily performed with a multichannel pipettor and a thermal cycler with a 96-well format (e.g., a GeneAmp PCR System 9600 thermal cycler; Perkin-Elmer).

2. Set the pipettor at 50 μl and pipette the arrayed bacterial cultures (from p. 185) up and down several times. Expel the cultures from the pipette tips. Rinse one of the pipette tips in each of the 50-μl PCR mixtures by pipetting up and down several times.

   Notes: The amount of bacteria that adheres to the sides of the pipette tip provides sufficient template DNA for the PCR assays. Adding larger volumes of the bacterial cultures inhibits the reactions.

   For thermal cyclers that do not have a top heating block, overlay each reaction mixture with several drops of light mineral oil and then cover the tube.

3. Preheat the thermal cycler to 94°C. Perform thermal cycling as follows for 30 cycles:

   94°C for 30 seconds
   59°C for 30 seconds
   72°C for 30 seconds

   At the end of the last cycle, hold the samples at 4°C.

   Note: These amplification conditions are optimal for a GeneAmp PCR System 9600 (Perkin-Elmer) thermal cycler.

4. Analyze 10 μl of each PCR product and the appropriate molecular-weight markers (e.g., 1-kb ladder) on a 1.5% agarose gel.

   Note: If a mineral oil overlay was used, avoid transferring the oil with the sample.

5. Detect the DNA by staining with **ethidium bromide**. Photograph the gel using a **UV** transilluminator.

   ethidium bromide, UV radiation (see Appendix for Caution)

6. Analyze the amplified inserts as described on pp. 187–189.

Confirming That cDNA Products Arise from Transcriptionally Active Genes in the Genomic Region of Interest

Direct cDNA selection yields a set of cDNA products that are highly enriched for genes from the genomic region used to perform the selection. This enrichment allows detection of mRNAs transcribed at levels that are difficult to detect by other means (e.g., by northern blots or by screening cDNA libraries). However, analysis of the isolated products is complicated by the fact that some segments of genes derived from other regions of the genome or sequences contaminating the starting cDNA material may be enriched as well. Some of these contaminating sequences, especially those composed of rRNA or repetitive elements, are relatively easy to identify; others are not. In general, the analysis of the cDNA products can be divided into two parts: (1) eliminating common artifacts and (2) proving that sequences in isolated clones are present in mature mRNA transcripts.

COMMON ARTIFACTS IN DIRECT cDNA SELECTION

The two most common artifacts seen in direct cDNA selection are due to repetitive sequences and sequences that are highly conserved between the cloning vector's host and the organism from which the cDNA is obtained for the procedure (the most common of which are rRNAs and mitochondrial mRNAs). The prehybridization step blocks repetitive sequences in the cDNA, but this reaction is not completely effective. Some repeat-containing clones will still be isolated. In particular, this prehybridization step does not block low-copy-number repeats or most medium-copy-number repeats. Even some high-copy-number repeats are not blocked. The strong evolutionary conservation of the rRNA genes across all species results in the isolation of rRNA genes in the selection procedure, particularly when YAC clones are used as the source of the cloned genomic DNA since YACs are harder to purify. Even with careful execution of the blocking step, artifactual clones usually comprise a significant percentage (5–50%) of the final products. Fortunately, in cases of artifactual clones composed of highly repetitive sequences and rRNA-derived cDNAs, labeled probes for these sequences can be used to eliminate these clones immediately after they are isolated and before additional analysis of the clones is performed.

More problematic are cDNA clones that are isolated because of hybridization to low-copy-number repeats or to host-cell-derived DNA sequences other than rRNA genes. The chromosomes of higher eukaryotes contain a heterogeneous population of sequences present at a level of 2–1000 copies per genome. These sequences include gene families, pseudogenes, locally repeated sequences, dispersed repeated sequences, and simple-sequence repeats. The presence of any of these sequences in a genomic DNA clone used for direct cDNA selection can lead to the isolation of artifactual cDNAs. For example, the presence of a pseudogene in a genomic DNA clone would be sufficient to cause the isolation of cDNAs derived from related genes, which may be located elsewhere in the genome.

Eliminating rRNA-derived cDNAs and Repetitive DNAs

The major artifacts in direct cDNA selection are the isolation of cDNAs arising from rRNA or mitochondrial mRNAs and cDNAs containing repetitive elements. These artifacts can easily be eliminated by screening the cloned PCR products with probes for these three types of sequences. An effective alternative is to hybridize to the YAC or other genomic clone used for direct cDNA selection and to a nonoverlapping, but similarly sized, YAC on duplicate Southern blots or dot blots. Only those inserts that hybridize to the YAC that was used for direct cDNA selection but not to the nonoverlapping YAC should be analyzed further.

Demonstrating That Products Represent Single-copy Sequences

A useful first test is to determine whether or not each remaining cDNA is present in a single, contiguous stretch of DNA in the cloned genomic DNA used to perform the selection. The cDNAs are labeled individually and used as probes for Southern hybridization of restriction-enzyme-digested DNA from all of the clones in a contig. Hybridization signals corresponding to a band or bands present in every lane (i.e., every clone in the contig) is a clear indication that a cDNA product recognizes host sequences and should be discarded. On the other hand, hybridization signals corresponding to a few bands in one or more contiguous genomic DNA clones would be expected if the cDNA is a single-copy sequence in the genomic region being tested. Hybridization to multiple products representing noncontiguous stretches of the genome should be treated with suspicion. This type of hybridization signal could result from a bona fide gene that has large introns, or it could be the result of repetitive sequences within the cDNA product used as a probe.

The next level of analysis determines whether or not any of the remaining cDNAs hybridize to sequences present elsewhere in the genome. This is most easily accomplished by using the cDNAs as probes for Southern hybridization of genomic DNA digested with restriction enzymes. If available, a somatic-cell hybrid containing only the chromosome or subchromosome region of interest should also be used (e.g., mapping panel #2 available from the Coriell cell repository; see Drwinga et al. 1993). Single-copy cDNA clones should produce one or a few bands that are identical in size to the band(s) seen with the cloned genomic DNA (with the exception that end fragments in the cloned inserts may be different sizes than their genomic counterparts are). cDNA clones that recognize low-copy-number repeats will recognize many bands and thus yield a pattern that cannot be explained by hybridization to the original genomic region.

Confirming That cDNA Products Represent Transcribed Sequences

The isolation of genomic DNA instead of cDNA is a potential artifact that sometimes occurs in direct cDNA selection. Although it is difficult to identify such an artifact, it is possible to show that a sequence is transcribed and therefore unlikely to represent such an artifact. Either amplification of cDNA by PCR or detection of hybridization signals on northern blots using cDNA clones as probes can be used to show that a sequence is transcribed. In both cases, some idea of the pattern of gene expression can be gained by using mRNA or cDNA from a variety of tissues in addition to the tissue source that was used as the starting material for direct cDNA selection.

If amplification by PCR is used as an assay, it is important to show rigorously that the amplification results from a cDNA and not from a genomic DNA contaminant. PCR primers on either side of an intron can be used to demonstrate that the cDNA is not a genomic contaminant because amplification will produce a larger PCR product from genomic DNA templates than they do from cDNA. In practice, this may be a difficult test to carry out, however, because it is not possible to easily predict the location of introns. In other words, the amplification of a similar size product from both cDNA and genomic templates does not necessarily indicate that the cDNA source is contaminated with genomic DNA.

Another test that is more predictable is to produce two cDNA templates. (The cDNA synthesis protocols on pp. 173–174 can be used for this purpose.) One of the templates is produced after first treating the RNA with DNase-free RNase (Sambrook et al. 1989). The second template is produced using the protocol as written. Both cDNA templates and a genomic DNA template are then used for amplification by PCR. If the same fragment is produced using all three templates, it is the result of genomic DNA contamination. If pretreatment with RNase prevents amplification from the cDNA template, it can be concluded that genomic contamination is not an issue for that particular direct selection cDNA product.

Detecting an mRNA on a northern blot by using a cDNA clone as a probe is good evidence that the cDNA product represents a transcribed sequence, although it is a less sensitive method of detection than amplification by PCR. Direct cDNA selection can detect sequences that are expressed at low levels in the starting material. Since the gene that is being analyzed may be expressed at higher levels in other tissues or cell lines, it is particularly important to include RNA from a variety of tissues on the northern blots that are used for this purpose.

The danger with this test is that genes transcribed at very low levels may be discarded as apparently representing artifacts. If a number of these clones are obtained, it is prudent to determine the DNA sequence of some of them. Computer programs such as BLAST and GRAIL can be used to analyze the sequence information and determine the likelihood that the clones encode proteins (see Baxevanis et al. 1997).

FURTHER ANALYSIS

Identifying Overlapping cDNA Clones

If direct cDNA selection has worked effectively, each gene will be represented by multiple overlapping cDNA clones. It is useful to identify overlapping cDNAs for two reasons. First, redundant steps in the analysis of the cDNA products can be avoided. For example, there is no need to hybridize to two different cDNA clones representing the same gene on northern blots. Second, with sufficient overlaps, large genes can be arranged into contigs as a set of overlapping cDNA clones. This is particularly useful if sequencing is a large part of the analysis since the sequence of entire cDNAs can be assembled from sufficiently redundant cDNA clones.

Overlaps can be identified by preparing probes from individual cloned PCR products and hybridizing them to a set of arrayed clones. This is most conveniently done by preparing multiple dot blots from the amplified PCR products so that different probes can be hybridized to the cDNA clones on several blots at the same time. After each round of hybridization, additional PCR products that have not been identified as overlapping clones by previous rounds of hybridization can be chosen for use as probes.

Sequence Analysis of cDNA Products

Sequence analysis of the cloned PCR products from direct cDNA selection can provide a great deal of information. Computer analysis of the sequence can often reveal a high degree of homology to a known gene or gene family. This can immediately provide functional and structural information about the newly isolated genes that cannot be obtained in any other fashion. (For discussions of sequence analysis and computer analysis, see Baxevanis et al. 1997; Kimmel et al. 1997; Wilson and Mardis 1997a,b.) Sequence analysis is also a prerequisite for synthesis of oligonucleotides that can be used for a PCR assay for expression of the putative gene.

REFERENCES

Baxevanis, A.D., M.S. Boguski, and B.F. Ouellette. 1997. Computational analysis of DNA and protein sequences. In *Genome analysis: A laboratory manual.* Vol. 1 *Analyzing DNA* (ed. B. Birren et al.), pp. 533–586. Cold Spring Harbor Laboratory Press, Cold Spring Harbor, New York.

Drwinga, H.L., L.H. Toji, C.H. Kim, A.E. Greene, and R.A. Mulivor. 1993. NIGMs human/rodent somatic cell hybrid mapping panels 1 and 2. *Genomics* **16:** 311–314.

Fanning, S. and R.A. Gibbs. 1997. PCR in genome analysis. In *Genome analysis: A laboratory manual.* Vol. 1 *Analyzing DNA* (ed. B. Birren et al.), pp. 249–299. Cold Spring Harbor Laboratory Press, Cold Spring Harbor, New York.

Feinberg, A.P. and B. Vogelstein. 1983. A technique for radiolabeling DNA restriction endonuclease fragments to high specific activity. *Anal. Biochem.* **132:** 6–13.

———. 1984. A technique for radiolabeling DNA restriction endonuclease fragments to high specific activity. *Anal. Biochem. (Addendum)* **137:** 266–267.

Green, E.D., P. Hieter, and F. Spencer. 1998. Yeast artificial chromosomes. In *Genome analysis: A laboratory manual.* Vol. 3 *Cloning systems* (ed. B. Birren et al.). Cold Spring Harbor Laboratory Press, Cold Spring Harbor, New York.

Kimmel, B.E., M.J. Palazzolo, C.H. Martin, J.D. Boeke, and S.E. Devine. 1997. Transposon-mediated DNA sequencing. In *Genome analysis: A laboratory manual.* Vol. 1 *Analyzing DNA* (ed. B. Birren et al.), pp. 455–532. Cold Spring Harbor Laboratory Press, Cold Spring Harbor, New York.

Korn, B., Z. Sedlacek, A. Manca, P. Kioschis, D. Konecki, H. Lehrach, and A. Poustka. 1992. A strategy for the selection of transcribed sequences in the Xq28 region. *Hum. Mol. Genet.* **1:** 235–242.

Lovett, M. 1994. Fishing for complements: Finding genes by direct selection. *Trends Genet.* **10:** 352–357. Lovett, M., J. Kere, and L.M. Hinton. 1991. Direct selection: A method for the isolation of cDNAs encoded by large genomic regions. *Proc. Natl. Acad. Sci.* **88:** 9628–9632.

Parimoo, S., R. Kolluri, and S.M. Weissman. 1993. cDNA selection from total yeast DNA containing YACs. *Nucleic Acids Res.* **21:** 4422–4423.

Parimoo, S., S.R. Patanjali, H. Shukla, D.D. Chaplin, and S.M. Weissman. 1991. cDNA selection: Efficient PCR approach for the selection of cDNAs encoded in large chromosomal DNA fragments. *Proc. Natl. Acad. Sci.* **88:** 9623–9627.

Sambrook, J., E.F. Fritsch, and T. Maniatis. 1989. *Molecular cloning: A laboratory manual,* 2nd edition. Cold Spring Harbor Laboratory Press, Cold Spring Harbor, New York.

Wilson, R.K. and E.R. Mardis. 1997a. Fluorescence-based DNA sequencing. In *Genome analysis: A laboratory manual.* Vol. 1 *Analyzing DNA* (ed. B. Birren et al.), pp. 301–395. Cold Spring Harbor Laboratory Press, Cold Spring Harbor, New York.

———. 1997b. Shotgun sequencing. In *Genome analysis: A laboratory manual.* Vol. 1 *Analyzing DNA* (ed. B. Birren et al.), pp. 397–434. Cold Spring Harbor Laboratory Press, Cold Spring Harbor, New York.

4

Exon Trapping

DAVID B. KRIZMAN

The fact that most eukaryotic genes are split into exons and introns makes the process of identifying those genomic regions that encode proteins a challenging one. To compound the problem, introns in higher eukaryotes are generally quite large, whereas exons are relatively small. For correct translation of proteins, the cell uses the mechanism of mRNA splicing to bring together the coding segments and remove the intervening noncoding sequences. The technique of exon trapping (sometimes called exon amplification) exploits this mechanism to isolate putative exons directly from genomic DNA via recognition of the splicing signals associated with exons.

A number of general strategies for exon trapping have been developed, each differing in the genomic target of interest. The original approach was designed to capture isolated 3' splice sites residing in fragments of genomic DNA (Duyk et al. 1990). More recently developed approaches capture either entire internal exons (Auch and Reth 1991; Buckler et al. 1991; Hamaguchi et al. 1992; Nehls et al. 1994) or entire 3'-terminal exons (Krizman and Berget 1993). These more recently developed approaches have proved to be more robust for isolating authentic exon sequences.

The substrate for trapping exons can be individual plasmid, bacteriophage λ, cosmid, bacteriophage P1, BAC, or YAC clones or pooled clones of any type containing higher eukaryotic DNA (Huntington's Disease Collaborative Research Group 1993; Krizman and Berget 1993; Church et al. 1994; Krizman et al. 1995). Experiments to date have shown that small numbers of distinct exons are typically trapped from individual plasmid, bacteriophage λ, cosmid, or bacteriophage P1 clones, whereas larger numbers are trapped by using pools of these clones. Exon trapping from YAC clones can be used for isolating exons from larger genomic regions, particularly in cases where smaller-insert clones spanning the region are not available.

Strategies for Trapping Intact Exons

Each of the two major strategies for exon trapping—internal exon trapping and 3′-terminal exon trapping—yields different products for subsequent analysis and is associated with distinct problems and limitations. Thus, the choice of which strategy to use is an important one. The major features of each strategy are discussed briefly below.

Internal exon trapping was designed to identify internal exons that have an average size of approximately 130 bp (range of 20–200 bp). An advantage of this strategy is that nearly all exons containing protein-coding information can potentially be recovered and sequenced. Subsequent comparison of the resulting sequences with DNA databases often reveals a significant degree of homology with known genes, thereby providing early insight about the function of the gene from which the exon was trapped. A disadvantage of internal exon trapping is that the probes generated from the trapped exons are often small and difficult to use for hybridization-based experiments (e.g., retrieving full-length cDNA clones or northern blot analysis). However, the resulting trapped sequences can be used for obtaining larger segments of the corresponding cDNAs by various methods, including 3′ and 5′ RACE (Frohman et al. 1988) as well as other oligonucleotide-based techniques (e.g., GeneTrapper; Life Technologies).

3′-Terminal exon trapping is designed to recover the last exons (i.e., exons at the 3′ ends) of genes. The recovered DNA segments are typically 200–2500 bp in size (averaging approximately 630 bp). Most of this DNA corresponds to the 3′ untranslated regions of genes and contains little, if any, protein-coding sequence. Thus, subsequent comparison of the resulting sequences with DNA databases often reveals no sequences or few sequences with a high degree of homology. However, the relatively large size of the recovered DNA segment and the unique nature of most 3′ untranslated regions of genes facilitate efforts to isolate corresponding cDNA clones by hybridization-based experiments and to identify the appropriate transcript by northern blot analysis. 3′-Terminal exon sequences can also be used for obtaining more full-length cDNA clones by the 5′ RACE technique.

INTERNAL EXON TRAPPING

General Principles

A typical internal exon is small (usually 20–200 bp, averaging approximately 130 bp) and is flanked on the 5′ end by a 3′ splice site (i.e., the splice acceptor) and on the 3′ end by a 5′ splice site (i.e., the splice donor). There is usually a single ORF spanning the entire sequence. Most authentic internal exons show some degree of homology to known protein-coding sequences in DNA databases.

To trap internal exons, restriction fragments derived from cloned genomic DNA are subcloned into a trapping vector engineered to produce vector-derived mRNA molecules after transfection into specific mammalian cells (e.g., COS-7). Internal exons present in the subcloned genomic fragments are incorporated into the vector's transcription unit and become trapped within the chimeric mRNA that results from splicing and processing the vector-derived transcripts. These trapped exons are amplified by RT-PCR, a method in which total RNA purified from the transfected mammalian cells is reverse transcribed and then used as the template for amplifying the exon sequences by PCR. The primers used in RT-PCR are specific for the mRNA species derived from the trapping vector. Thus, only vector-derived products—and not endogenous-mRNA-derived products—are amplified. The resulting PCR products are then cloned and typically sequenced.

Vectors for Internal Exon Trapping

There are currently a small number of vectors used for internal exon trapping. All but one of these vectors are plasmid-based and contain amp^r or cam^r and a bacterial origin of replication (*ori*) for propagation in *E. coli*. All of the vectors contain a trapping cassette with a eukaryotic enhancer/promoter that drives the transcription of a two-exon transcription unit containing a multiple cloning site between the two exons. The first of these exons functions as a 5′-terminal exon, whereas the second exon functions as a 3′-terminal exon capable of directing polyadenylation. In addition to the common features of these vectors, each has unique characteristics.

• *pSPL3:* This plasmid (Figure 1) is a newer version of pSPL1, the original vector used for inter-

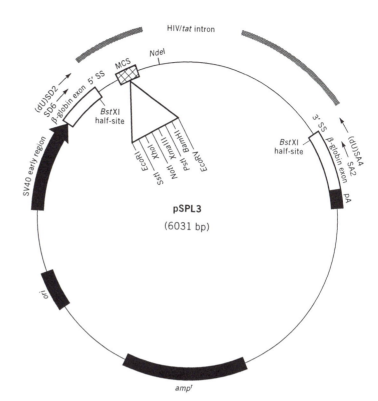

Figure 1 Vector pSPL3 for internal exon trapping. SS denotes a splice site, MCS denotes the multiple cloning site, pA denotes the polyadenylation signal, *ori* denotes the ColE1 origin of replication, and *amp*^r denotes the gene conferring ampicillin resistance. The positions of primers SD6, (dU)SD2, (dU)SA4, and SA2 are indicated. For further description, see text.

nal exon trapping (Buckler et al. 1991; Church et al. 1994). The trapping cassette contains the SV40 early region, rabbit β-globin exonic sequences, HIV/*tat* 5′ and 3′ splice sites, the HIV/*tat* intron, and the SV40 polyadenylation signal. The SV40 early region directs transcription and replication in the African green monkey cell line COS-7. This cell line harbors a replication-defective mutant SV40 that expresses the large T antigen, which initiates both replication and transcription from the SV40 early region. Exon 1 of pSPL3 is a chimera constructed from rabbit β-globin exonic sequences and the HIV/*tat* 5′ splice site. The intron corresponds to HIV/*tat* intronic sequences. Exon 2 begins with the HIV/*tat* 3′ splice site followed by rabbit β-globin exonic sequence and ends with the SV40 polyadenylation signal sequence that directs correct cleavage and polyadenylation of the resulting mRNA. *Bst*XI half-sites are present on either side of the intron. After the RNA is processed and amplified by PCR, digestion with *Bst*XI or *Nde*I eliminates vector-only background products (see pp. 205-206 and Figure 2), which occur when there is no internal exon present in the foreign DNA fragment, when RNA is produced from nonrecombinant vector, or when

splicing occurs involving a known HIV/*tat* intronic cryptic splice donor. The multiple cloning site in the intron contains sites for *Eco*RI, *Sst*I, *Xho*I, *Not*I, *Xma*III, *Pst*I, *Bam*HI, and *Eco*RV for cloning the target DNA. pSPL3 is the most widely used vector for internal exon trapping.

• *pSPL3B* and *pSPL3Bcam:* These vectors were derived from pSPL3. A cryptic splice site was identified in the HIV/*tat* intron of pSPL3 that led to a significant amount of vector-derived background. This cryptic splice site was removed to create pSPL3B. Subsequent substitution of *cam*^r for *amp*^r yielded pSPL3Bcam (Burn et al. 1995).

• *pL53In:* The trapping cassette of this plasmid vector contains the Rous sarcoma virus LTR, which drives transcription. The first exon of the cassette is derived from the human phosphatase gene, whereas the second exon and the intervening intron are derivatives of the rat preproinsulin gene. A unique *Kpn*I site serves as the cloning site within the intron (Auch and Reth 1991).

• *pMHC2:* Transcription from this plasmid vector is driven by the SV40 early region. The trapping cassette consists of part of exon 10, intron 10, and part of exon 11 from the human *p53* gene

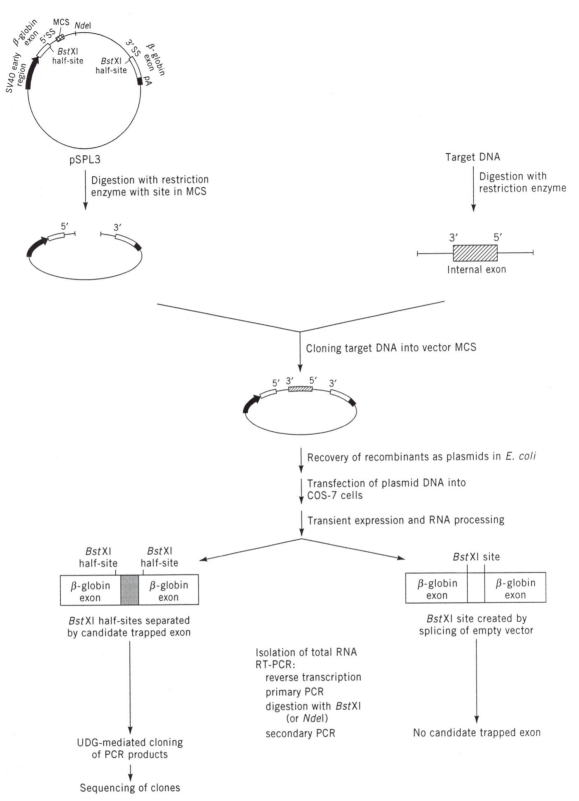

Figure 2 Overview of internal exon trapping using the vector pSPL3. Any of the restriction enzymes with a site in the multiple cloning site of the vector can be used for digestion of the vector and target DNA. SS denotes a splice site, MCS denotes the multiple cloning site, and pA denotes the polyadenylation signal. For details, see text.

plus the SV40 polyadenylation signal. A unique *Bgl*II site serves as the cloning site within the intron (Hamaguchi et al. 1992).

- *LambdaGET:* This bacteriophage-based vector is designed to clone and analyze fragments of genomic DNA larger than those that can be handled with the plasmid-based trapping vectors (i.e., approximately 20 kb versus approximately 8 kb) (Nehls et al. 1994). It is derived from pL53In (see above) and has the same trapping cassette. Cloning of foreign target DNA is highly efficient since a bacteriophage packaging system is used.

3'-TERMINAL EXON TRAPPING

General Principles

A typical 3'-terminal exon (usually 200–2500 bp, averaging approximately 630 bp) is larger than a typical internal exon and consists mostly of a 3' untranslated region flanked on the 5' end by a 3' splice site (i.e., the splice acceptor) and on the 3' end by a consensus polyadenylation signal AATAAA or ATTAAA. Computational analysis of a 3'-terminal exon typically reveals little protein-coding sequence and little homology to DNA sequences in the databases. In fact, the presence of

multiple stop codons in all frames is suggestive of a 3'-terminal exon.

To trap 3'-terminal exons, restriction fragments derived from cloned genomic DNA are ligated to the trapping vector pTAG4. The resulting ligation products (not the recombinant plasmids derived from transformed *E. coli* that are used for internal exon trapping) are transfected directly into COS-7 cells. pTAG4 contains a trapping cassette with a transcription unit that lacks a last exon and the polyadenylation signal. The donation of a 3'-terminal exon to the vector by an exogenous DNA fragment in essence completes the three-exon transcription unit, allowing the generation of a stable vector-derived mRNA molecule after transfection of the DNA into COS-7 cells. Exons trapped in this manner are amplified by a modification of the RT-PCR method used for internal exons. This modification is called 3' RACE (Frohman et al. 1988) and was originally designed to amplify mRNA species from the 3' end of the molecule by using the poly(A) tail as an anchor.

pTAG4 Vector for 3'-Terminal Exon Trapping

Only one vector for 3'-terminal exon trapping has been developed (Figure 3). pTAG4 contains *amp*[r] and a bacterial origin of replication for propagation

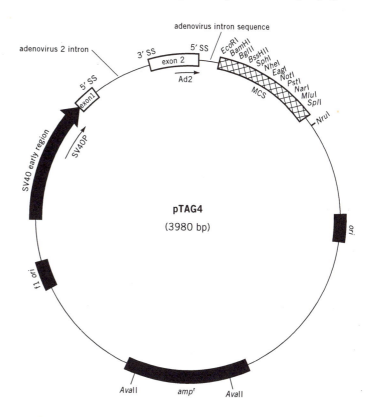

Figure 3 Vector pTAG4 for 3'-terminal exon trapping. SS denotes a splice site, MCS denotes the multiple cloning site, *ori* denotes the ColE1 origin of replication, f1 *ori* denotes the filamentous bacteriophage origin of replication, and *amp*[r] denotes the gene conferring ampicillin resistance. The positions of primers SV40P and Ad2 are indicated. For further description, see text.

in *E. coli* (*ori*) as well as bacteriophage f1 *ori*. The trapping cassette contains the SV40 early region, which drives transcription after transfection of the DNA into COS-7 cells. Exons 1 and 2, the intervening intron, and the splice sites were derived from leader exon/intron sequences of the human adenovirus 2 genome. pTAG4 lacks a last exon containing a polyadenylation signal; thus, no mature polyadenylated mRNA is produced from the vector after it is transfected into COS-7 cells. A multiple cloning site downstream from the 5′ splice site of exon 2 contains the following unique restriction sites: *Eco*RI, *Bam*HI, *Bgl*II, *Bss*HII, *Sph*I, *Nhe*I, *Eag*I, *Not*I, *Pst*I, *Nar*I, *Mlu*I, and *Spl*I.

PROTOCOLS FOR TRAPPING INTACT EXONS

The starting material for exon trapping is genomic DNA cloned into any one of a number of vectors (e.g., plasmid, bacteriophage λ, cosmid, bacteriophage P1, BAC, or YAC vectors). Successful exon trapping requires that the cloned DNA be sufficiently purified from the endogenous *E. coli* or yeast DNA to minimize the recovery of nonspecific (background) products. It is not necessary to purify the DNA insert from the cloning vector, since the vectors do not generally contain exonic or intronic sequences. The target DNA prepared on p. 198 can be used for either internal exon trapping (see pp. 199–206) or 3'-terminal exon trapping (see pp. 207–214).

PROTOCOL

Preparation of Target DNA

1. Purify cloned target DNA free of endogenous host DNA according to standard protocols (e.g., see Wilson and Mardis 1997b).

2. Digest to completion at least 1 μg of target DNA in a 1.5-ml microcentrifuge tube. For each microgram of DNA, use 1 unit of an appropriate restriction enzyme.

   Notes: For internal exon trapping using pSPL3, use *Eco*RI, *Sst*I, *Xho*I, *Not*I, *Xma*III, *Pst*I, *Bam*HI, or *Eco*RV. For 3'-terminal exon trapping using pTAG4, use *Eco*RI, *Bam*HI, *Bgl*II, *Bss*HII, *Sph*I, *Nhe*I, *Eag*I, *Not*I, *Pst*I, *Nar*I, *Mlu*I, or *Spl*I. If *Eco*RV is used to subclone insert DNA into pSPL3, the standard digestion with *Bst*XI cannot be used because of disruption of the *Bst*XI site (see pp. 205–206, step 12).

   The starting amount of target DNA is likely to depend on the source of the cloned DNA. For example, cosmid DNA can be purified more readily than YAC DNA, which requires purification by PFGE (see Wilson and Mardis 1997b). A typical target DNA preparation using cosmids may therefore start with 2–4 μg of purified DNA, whereas a preparation using YACs may start with as little as 1 μg.

3. Extract with **phenol:chloroform**:isoamyl alcohol (25:24:1) (see Appendix). Use phenol saturated with TE (pH 8.0).

   phenol, chloroform (see Appendix for Caution)

4. Precipitate the target DNA as follows:

   a. Add 0.5 volume of 7.5 M ammonium acetate and 2.5 volumes of ice-cold absolute ethanol to the aqueous phase and mix. Place at –20ºC for at least 15 minutes.

   b. Centrifuge in a microcentrifuge at 12,000g at 4ºC for 15 minutes. Discard the supernatant.

   c. Wash the pellet by adding 1 ml of ice-cold 70% ethanol, recentrifuging for 5 minutes, and then carefully removing and discarding the ethanol.

   d. Dry the DNA pellet under vacuum.

5. Dissolve the target DNA in TE (pH 8.0) at a final concentration of 250 ng/μl.

   Notes: The DNA can be stored at 4ºC indefinitely.

   If there is a sufficient amount of target DNA, analyze small aliquots of digested and undigested DNA samples and the appropriate molecular-weight markers on a 1% agarose gel to confirm that the digestion was complete. The resulting number of bands in the digested sample will vary with the complexity of the target DNA. Proceed with the exon trapping protocol only if there is convincing evidence that the digestion was successful. If the DNA was not completely digested, add more enzyme and incubate for an additional period of time.

6. Use the digested target DNA in the exon trapping protocol on pp. 201–202 or pp. 210–211.

INTERNAL EXON TRAPPING USING pSPL3

The basic protocol for exon trapping using pSPL3 (Church et al. 1994) is provided on pp. 200–206 and shown in Figure 2. pSPL3B and pSPL3Bcam can be used in the identical fashion outlined for pSPL3. This general protocol can also be applied to other non-pSPL3-derived vectors for internal exon trapping. A complete kit for performing internal exon trapping is also available from Life Technologies. Restriction fragments of genomic DNA from any source (e.g., individual cosmid, bacteriophage P1, plasmid, bacteriophage λ, BAC, or YAC clones or pools of these clones) are first subcloned into an appropriate vector. Recombinant DNA molecules are then recovered as plasmids in *E. coli*. The resulting plasmids are transfected en masse into COS-7 cells. After transient expression for 16–24 hours, total RNA is purified and reverse transcribed by using a vector-specific oligonucleotide to yield first-strand cDNA. A primary round of PCR is performed by using two vector-specific primers. The resulting product (except for *Eco*RV-subcloned targets) is digested with *Bst*XI or *Nde*I, which removes products of splicing events that only contain vector sequences and thereby reduces the fraction of final products that lack a candidate trapped exon. A secondary round of PCR is then performed by using nested vector-specific primers. By including a uracil-containing 12-base tail on the secondary PCR primers, the final products can be efficiently cloned by using a UDG-mediated cloning strategy (see Nisson et al. 1991; see also Fanning and Gibbs 1997).

PROTOCOL

Preparation of pSPL3

1. Purify the exon trapping vector pSPL3 (see Figure 1) by ultracentrifugation on a CsCl gradient (see Sambrook et al. 1989).

2. In a 1.5-ml microcentrifuge tube, digest 5 μg of purified pSPL3 with the same restriction enzyme used to digest the target DNA on p. 198.

   Note: Alternatively, two different restriction enzymes can be used in combination to digest both the vector and the target DNA. This will make the CIP treatment of pSPL3 in step 4 unnecessary. Any combination of two enzymes with sites in the vector's multiple cloning site can be used except *Not*I plus *Xma*III, which have recognition sites that overlap in the multiple cloning site.

3. Analyze 100 ng of the digested vector DNA and the appropriate molecular-weight markers on a 1% agarose gel. Keep the remainder of the DNA on ice or at 4°C during this step. Confirm that the vector is linearized (6031 bp in length) and then proceed with step 4.

   Notes: If the vector is not completely linearized, add more enzyme and incubate for an additional period of time. Alternatively, digest another 5-μg aliquot of purified vector from the same preparation or a new preparation.

   If two different enzymes were used in combination to digest both pSPL3 and the target DNA, gel purify the linearized vector (along with the appropriate molecular-weight markers) on a 0.8% agarose gel. Extract the vector DNA from the agarose as specified by the manufacturer of one of the kits for purification of DNA (e.g., GENECLEAN [BIO 101], Prep-A-Gene [Bio-Rad], QIAEX [Qiagen], or BandPrep [Pharmacia]). This removes the sequences liberated from the multiple cloning site by digestion with two restriction enzymes. Proceed with step 5 using the gel-purified vector.

4. Dephosphorylate the linearized vector with CIP as specified by the enzyme's manufacturer.

   Note: CIP can generally be added directly to the restriction enzyme reaction mixture once the vector has been determined to be linear.

5. Extract with **phenol:chloroform**:isoamyl alcohol (25:24:1) (see Appendix). Use phenol saturated with TE (pH 8.0).

   phenol, chloroform (see Appendix for Caution)

6. Precipitate the linearized pSPL3 with 0.5 volume of 7.5 M ammonium acetate and 2.5 volumes of ice-cold absolute ethanol and dry the pellet under vacuum as described on p. 198, step 4.

7. Dissolve the pSPL3 DNA in TE (pH 8.0) at a final concentration of 250 ng/μl.

   Note: The DNA can be stored at 4°C indefinitely.

8. Subclone the target DNA into the prepared vector DNA as described on pp. 201–202.

PROTOCOL

Subcloning Target DNA into pSPL3 and Transforming *E. coli*

1. Prepare a ligation mixture by gently combining the components below in a 0.5-ml microcentrifuge tube. Also prepare a control reaction mixture containing all of the components except the target DNA (i.e., a vector-only control).

digested pSPL3 (250 ng/μl; see p. 200)	1 μl
digested target DNA (250 ng/μl; see p. 198)	1–4 μl
5x exon-trapping ligation buffer	2 μl
bacteriophage T4 DNA ligase (1 unit/μl; Life Technologies)	1 μl
H$_2$O	to a final volume of 10 μl

   Note: The control reaction is useful for monitoring the extent of self-ligation of the vector (i.e., ligations devoid of target DNA) (see step 4). Use this control in parallel reactions throughout the remainder of the trapping protocol.

   5x Exon-trapping ligation buffer

Component and final concentration	Amount to add per 10 ml
250 mM Tris-Cl	2.5 ml of 1 M (pH 7.6)
50 mM MgCl$_2$	0.5 ml of 1 M
5 mM ATP	100 μl of 0.5 M
5 mM DTT	50 μl of 1 M
25% (w/v) PEG 8000	2.5 g
H$_2$O	to make 10 ml

   Store at –20°C for at least 6 months.

2. Incubate at room temperature for 1 hour or at 15°C overnight.

3. Transform *E. coli* with 1 μl of each ligation product (ligated vector/target DNA and vector-only control). Use a standard protocol for either transformation by electroporation or CaCl$_2$-mediated (i.e., chemical) transformation (Sambrook et al. 1989), but plate the transformation mixture as described in step 4 below.

   Note: Various strains of *E. coli* (e.g., XL1-Blue, JM109, DH10B, DH5α, or HB101) can be used. These are available in transformation-competent form from a number of suppliers (e.g., Stratagene, Promega, or Life Technologies).

4. Spread 1/10 and 1/100 of the volume of each transformation mixture on separate LB agar plates containing ampicillin (80 μg/ml) and incubate the plates at 37°C overnight. Inoculate 5 ml of LB medium containing ampicillin (80 μg/ml) with half of the original volume of the transformation mix-

ture prepared with the ligated vector/target DNA and incubate at 37ºC with agitation at 200 rpm overnight.

Note: The plates should reveal the numbers of transformants obtained in the presence and absence of target DNA, thereby indicating the extent of vector-only ligation. A high percentage of nonrecombinant vector in the vector/target DNA ligation indicates that the CIP treatment during the preparation of the vector did not work effectively. If the number of colonies produced on the plates with the control ligation exceeds 10% of the number produced in the presence of target DNA, start the procedure over with vector prepared again as described on p. 200, steps 2–7. If the number of colonies produced on the plates with the control ligation is less than 10% of the number produced in the presence of target DNA, proceed with step 5.

5. Purify plasmid DNA from the 5-ml culture (prepared in step 4) according to a standard alkaline lysis procedure (see Wilson and Mardis 1997a).

6. Dissolve the purified DNA in TE (pH 8.0) at a final concentration of 0.5–1 μg/μl.

   Note: The DNA can be stored at 4ºC indefinitely.

7. Use the purified plasmid DNA to transfect COS-7 cells as described on p. 203.

PROTOCOL

Transfection of Recombinant Plasmids into COS-7 Cells

1. For each transfection, transfer approximately 300,000 COS-7 cells into a well of a 3.5-cm 6-well tissue-culture plate. Incubate at 37°C for 16 hours.

   Note: Maintain and passage (Jakoby and Pastan 1979) a single T-75 flask of COS-7 cells (ATCC accession number CRL1651) for use in transfections. Grow the cells at 37°C in DMEM supplemented with 10% FCS.

2. Transfect each well of COS-7 cells with 1 μg of the purified plasmid DNA prepared on pp. 201–202.

   Note: Cationic lipid-mediated transfection can be used to introduce the DNA into the cells. This method is more efficient than transfection mediated by calcium phosphate and is more gentle to the cells than electroporation. The necessary reagents are available from several good suppliers and should be used as specified by the manufacturer.

3. Incubate the transfected cells at 37°C for 16–24 hours.

4. Purify total RNA from the transfected cells by extraction with acidic **phenol** and guanidinium thiocyanate.

   Note: Kits for performing rapid and efficient isolation of high-quality total RNA are available from several suppliers (e.g., Life Technologies, Promega, or Pharmacia) and should be used as specified by the manufacturer. Working with RNA requires the use of appropriate precautions to maintain RNase-free conditions. These precautions include using **DEPC**-treated H_2O for preparing solutions (except those containing Tris), wearing gloves, and using sterile plasticware instead of glassware.

   phenol, DEPC (see Appendix for Caution)

5. Dissolve the purified total RNA in 20 μl of TE (pH 8.0). Typically, the resulting RNA concentration is 1–2 μg/μl.

   Note: The RNA can be stored at –80°C indefinitely.

6. Use the purified total RNA to synthesize cDNA as described on pp. 204–206.

PROTOCOL

Synthesis of a cDNA Pool and Amplification of Trapped Exons by RT-PCR

1. Combine the following in a 0.5-ml microcentrifuge tube:

20 µM oligonucleotide SA2	1 µl
purified total RNA (see p. 203)	1–3 µg
H_2O	to a final volume of 12 µl

   Notes: The SA2 primer is 5′ATCTCAGTGGTATTTGTGAGC3′.
   DEPC-treated H_2O should be used in all steps and in preparing all solutions (except those containing Tris). Other appropriate precautions to maintain RNase-free conditions include wearing gloves and using sterile plasticware instead of glassware.

   DEPC (see Appendix for Caution)

2. Incubate at 70°C for 5 minutes, and then cool on ice.

3. Centrifuge briefly to collect the liquid at the bottom of the tube, add the following components equilibrated to room temperature, and mix gently:

5x first-strand cDNA buffer	4 µl
100 mM DTT	2 µl
10 mM dNTP mixture (see Appendix)	1 µl

   5x First-strand cDNA buffer

Component and final concentration	Amount to add per 100 ml
250 mM Tris-Cl	25 ml of 1 M (pH 8.3)
375 mM KCl	37.5 ml of 1 M
15 mM $MgCl_2$	1.5 ml of 1 M
H_2O	36 ml

   Store at –20°C indefinitely.

4. Centrifuge briefly to collect the liquid at the bottom of the tube, and then incubate at 42°C for 5 minutes.

5. Add 1 µl of reverse transcriptase (200 units/µl) and mix gently. Incubate at 42°C for 30 minutes.

6. Incubate at 55°C for 5 minutes, and then add 1 µl of DNase-free RNase H (2 units/µl; e.g., Life Technologies) and mix gently. Incubate at 55°C for an additional 10 minutes.

   Note: RNase H digests the RNA strand of a DNA:RNA hybrid. The presence of RNA has been shown to inhibit DNA amplification by PCR. Digestion with RNase H is carried out at 55°C to eliminate the possibility of snapback structures and second-strand products resulting from residual reverse transcriptase activity. Reverse transcriptase is inactive at this temperature.

7. Centrifuge briefly to collect the liquid at the bottom of the tube, and then cool on ice.

   Note: This product is essentially a cDNA pool. It can be stored at −20°C indefinitely.

8. Prepare the primary PCR mixture by gently combining the following in a PCR tube:

cDNA pool	5 μl
10x amplification buffer	5 μl
50 mM MgCl$_2$	1.5 μl
10 mM dNTP mixture (see Appendix)	1 μl
20 μM oligonucleotide SA2	2.5 μl
20 μM oligonucleotide SD6	2.5 μl
H$_2$O	29.5 μl

   Notes: The SA2 primer is 5'ATCTCAGTGGTATTTGTGAGC3'. The SD6 primer is 5'TCTGAGTCACCTGGACAACC3'.

   For thermal cyclers that do not have a top heating block, overlay each reaction mixture with 50 μl of light mineral oil and then cover the tube.

   10x Amplification buffer

Component and final concentration	Amount to add per 100 ml
0.5 M KCl	50 ml of 1 M
100 mM Tris-Cl	10 ml of 1 M (pH 8.3)
H$_2$O	40 ml

   Store at −20°C indefinitely.

9. Preheat the thermal cycler to 94°C. Incubate the PCR mixture at 94°C for 5 minutes.

10. Reduce the temperature of the thermal cycler to 80°C and add diluted *Taq* DNA polymerase (2.5 units diluted with H$_2$O to a final volume of 3 μl).

11. Perform thermal cycling as follows for 6 cycles:

    94°C for 1 minute
    60°C for 1 minute
    72°C for 5 minutes

    Perform an additional cycle of 72°C for 10 minutes, and then hold the samples at 4°C.

12. Unless the target DNA was subcloned into either the *Eco*RI or the *Eco*RV site, add 25 units of *Bst*XI to the PCR product and incubate at 55°C overnight. If the target DNA was subcloned into either the *Eco*RI or the *Eco*RV site, add 25 units of *Nde*I to the PCR product and incubate at 37°C overnight.

    Note: Digestion with *Bst*XI is used to reduce background caused by vector-vector splicing and splicing at cryptic splice sites in the vector and in the cloned target DNA. Subcloning the target DNA into either the *Eco*RI or the *Eco*RV site destroys a *Bst*XI recognition site, making it necessary to use *Nde*I to reduce this type of background.

13. Add an additional 5 units of the appropriate restriction enzyme (*Bst*XI or *Nde*I) and incubate at 55°C or 37°C, respectively, for 2 hours.

14. Prepare the secondary PCR mixture by gently combining the following in a new PCR tube:

digested primary PCR product	5 μl
10x amplification buffer	5 μl
50 mM MgCl$_2$	1.5 μl
10 mM dNTP mixture (see Appendix)	1 μl
20 μM oligonucleotide (dU)SA4	1 μl
20 μM oligonucleotide (dU)SD2	1 μl
H$_2$O	32.5 μl

Notes: The (dU)SA4 primer is 5′CUACUACUACUACACCTGAGGAGTGAATTGGTCG3′. The (dU)SD2 primer is 5′CUACUACUACUAGTGAACTGCACTGTGACAAGCTGC3′. The presence of the triplet repeats CUA on the primers facilitates the cloning of the resulting PCR product by the UDG-mediated cloning strategy.

For thermal cyclers that do not have a top heating block, overlay each reaction mixture with 50 μl of light mineral oil and then cover the tube.

The remainder of the digested primary PCR product can be stored at –20°C indefinitely.

15. Repeat steps 9–10.

16. Perform thermal cycling as follows for 30 cycles:

94°C for 30 seconds
60°C for 30 seconds
72°C for 2 minutes

Perform an additional cycle of 72°C for 10 minutes, and then hold the samples at 4°C.

17. Analyze 10 μl of the secondary PCR product and the appropriate molecular-weight markers on a 1.2% agarose gel and detect the DNA by staining with **ethidium bromide**.

Note: Analysis of the vector-only control reaction should help identify vector-derived PCR products. A major PCR product of 177 bp and several larger, but less abundant, products will be seen even after digestion with *Bst*XI or *Nde*I. These products are derived from pSPL3 and are the result of both correct splicing of nonrecombinant vector and cryptic splicing of recombinant vector at the cryptic splice sites within the HIV/*tat* intron of pSPL3.

ethidium bromide (see Appendix for Caution)

18. Clone the secondary PCR products and analyze the resulting isolates as described on p. 215.

Note: PCR products can be cloned directly or after gel purification. If the secondary PCR is performed on products that were not treated with *Bst*XI or *Nde*I, a 177-bp vector-derived product will be seen in a large fraction of the clones. In general, any PCR product larger than 177 bp should be assumed to contain a putative trapped exon.

3′-TERMINAL EXON TRAPPING USING pTAG4

The protocol for 3′-terminal exon trapping using pTAG4 is provided on pp. 209–214 (Krizman and Berget 1993) and is shown in Figure 4. A complete kit for 3′-terminal exon trapping is also available from Life Technologies. The target genomic DNA is first digested to completion with one of the restriction enzymes that has a site in the multiple cloning site (e.g., *Eco*RI, as shown in Figure 4). Digestion of pTAG4 with *Ava*II and the same restriction enzyme used to digest the target DNA yields a linearized molecule with a sticky end downstream from the second exon. Digestion with *Ava*II functions to remove *amp*^{r} from pTAG4. This gene has been shown to give a small amount of background in this protocol. The digested target DNA and pTAG4 are ligated to form linear concatemers consisting of target DNA fragments flanked by pTAG4 sequences. The product of the ligation is directly transfected into COS-7 cells, and transcription occurs from the linear concatemers. After transient expression, poly(A)$^+$ mRNA is purified and reverse transcribed by using an oligo(dT)-based primer-adapter. Trapped exons are then amplified by a seminesting PCR approach. (This modification of the RT-PCR method is called 3′RACE.) The primary PCR uses one primer specific for the 5′tail sequence on the reverse transcription primer-adapter and one primer specific for the SV40 promoter in pTAG4. The resulting PCR products are then digested with the same restriction enzyme used for preparative digestion of the target DNA and pTAG4. This step helps remove background products resulting from reverse transcription of unspliced precursor RNA or residual contaminating DNA in the mRNA preparation. The secondary PCR uses the same primer specific for the 5′tail sequence on the reverse transcription primer-adapter and a nested primer specific for the second exon in pTAG4. Both of these secondary PCR primers have uracil-containing 12-base tails that allow UDG-mediated cloning of the final product (see Fanning and Gibbs 1997).

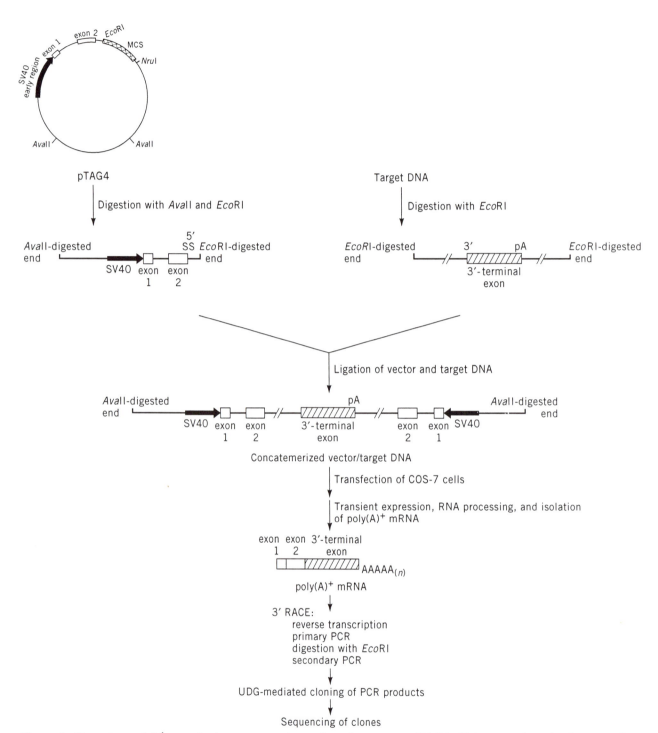

Figure 4 Overview of 3′-terminal exon trapping using the vector pTAG4. This procedure is shown using *Eco*RI, but any of the restriction enzymes with a site in the multiple cloning site of the vector can be used. SS denotes a splice site, MCS denotes the multiple cloning site, and pA denotes the polyadenylation signal. For details, see text.

PROTOCOL

Preparation of pTAG4

1. Purify the exon trapping vector pTAG4 (see Figure 3) by ultracentrifugation on a CsCl gradient (Sambrook et al. 1989).

2. In a 1.5-ml microcentrifuge tube, digest 5 µg of purified pTAG4 with *Ava*II and the same restriction enzyme used to digest the target DNA on p. 198.

3. Gel purify the largest linearized vector band (along with the appropriate molecular-weight markers) on a 0.8% agarose gel. Extract the vector DNA from the agarose as specified by the manufacturer of one of the kits for purification of DNA (e.g., GENECLEAN [BIO 101], Prep-A-Gene [Bio-Rad], QIAEX [Qiagen], or BandPrep [Pharmacia]).

   Notes: After digestion with both *Ava*II and the enzyme chosen for the target DNA, three bands will be visible on the gel. The vector band to be used for trapping is the largest of the three and ranges in size from 2340 bp to 2580 bp, depending on which enzyme is used in combination with *Ava*II. This step removes the small DNA fragment liberated from the multiple cloning site by digestion with two restriction enzymes.

   In general, the linearized vector will be 3600–3900 bp in length, depending on which restriction enzyme besides *Ava*II was used for the digestion. If the vector is not linearized, add more enzyme and incubate for an additional period of time. Alternatively, digest another 5-µg aliquot of purified vector from the same preparation or a new preparation.

4. Dissolve the purified, linearized pTAG4 in TE (pH 8.0) at a final concentration of 500 ng/µl.

   Note: The DNA can be stored at 4°C indefinitely.

5. Ligate the target DNA with the prepared vector DNA as described on pp. 210–211.

PROTOCOL

Ligating Target DNA with pTAG4 and Transfecting into COS-7 Cells

1. Prepare a ligation mixture by gently combining the components below in a 0.5-ml microcentrifuge tube. Also prepare a control reaction mixture containing all of the components except the target DNA (i.e., a vector-only control).

digested pTAG4 (500 ng/µl; see p. 209)	1 µl
digested target DNA (500 ng/µl; see p. 198)	2 µl
5x exon-trapping ligation buffer (for preparation, see p. 201)	1 µl
bacteriophage T4 DNA ligase (1 unit/µl; Life Technologies)	1 µl

   Note: To induce concatemerization, the total concentration of vector DNA plus target DNA should be approximately 200 ng/µl.

2. Incubate at 15°C overnight.

3. For each transfection, transfer approximately 300,000 COS-7 cells into a well of a 3.5-cm 6-well tissue-culture plate. Incubate at 37°C for 16 hours.

   Note: Maintain and passage (Jakoby and Pastan 1979) a single T-75 flask of COS-7 cells (ATCC accession number CRL1651) for use in transfections. Grow the cells at 37°C in DMEM supplemented with 10% FCS.

4. Transfect each well of COS-7 cells with the entire 5-µl ligation mixture from step 2.

   Note: Cationic lipid-mediated transfection can be used to introduce the DNA into the cells. This method is more efficient than transfection mediated by calcium phosphate and is more gentle to the cells than electroporation. The necessary reagents are available from several good suppliers and should be used as specified by the manufacturer.

5. Incubate the transfected cells at 37°C for 16–24 hours.

6. Purify poly(A)$^+$ mRNA from the transfected cells.

   Notes: The use of poly(A)$^+$ mRNA is critical for 3′-terminal exon trapping since the subsequent 3′RACE steps use an oligo(dT)-based primer for reverse transcription.

   Kits for performing rapid and efficient purification of poly(A)$^+$ mRNA are available from several suppliers (e.g., Invitrogen or Pharmacia) and should be used as specified by the manufacturer. Approximately 1 µg of poly(A)$^+$ mRNA should be obtained from each transfection. Working with RNA requires the use of appropriate precautions to maintain RNase-free conditions. These precautions include using **DEPC**-treated H_2O for preparing solutions (except those containing Tris), wearing gloves, and using sterile plasticware instead of glassware.

   DEPC (see Appendix for Caution)

7. Dissolve the purified poly(A)+ mRNA in 10 µl of TE (pH 8.0).

   Note: The poly(A)+ mRNA can be stored at −80°C for at least 6 months.

8. Use the purified poly(A)+ mRNA to synthesize cDNA as described on pp. 212–214.

PROTOCOL

Synthesis of a cDNA Pool and Amplification of Trapped 3'-Terminal Exons

1. Combine the following in a 0.5-ml microcentrifuge tube:

50 μM oligonucleotide AP	1 μl
purified poly(A)$^+$ mRNA	
(see pp. 210–211)	500 ng
20 mM EDTA (pH 8.0)	1 μl
H$_2$O	to a final volume of 20 μl

   Notes: The AP primer-adapter is 5'AAGGATCCGTCGACATCGATAATACGAC(T)$_{17}$3'. It is 45 nucleotides in length and has 17 T residues at the 3'end that bind to poly(A) tails of mRNA species. The remainder of the primer-adapter is an engineered sequence that does not appear to anneal to endogenous sequences in COS-7 cells. The synthesized cDNA can thus be amplified by PCR by using one primer specific for the pTAG4 sequence and one primer specific for the engineered 5'tail sequence of the AP primer-adapter. This is the basis for the 3'RACE technique.

   DEPC-treated H$_2$O should be used in all steps and in preparing all solutions (except those containing Tris). Other appropriate precautions to maintain RNase-free conditions include wearing gloves and using sterile plasticware instead of glassware.

   DEPC (see Appendix for Caution)

2. Incubate at 70°C for 5 minutes and then allow to cool to 42°C.

3. Combine the components below (in the order listed) and preheat to 42°C. Centrifuge the tube from step 2 briefly to collect the liquid at the bottom of the tube, add the preheated mixture, and mix gently.

H$_2$O	10.5 μl
5x first-strand cDNA buffer (for preparation,	
see p. 204)	10 μl
100 mM DTT	5 μl
10 mM dNTP mixture (see Appendix)	2.5 μl
reverse transcriptase (200 units/μl)	2 μl

4. Incubate at 42°C for 30 minutes.

5. Incubate at 55°C for 5 minutes, and then add 1 μl of DNase-free RNase H (2 units/μl; e.g., Life Technologies) and mix gently. Incubate at 55°C for an additional 10 minutes.

   Note: RNase H digests the RNA strand of a DNA:RNA hybrid. The presence of RNA has been shown to inhibit DNA amplification by PCR. Digestion with RNase H is carried out at 55°C to eliminate the possibility of snapback structures and second-strand products resulting from residual reverse transcriptase activity. Reverse transcriptase is inactive at this temperature.

6. Centrifuge briefly to collect the liquid at the bottom of the tube, and then cool on ice.

   Note: This product is essentially a cDNA pool. It can be stored at –20°C indefinitely.

7. Prepare the primary PCR mixture by gently combining the following (in the order listed) in a PCR tube:

cDNA pool	5 µl
H_2O	72.5 µl
10x amplification buffer (see p. 205)	10 µl
25 mM $MgCl_2$	5 µl
10 mM dNTP mixture (see Appendix)	2.5 µl
50 µM oligonucleotide SV40P	1 µl
50 µM oligonucleotide UAP	1 µl

   Notes: The SV40P primer is 5′AGCTATTCCAGAAGTAGTGA3′. The UAP primer is 5′CUACUACUACUAGUCGACATCGATAATACGAC3′. SV40P anneals to the cDNA derived from the SV40 promoter of pTAG4, whereas UAP anneals to the 5′ tail region of the AP primer-adapter.

   For thermal cyclers that do not have a top heating block, overlay each reaction mixture with 100 µl of light mineral oil and then cover the tube.

8. Preheat the thermal cycler to 94°C. Incubate the PCR mixture at 94°C for 5 minutes.

9. Reduce the temperature of the thermal cycler to 80°C and add diluted *Taq* DNA polymerase (2.5 units diluted with H_2O to a final volume of 3 µl).

10. Perform thermal cycling as follows for 20 cycles:

    94°C for 30 seconds
    55°C for 30 seconds
    72°C for 2 minutes

    Perform an additional cycle of 72°C for 5 minutes, and then hold the sample at 4°C.

11. Digest the primary PCR product by gently combining the components below and incubating at 37°C for 1 hour. Include the same restriction enzyme used to digest pTAG4 and the target DNA.

primary PCR product	17 µl
10x restriction enzyme buffer (supplied with the enzyme)	2 µl
restriction enzyme (10 units/µl)	1 µl

    Notes: This step removes background products resulting from unspliced precursor RNA or from residual DNA that is reverse transcribed at AT-rich regions by using the AP primer-adapter.

    The remainder of the primary PCR product can be stored at –20°C indefinitely.

12. Add 150 µl of H_2O to the digested primary PCR product.

13. Prepare the secondary PCR mixture by gently combining the following (in the order listed) in a new PCR tube:

digested primary PCR product	1 µl
H$_2$O	76.5 µl
10x amplification buffer	10 µl
25 mM MgCl$_2$	5 µl
10 mM dNTP mixture (see Appendix)	2.5 µl
50 µM oligonucleotide Ad2	1 µl
50 µM oligonucleotide UAP	1 µl

Notes: The Ad2 primer is 5'CAUCAUCAUCAUCAGTACTCTTGGATCGGA3'. The UAP primer is 5'CUACUACUACUAGTCGACATCGATAATACGAC3'. The presence of the triplet repeats CAU on the Ad2 primer and CUA on the UAP primer facilitates the cloning of the resulting PCR product by the UDG-mediated cloning strategy.

For thermal cyclers that do not have a top heating block, overlay each reaction mixture with 100 µl of light mineral oil and then cover the tube.

The remainder of the digested primary PCR product can be stored at –20°C indefinitely.

14. Repeat steps 8–9.

15. Perform thermal cycling as follows for 30 cycles:

> 94°C for 30 seconds
> 55°C for 30 seconds
> 72°C for 2 minutes

Perform an additional cycle of 72°C for 5 minutes, and then hold the samples at 4°C.

16. Analyze 10 µl of the secondary PCR product and the appropriate molecular-weight markers on a 1.2% agarose gel and detect the DNA by staining with **ethidium bromide.**

Note: In general, the resulting trapped 3'-terminal exons range in size from 200 bp to 1000 bp.

ethidium bromide (see Appendix for Caution)

17. Clone the secondary PCR products and analyze the resulting isolates as described on p. 215.

Analysis and Identification of Trapped Exons

CLONING PCR PRODUCTS

The number of unique PCR products generated by the secondary PCR in internal exon trapping or 3'-terminal exon trapping is typically proportional to the complexity of the target DNA. For example, gel analysis typically reveals one to three unique products from a trapping experiment with a single cosmid or a bacteriophage P1 or λ clone. A larger number of products is usually encountered in a trapping experiment with a YAC clone or a pool of bacterial clones. Cloning of the resulting PCR products can involve either excision of individual bands from the gel or a "shotgun" approach using the entire set of PCR products.

The secondary PCR products can be cloned most conveniently by using either a UDG-mediated cloning strategy (Nisson et al. 1991) or a TA cloning strategy (Mead et al. 1991). Vectors for each type of strategy are commercially available and should be used as specified by the manufacturer (see also Fanning and Gibbs 1997). In the secondary PCR assays in both of the exon trapping protocols provided in this chapter, primers that allow subsequent UDG-mediated cloning of the PCR products are used. UDG-mediated cloning is recommended because of its high efficiency and low background of nonrecombinant clones. However, TA cloning is also a viable alternative that provides sufficient numbers of clones for subsequent analysis.

SEQUENCE ANALYSIS OF CANDIDATE TRAPPED EXONS

The typical initial step in the preliminary characterization of cloned candidate trapped exons is DNA sequencing. The results obtained differ between the two exon trapping strategies. Internal exon trapping yields small products (usually 25–200 bp). Each putative internal exon should show evidence of splice events on both sides of the exonic sequence (as reflected by the removal of vector intron sequences via splicing) and an ORF all the way through the sequence in at least one frame. Each unique sequence should be compared with existing DNA sequence databases by using an appropriate computer program (e.g., BLAST [see Baxevanis et al. 1997]). In contrast, 3'-terminal exon trapping yields candidate exons that are typically 200–2500 bp in size. Authentic trapped 3'-terminal exons should show evidence of a splice event at the 5' end of the sequence and a poly(A) tail at the 3' end. In approximately 90% of the cases, a consensus polyadenylation signal consisting of the hexanucleotide AATAAA or ATTAAA is present 12–30 bases upstream of the poly(A) tail. Computational analysis typically reveals multiple stop codons in most frames. The great majority of 3'-terminal exon sequence is within the 3' untranslated region, and thus very little protein-coding sequence is typically encountered.

PCR ASSAYS AND HYBRIDIZATION STUDIES

It is often useful to develop a PCR assay for each candidate exon. Such assays can be used to establish the genomic map position of the candidate exon, to study gene expression by using RT-PCR, or to generate a hybridization probe for screening cDNA libraries or for analysis of Southern and northern blots. The entire cloned candidate exon can also be used as a hybridization probe. Since trapped 3'-terminal exons are larger than trapped internal exons, 3'-terminal exons tend to be better hybridization probes for some hybridization experiments (e.g., isolation of full-length cDNA clones). The small size of many trapped internal exons makes the use of PCR assays more appropriate for many studies.

REFERENCES

Auch, D. and M. Reth. 1991. Exon trap cloning: Using PCR to rapidly detect and clone exons from genomic DNA fragments. *Nucleic Acids Res.* **18:** 6743–6744.

Baxevanis, A.D., M.S. Boguski, and B.F. Ouellette. 1997. Computational analysis of DNA and protein sequences. In *Genome analysis: A laboratory manual.* Vol. 1 *Analyzing DNA* (ed. B. Birren et al.), pp. 533–586. Cold Spring Harbor Laboratory Press, Cold Spring Harbor, New York.

Buckler, A.J., D.D. Chang, S.L. Graw, J.D. Brook, D.A. Haber, P.A. Sharp, and D.E. Housman. 1991. Exon amplification: A strategy to isolate mammalian genes based on RNA splicing. *Proc. Natl. Acad. Sci.* **88:** 4005–4009.

Burn, T.C., T.D. Connors, K.W. Klinger, and G.M. Landes. 1995. Increased exon-trapping efficiency through modifications to the pSPL3 splicing vector. *Gene* **161:** 183–187.

Church, D.M., C.J. Stotler, J.L. Rutter, J.R. Murrell, J.A. Trofatter, and A.J. Buckler. 1994. Isolation of genes from complex sources of mammalian genomic DNA using exon amplification. *Nat. Genet.* **6:** 98–105.

Duyk, G.M., S. Kim, R.M. Myers, and D.R. Cox. 1990. Exon trapping: A genetic screen to identify candidate transcribed sequences in cloned mammalian genomic DNA. *Proc. Natl. Acad. Sci.* **87:** 8995–8999.

Fanning, S. and R.A. Gibbs. 1997. PCR in genome analysis. In *Genome analysis: A laboratory manual.* Vol. 1 *Analyzing DNA* (ed. B. Birren et al.), pp. 249–299. Cold Spring Harbor Laboratory Press, Cold Spring Harbor, New York.

Frohman, M.A., M.K. Dush, and G.R. Martin. 1988. Rapid production of full-length cDNAs from rare transcripts: Amplification using a single gene-specific oligonucleotide primer. *Proc. Natl. Acad. Sci.* **85:** 8998–9002.

Hamaguchi, M., H. Sakamoto, H. Tsuruta, H. Sasaki, T. Muto, T. Sugimura, and M. Terada. 1992. Establishment of a highly sensitive and specific exon-trapping system. *Proc. Natl. Acad. Sci.* **89:** 9779–9783.

Huntington's Disease Collaborative Research Group. 1993. A novel gene containing a trinucleotide repeat that is expanded and unstable on Huntington's disease chromosomes. *Cell* **72:** 971–983.

Jakoby, W. and I. Pastan, eds. 1979. *Methods in enzymology: Cell culture*, vol. 58. Academic Press, New York.

Krizman, D.B. and S.M. Berget. 1993. Efficient selection of 3'-terminal exons from vertebrate DNA. *Nucleic Acids Res.* **21:** 5198–5202.

Krizman, D.B., T.A. Hofmann, U. DeSilva, E.D. Green, P.S. Meltzer, and J.M. Trent. 1995. Identification of 3' terminal exons from yeast artificial chromosomes. *PCR Methods Appl.* **4:** 322–326.

Mead, D.A., N.K. Pey, C. Herrnstadt, R.A. Marcil, and L.M. Smith. 1991. A universal method for the direct cloning of PCR amplified nucleic acid. *Bio/Technology* **9:** 657–663.

Nehls, M., D. Pfeifer, and T. Boehm. 1994. Exon amplification from complete libraries of genomic DNA using a novel phage vector with automatic plasmid excision facility: Application to mouse neurofibromatosis-1 locus. *Oncogene* **9:** 2169–2175.

Nisson, P.E., A. Rashtchian, and P.C. Watkins. 1991. Rapid and efficient cloning of *Alu*-PCR products using uracil DNA glycosylase. *PCR Methods Appl.* **1:** 120–123.

Sambrook, J., E.F. Fritsch, and T. Maniatis. 1989. *Molecular cloning: A laboratory manual*, 2nd edition, Chapter 1. Cold Spring Harbor Laboratory Press, Cold Spring Harbor, New York.

Wilson, R.K. and E.R. Mardis. 1997. Fluorescence-based DNA sequencing. In *Genome analysis: A laboratory manual.* Vol. 1 *Analyzing DNA* (ed. B. Birren et al.), pp. 301–395. Cold Spring Harbor Laboratory Press, Cold Spring Harbor, New York.

5

Gene Detection by the Identification of CpG Islands

ROSALIND M. JOHN AND SALLY H. CROSS

The isolation of CpG islands is a robust method for gene identification in vertebrate genomes, particularly when used in conjunction with a complementary approach, such as direct cDNA selection (see Chapter 3) or exon trapping (see Chapter 4). This chapter describes two strategies for the isolation of CpG-rich sequences from genomic DNA. In the first strategy, identification of CpG-rich DNA from cloned genomic sequences takes advantage of restriction enzymes that cut more frequently within CpG islands than elsewhere in the genome. The second strategy involves the purification of differentially methylated CpG island DNA from uncloned genomic DNA by binding to a column.

Comparison of CpG Islands and Vertebrate Genomic DNA

CpG islands were originally identified because of their unusual sequence content and methylation status which distinguished them from genomic DNA (Tykocinski and Max 1984; Bird et al. 1985). Most vertebrate genomic DNA is relatively AT-rich (~40% G + C) and heavily methylated. Within vertebrate genomic DNA, the dinucleotide CpG occurs at only 20% of the statistically expected frequency (CpG << GpC); 60–90% of these CpG dinucleotides in genomic DNA also contain a cytosine that is methylated (for review, see Bird 1987). In contrast, CpG islands, which are discrete 1–2-kb DNA segments found at the 5′ ends of many vertebrate genes, are relatively GC-rich (60–70% G + C) and contain clusters of multiple unmethylated CpG dinucleotides at the predicted frequency (where CpG = GpC) (Bird 1987). The majority of CpG dinucleotides outside CpG islands are symmetrically methylated at C5 on their cytosine rings.

The following are features that distinguish CpG island DNA from genomic DNA in vertebrates:

- CpG islands are GC-rich. The G + C content of CpG islands is 60–70%, whereas the overall G + C content of vertebrate genomic DNA is 40%.
- CpG islands show the expected frequency of one CpG dinucleotide per 10 bp. In contrast, outside CpG islands, approximately one CpG occurs per 100 bp (Bird 1987). This difference occurs because CpG in non-CpG island genomic DNA is underrepresented.
- CpG islands are unmethylated in all vertebrate tissues, including germ cells, with three exceptions: CpG islands on the inactive X chromosome (Toniolo et al. 1984; Wolf et al. 1984; Yen et al. 1984; Bird 1987); those associated with nonessential genes in tissue-culture cell lines (e.g., α-globin and retinol-binding protein in HeLa cells, Thy-1, and major histocompatibility complex [H-2K] genes in NIH 3T3 cells [Antequera et al. 1990]); and those associated with certain imprinted genes (e.g., the *Igf2r* gene in mice [Stoger et al. 1993] and H19 and SNRPN [Razin and Cedar 1994]).

CpG Islands as Markers for Genes

In the human genome, there are an estimated 45,000 CpG islands (Antequera and Bird 1993). All currently identified "housekeeping" genes and 40% of genes with a tissue-restricted pattern of expression in humans (in total, ~60% of all identified human genes) contain CpG islands (Gardiner-Garden and Frommer 1987; Larsen et al. 1992). CpG islands may contain both promoter and transcribed sequences, with the majority of CpG islands analyzed to date containing at least part of one exon from the gene with which they are associated (Larsen et al. 1992). Examples of typical genes associated with CpG islands are shown in Figure 1. In each case, the dense 1–2-kb cluster of CpG dinucleotides defines the position of the CpG island, and the first exon and part of the promoter are within the CpG island. The association of CpG islands with the 5′ end of genes provides the basis for using CpG-rich DNA for gene isolation. CpG island DNA can subsequently be used as probes to isolate full-length cDNAs and place genes on transcription maps.

Evolution of Approaches for Identifying CpG Islands

Many of the procedures for isolating genes by their association with CpG islands use a set of restriction enzymes that cut more frequently within CpG islands than elsewhere in the genome (e.g., see the strategy on pp. 224–244). These rare-cutting enzymes, known as RCREs, frequently recognize sequences that are composed of C and G residues and that include one or more CpG dinucleotide (Table 1). The majority of RCRE recognition sites in genomic DNA occur within CpG islands. In addition, CpG dinucleotides located outside islands are generally methylated and therefore resistant to cleavage by this class of enzymes. RCREs can thus be used both to define CpG islands and to isolate DNA from these regions (Bickmore and Bird 1992). Several strategies for finding genes within long stretches of DNA have been devised to take advantage of the properties of RCREs and their affinity for CpG islands.

One early approach for identifying CpG islands involved generating a plasmid library enriched for CpG island DNA (Gao et al. 1991; Triboli et al. 1992). The library is made by first digesting genomic DNA with a RCRE and then digesting with a second enzyme that cuts randomly (at ~3–7-kb intervals) within genomic DNA. The resulting digestion products are then cloned into a plasmid vector. The cloned sequences are then

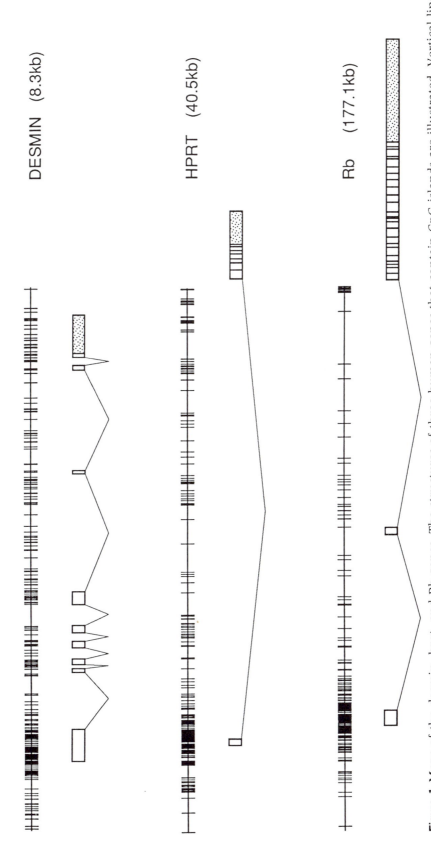

Figure 1 Maps of the desmin, hprt, and Rb genes. The structures of three human genes that contain CpG islands are illustrated. Vertical lines show the positions of CpG dinucleotides in the first 10 kb of the desmin gene (EMBL hsdes01), the hypoxanthine phosphoribosyl transferase gene (hprt; EMBL hshprt8a), and the retinoblastoma gene (Rb; EMBL L11910). (*Open boxes*) Exons; (*stippled boxes*) 3′ untranslated regions. Any exons not present in the first 10 kb of genomic DNA are shown fused together to the right. The genomic length (in kb) of each gene is given in parentheses. (Modified, with permission, from Cross and Bird 1995 and Cross et al. 1994.)

Table 1 Occurrence of Commonly Used RCREs in CpG Islands in Human DNA

RCRE	Recognition site[a]	Number of sites per island[b] expected	observed	Estimated % of sites that occur in islands[c]
*Not*I	GCGGCCGC	0.14	0.35	93
*Asc*I	GGCGCGCC	0.14	0.27	93
*Bss*HII	GCGCGC	1.68	2.11	76
*Eag*I	CGGCCG	1.68	1.70	76
*Sac*II	CCGCGG	1.68	1.89	76

These data summarize those of Bickmore and Bird (1992).
[a]The enzymes listed recognize sites that are entirely composed of Gs and Cs and that contain two CpG dinucleotides. Methylation of the site blocks digestion.
[b]This number assumes that the average size of a CpG island is 1.4 kb (based on a survey of 37 human CpG islands).
[c]This percentage is based on the observed frequency of recognition sites for the enzyme in CpG islands relative to the rest of the genome.

evaluated to ascertain whether or not they originated from CpG islands by determining their overall CpG content by sequencing and/or by testing for the presence of sites for restriction enzymes such as *Bst*UI, which are found frequently within CpG islands but rarely elsewhere (Cross et al. 1994). It is important to note that methylation of cytosine residues is not maintained when the cloned DNA is propagated in *E. coli*. It is therefore important to ascertain their original methylation status in the corresponding segment of genomic DNA. One way of doing this is to use them as probes on Southern blots of genomic DNA cleaved with methylation-sensitive restriction enzymes.

*Not*I is the RCRE used for this approach in many studies because of its relative specificity for CpG islands; fewer than 10% of *Not*I sites occur outside CpG islands (Table 1). However, because more than two thirds of CpG islands that have been identified do not contain a *Not*I site (Bickmore and Bird 1992), many CpG islands are not detected by this approach (use of or suggestions for alternative RCREs to reveal the remaining CpG islands in the genome are described in the protocol). A more significant disadvantage of this approach is that it is necessary to use uncloned genomic DNA, instead of cloned DNA, as the starting material for preparation of the library to preserve the native methylation status of the DNA. This could be a problem if the genomic DNA is scarce, but more importantly, CpG islands from an entire genome are also included in a library prepared in this manner and the majority of these will not contain the region of interest. Clones that do not lie within the region of interest must be eliminated to avoid time-consum-

ing analysis. This can be done in two ways—either by mapping the clone using a chromosome mapping panel or by hybridizing all of the subclones to a cloned contig of DNA from the region being studied. Furthermore, if a specific chromosomal region is being targeted, the use of human–rodent hybrid cell lines is required (Tribioli et al. 1992), entailing additional analysis to identify human-derived clones. The library will still contain clones outside the region, but these will be derived from rodent DNA, which can be eliminated by hybridization with rodent genomic DNA. There is also the problem that the methylation pattern in the hybrid cell might not accurately reflect that in the animal. In addition, both of the elimination strategies outlined above require the initial analysis of a large number of clones, most of which have no future purpose.

To locate CpG islands in defined regions of the genome, a different approach using cloned genomic DNA contigs has been developed (see pp. 224–244). Instead of relying on the presence of a single unmethylated RCRE site in genomic DNA to indicate the presence of a CpG island, the cloned DNA (which is no longer methylated because it has been cloned) is digested with multiple RCREs (Table 1), and only regions containing clusters of sites (which are smaller than 1 kb after digestion with multiple RCREs) are analyzed further (Lindsay and Bird 1987; Weber et al. 1991). Another approach, using the differential affinity of methylated and unmethylated genomic DNA for a binding matrix, has allowed the separation of CpG-island-rich regions of DNA from uncloned genomic DNA (pp. 245–283).

Recently, the traditional approaches for gene isolation by identification of CpG islands (using individual RCREs to identify and clone CpG islands) were superseded by other techniques for gene detection, including direct cDNA selection (Lovett et al. 1991; Parimoo et al. 1991; see Chapter 3) and exon trapping/amplification (Duyk et al. 1990; Buckler et al. 1991; see Chapter 4). Although these two approaches are powerful and have been applied successfully to clone human disease genes, they are technically demanding. To have some degree of completeness, they also require the analysis of larger numbers of clones than either of the gene identification strategies in this chapter (depending on the stringency of the screen, up to ten times more clones may have to be analyzed to identify the desired gene).

Recently Developed Techniques for Identifying CpG Islands

A method based on PCR (see Valdes et al. 1994) and one based on DGGE (see Shiraishi et al. 1995) are two recently developed strategies with similar applications. The PCR-based method is called island rescue PCR. In this method, YAC clones containing human genomic DNA are digested with individual RCREs and adapters are ligated to the cleaved ends (the adapters have the appropriate restriction site overhang for each of the enzymes used). The regions adjacent to the RCRE site are amplified by PCR using one PCR primer that matches the adapter sequence and a second PCR primer that matches the human *Alu* sequence, a repetitive DNA element that occurs frequently in the human genome. This method is useful when applied to genomic DNA cloned into YACs, since the use of the *Alu*-specific primer excludes contaminating yeast DNA. However, genes that lack an *Alu* repeat within approximately 2 kb of a CpG island are less likely to be isolated.

DGGE is an electrophoretic technique that allows the separation of DNA fragments on the basis of their nucleotide sequence content and their length. Shiraishi et al. (1995) use DGGE in a manner that results in the retention on the gel of long, GC-rich DNA sequences, thus allowing their purification. Cosmid DNA is digested with four restriction enzymes predicted to cut frequently outside CpG island regions but leave the CpG island DNA relatively intact. When this DNA is separated on a

denaturing gradient gel, approximately half of the bands with reduced mobility meet criteria compatible with CpG island DNA. An advantage of using this method is that it does not depend on the existence of RCRE sites in the islands (see pp. 218–220).

Advantages of Isolating CpG Islands as an Approach to Gene Identification

Methods for isolating genes on the basis of their association with CpG islands have several advantages over other approaches:

- In general, only one CpG island is found associated with one gene (although occasionally two bidirectionally transcribed genes are associated with one CpG island [Lavia et al. 1987]), which means that only one or two clones are required to identify each gene. In the case of exon trapping or direct cDNA selection, many different clones may be derived from different parts of a single gene, thus increasing the level of redundancy in the analysis.

  For example, the Huntington's disease gene is associated with one CpG island, covers 200 kb, contains 67 exons, and produces two transcripts of 10.5 kb and 13.5 kb. Theoretically, a CpG island library could contain two clones associated with the Huntington's disease CpG island, an exon-trapping library could contain 67 or more exon-trapped fragments, and a cDNA selection library could contain many more. All of these clones would require analysis.

- CpG islands generally lie at the 5′ end of the gene, a region that is typically difficult to isolate using other methods. (Because cDNA is commonly primed with oligo[dT] and the reverse transcriptase enzyme may not have sufficient processivity to reach the 5′ end, the 5′ end will be missing from a cDNA library. The first exon of a gene only contains a splice donor site and therefore will not be trapped in a vector relying on a splice acceptor site.) CpG island clones can furthermore be used as probes to obtain full-length cDNAs or the 5′ end of a cDNA clone.

- CpG island DNA is usually single-copy DNA and does not contain repetitive DNA sequences, rendering it a convenient probe for genetic

Table 2 Comparison of Strategies for Identifying CpG Islands

Advantages	Disadvantages
Identification of CpG-rich sequences from cloned genomic DNA using RCREs	
It can be applied early in a positional cloning project since all that is required is the existence of genomic clones.[a]	CpG islands lacking RCRE sites or a nearby site for the non-RCRE used in the cloning steps will be missed.
Only standard techniques (cloning and sequencing) are required and results can be achieved rapidly.	The success of this approach depends on the density of CpG islands in a specific region.[b]
Sequence information is obtained early in the procedure, allowing rapid identification of genes.	
Identification of CpG islands from uncloned genomic DNA using differential methylation	
Mainly CpG island fragments are obtained, since other GC-rich fragments, that are heavily methylated in genomic DNA are eliminated in the first stripping step.	The construction of the HMBD column requires specialized reagents (i.e., the cloned MBD gene and the nickel-agarose resin).
Intact CpG islands are isolated.	This technique cannot be readily applied to cloned DNA.[c]

[a]Exon trapping/amplification requires the purchase of the exon trapping vector and the appropriate mammalian cell line. Direct cDNA selection requires the production of high-quality cDNA from a variety of tissues and ligation to adapters as well as the purchase of streptavidin-biotin labeling and separating reagents.

[b]If the region is deficient in CpG islands, the background of non-CpG-island clones will be high since the relative number of RCRE sites outside islands is likely to be constant.

[c]The native methylation pattern is lost when genomic DNA is cloned into *E. coli* or yeast. Thus, the stripping step that removes methylated GC-rich DNA cannot be effectively carried out. Consequently, any non-CpG-island GC-rich DNA (e.g., from contaminating yeast or bacterial sources) will copurify with the CpG island DNA. As an alternative, CpG island libraries could be used in an analogous way to cDNA for direct selection of CpG islands from cloned DNA (see Chapter 3). However, the method has been successfully applied to genomic DNA cloned into both cosmid and PAC vectors (S.H. Cross, unpubl.).

Table 3 Applications of Strategies for the Identification of CpG Islands

Identification of CpG-rich sequences from cloned genomic DNA using RCREs	Identification of CpG islands from uncloned genomic DNA using differential methylation
Isolation of candidate genes in a positional cloning project	Isolation of large number of genes in the assembly of transcription maps
Isolation of single-copy DNA sequences for the development of STS markers	Isolation of single-copy DNA sequences for the development of STS markers
	Differential isolation of CpG islands

mapping and for isolating the full-length cDNA associated with a gene.

- All CpG islands are represented at an equal level (i.e., each island is present once per haploid genome, regardless of the pattern and level of expression of the associated transcript). This is in contrast to cDNA libraries, where the representation of each clone is dependent on the temporal and spatial level of expression of the associated gene.

Choosing a Strategy for Isolating Genes Associated with CpG Islands

The two strategies in this chapter assume the association of a large number of vertebrate genes with CpG islands. The choice of whether or not to use a CpG-island-based strategy for the isolation of genes depends on the advantages and disadvantages of this approach when applied to a specific situation. The advantages and disadvantages of the two strategies presented here are very different with respect to the isolation of specific genes and the assembly of transcription maps (Table 2).

Applications of the first strategy in this chapter include the isolation of candidate genes in a positional cloning project (this method was as effective as direct cDNA selection for identifying genes in the Huntington's disease region on human chromosome 4 [John et al. 1994]) and in the development of STSs (Table 3). STSs are short stretches of DNA that can be specifically detected by PCR (Green et al. 1991) and that are used in genome mapping projects to detect and localize genomic clones in the development of long-range physical maps. Sequences adjacent to RCRE sites are frequently single-copy sequences (since they often contain an exon of a gene) and thus make ideal STS markers.

A typical application of the second strategy in this chapter is the development of a transcript map of a whole genome or a specific chromosome (Table 3). Another application is the identification of differentially methylated regions by comparing the CpG islands obtained from different genomic sources (e.g., normal versus malignant tissue, tissue versus cell line). Since gene silencing by CpG methylation appears to be important in such processes as X-chromosome inactivation, imprinting, and, perhaps, cancer (Cross and Bird 1995), this strategy could potentially be used to isolate the CpG islands involved in these processes. To identify CpG islands in cloned genomic DNA, CpG island libraries can be used in direct-selection protocols in a fashion similar to that in which cDNA libraries are used (see Chapter 3). As with the first strategy, this strategy can be used to generate STS markers.

Combining Approaches for Gene Isolation

It is important to recognize that no single technique for the isolation of genes is capable of detecting all genes in a region. Any method that relies on the identification of CpG islands will miss genes without CpG islands (~40% of human genes). However, methods that rely on gene expression (e.g., direct cDNA selection) will miss genes that are not expressed in the tissue from which the target cDNA is generated, and methods that rely on functional identification of gene sequences (e.g., exon trapping/amplification) will miss genes that lack splice acceptor/donor sequences such as intronless genes or that have unusual splice acceptor/donor sequences. The most comprehensive approach is to apply more than one strategy to identify genes within a particular region. Because the expression of a gene is not required for its detection by a CpG-island-based approach, it works well in combination with a second approach (e.g., direct cDNA selection) that does require the presence of a transcript. The completion of the human genome sequencing project in the next 5–10 years should enable the use of more computer-based searching methods to identify CpG islands. Currently, however, the techniques outlined here will assist in the identification of genes in all vertebrate genomes. Combining all of the gene identification techniques outlined in this manual currently provides an estimated 70% success rate in predicting the positions of genes within genomic DNA.

USING RCREs TO IDENTIFY CpG-RICH SEQUENCES IN CLONED GENOMIC DNA

The strategy on pp. 229–234 (Figure 2) is a technically simple and robust method for isolating and evaluating the coding potential of CpG-rich segments derived from cloned genomic DNA (John et al. 1994). In these protocols, 100-kb segments of genomic DNA (or single clones) are digested with two enzymes, one that cuts within the CpG island (the RCRE) and one that cuts at sites represented randomly throughout the genome (the non-RCRE) (see Table 1). This liberates DNA adjacent to the RCRE sites, which can then be directionally cloned into a plasmid vector generating a library of subclones enriched for CpG island DNA. The plasmid vector pBluescript II (Short et al. 1988; Stratagene), which contains sites for *Bss*HII, *Eag*I, *Not*I, and *Sac*II and allows a blue/white selection for the presence of an insert, is used here, but any plasmid vector with suitable restriction enzyme recognition sequences within its cloning site may be substituted.

Generating the CpG island library requires the following steps:

- *Preparation of the pBluescript II vector (pp. 229–231).* High-quality preparations of the plasmid DNA are required to ensure the integrity of the prepared vector, the absence of contaminating *E. coli* DNA, and the ability to digest the DNA to completion. The DNA is prepared according to the CsCl method of Sambrook et al. (1989) or kits may be purchased from Qiagen (Plasmid Midi Kit 12143) or Promega (Wizard Midipreps A7640). The vector is first cut with the RCRE. Before it is digested with the non-RCRE, the linearized vector is isolated on an agarose gel; this decreases the likelihood of unproductive ligations resulting from incomplete digestion.

- *Estimation of the density of sites recognized by the RCRE in individual clones (p. 232).* The density of RCRE sites in the region of interest must be estimated since the density dictates the number of subclones to be analyzed and since unique subclones often cannot be identified on the basis of the size of the insert alone, particularly if the density of RCRE sites in a region is high or if the genomic clones have insert sizes much greater than 100 kb (see pp. 241–242).

- *Digestion of the cloned genomic DNA with the RCRE and the non-RCRE (p. 233).*

- *Ligation of the digested genomic DNA to the vector DNA and transformation of* E. coli *with the ligated DNAs (p. 234).*

Approximately 20 individual clones from this library are analyzed, first by size and then by fingerprinting (see pp. 241–242). Fingerprinting is a diagnostic analysis of a segment of DNA using restriction enzymes that cut frequently (4-base recognition site) to give a unique digestion pattern. Unique clones are sequenced from the RCRE-derived end of the subcloned DNA (Figure 2). The sequences are then analyzed by using the BLAST (Altschul et al. 1990) and GRAIL (Uberbacher and Mural 1991) computer programs to assess the coding potential of the sequences and identify the 60–70% that correspond to genes.

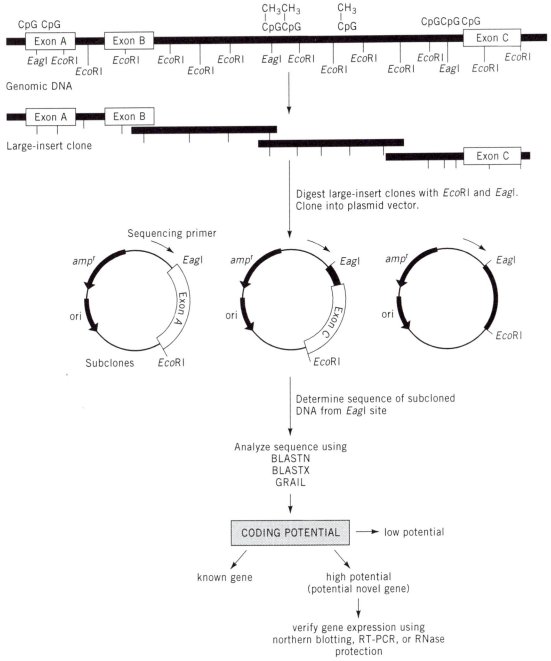

Figure 2 Strategy for isolating and evaluating the coding potential of CpG-rich segments from cloned genomic DNA. This strategy uses two enzymes to digest 100-kb segments of genomic DNA (or single clones), one enzyme that cuts within the CpG island (the RCRE) and one that cuts at sites represented randomly throughout the genome (the non-RCRE). See text for details.

Starting Material for Isolating CpG-rich DNA

The positional cloning approach to gene identification (see Chapter 1) requires the generation of a collection of large-insert clones so that the genomic region of interest can be physically mapped by arranging overlapping genomic clones into a series of contigs. These contigs can cover several megabases of DNA and may consist of YAC (0.2–2 Mb), BAC (100–250 kb), bacteriophage P1 (70–100 kb), and/or cosmid (35–45 kb) clones. In such clones, the native methylation state of the genomic DNA is erased. (The G + C composition and CpG-dinucleotide content of the cloned segment and/or the methylation status of the RCRE site selected within the genomic DNA of origin can be used as criteria to distinguish between clones originating from CpG islands and those originating from non-island sites before any associated genes are identified.)

The total size of the genomic DNA region represented by the cloned DNA is an important factor in determining the complexity of the subsequent steps for identifying CpG-rich sequences. Analysis can be simplified by dividing the region of interest into smaller segments. Not only are smaller regions more manageable, but CpG-rich subclones obtained from each segment can also be compared within a small group. This allows the elimination of clones of a similar size (which are likely to be identical), which, in turn, substantially reduces the amount of work involved. Segments of 100 kb have worked well (John et al. 1994), but factors such as the density of potential CpG islands should also be considered. Since segments of 100 kb are preferred, the type of vector used in cloning the genomic DNA is important in determining the insert size (e.g., inserts may be larger than 100 kb for most YAC clones and some BAC and bacteriophage P1 clones).

In the strategy presented here, CpG-rich DNA is isolated from pools of overlapping cosmid clones with a combined coverage of approximately 100 kb. As alternative starting DNA, large DNA clones with average insert sizes of at least 100 kb (e.g., YAC, BAC, or bacteriophage P1 vectors) can be used and each clone analyzed individually. The major difference is that insert size of the CpG subclones cannot be used as the sole criterion for identifying unique subclones if the genomic region covered by the large-insert clones is significantly larger than 100 kb, since there is an increased likelihood that two different subclones will be indistinguishably close in size. Likewise, fingerprint comparisons (see pp. 241–242) become more complicated when larger numbers of subclones need to be analyzed to ensure adequate coverage of a region.

USE OF INDIVIDUAL GENOMIC CLONES WITH INSERTS OF AT LEAST 100 KB

Since most YAC (0.2–2 Mb) and some BAC (100–250 kb) and bacteriophage P1 (70–100 kb) genomic clones cover 100 kb of DNA or more, single clones can be used for the identification of CpG-rich sequences. High-quality preparations of the purified cloned DNA samples are required to ensure complete digestion of the DNA and the absence of contaminating host chromosomal DNA (i.e., *E. coli* DNA in the case of BAC and bacteriophage P1 clones or yeast DNA in the case of YACs). Cloning systems with bacterial hosts are preferable because large quantities of cloned DNA, uncontaminated with host chromosomal DNA, can be obtained quickly and easily. The isolation of pure cloned DNA is complicated by the requirement to separate the YAC DNA away from the host yeast DNA. After DNA is prepared from the yeast, it must then be run on a PFG so that the YAC is separated from the yeast chromosomes by size (see Riethman et al. 1997). It then must be purified away from the agarose before it can be used in the cloning step. These additional steps may result in a low yield of pure DNA, and if the YAC is of a similar size to a yeast chromosome (*S. cerevisiae* has 16 chromosomes that range in size from 225 to 1900 kb), the final preparation will be substantially contaminated with yeast DNA. If a physical contig is already established in overlapping YAC clones, the smallest clones should be used (<250 kb). For larger clones, it is advisable to subclone the YAC insert into cosmids or to use the YAC DNA to screen gridded cosmid libraries to identify cosmid clones covering the same region (Zuo et al. 1993).

USE OF POOLED GENOMIC CLONES

The average insert of a cosmid clone is 35–45 kb. If the genomic region is represented by contigs of overlapping cosmid clones, more than one clone will be required to cover 100 kb of DNA. A minimum of three cosmids could cover 100 kb, but the overlap between cosmids will generally vary so that perhaps as many as six cosmid clones may be required. It is important that each clone be represented equally so that subclones from each cosmid are represented equally in the ligation. For example, for three cosmids, 33% of the DNA should come from each cosmid (~0.33 µg), but for six cosmids, 15% of the DNA should come from each cosmid (~0.15 µg).

The separate analysis of each 100-kb segment of a larger contig allows the identification of the majority of unique CpG-rich subclones by comparing the sizes of their inserts (which should be 0.4–7 kb [John et al. 1994]). Since there is a low probability that two different CpG islands will produce fragments of the same size within one segment of 100 kb, it is reasonable to distinguish identical clones on the basis of size; however, if the segment were 1000 kb, there would be ten times the number of islands, which would increase the probability that two different fragments might coincidentally be of a similar size and make it risky to distinguish them purely on the basis of size.

Choosing the Restriction Enzymes

Table 1 lists some of the enzymes that can be used to isolate CpG-rich DNA. *Not*I most accurately defines a CpG island because 93% of all *Not*I sites occur within CpG island regions. However, *Not*I, which has an 8-base recognition site, cuts infrequently in the human genome and only 35% of CpG islands contain a *Not*I site. The more useful enzymes for this procedure are *Bss*HII, *Eag*I, and *Sac*II, each of which has a 6-base recognition sequence; although 24% of these sites lie outside islands, it has been observed that *all islands* contain at least one of each of these sites (Bickmore and Bird 1992). It is also convenient to use *Eag*I and *Sac*II since many commonly used plasmid vectors contain sites for these enzymes in their multiple cloning sites. Although *Bss*HII would seem to be a good alternative, most cloning vectors lack *Bss*HII sites and *Bss*HII requires an incubation temperature of 50ºC instead of 37ºC, which means that two separate digestions must be performed to prepare the genomic DNA for subcloning.

A second enzyme without a bias for cutting within CpG islands is used to obtain DNA fragments of the appropriate size for cloning from either side of the CpG island. The use of an RCRE alone is not recommended because few CpG islands contain more than one recognition site for a given RCRE (see Table 1), and the DNA fragments generated by digestion of CpG islands with multiple recognition sites for a particular RCRE are likely to be small (<1 kb) and therefore less useful in later analytical steps (e.g., identification of cDNA or mapping to a specific locus). The only requirements for the second restriction enzyme are that it should cut randomly throughout the genome, it should not contain a CpG dinucleotide in its recognition site, and it should generate DNA fragments in the 3–7 kb range (see Table 4).

It is advantageous to use two sets of double digestions (e.g., *Eag*I/*Hin*dIII plus *Sac*II/*Eco*RI; see Table 4) on the same starting DNA to increase the percentage of CpG islands that are subcloned with this method since there will be some DNA regions that lack the recognition sequence for a given individual RCRE. Subclones that arise from the same CpG island can later be identified when the subclones are mapped to the contig.

Table 4 Recommended Restriction Enzymes for Subcloning CpG-rich Sequences

RCRE[a]	Non-RCRE[b]
*Eag*I	*Bam*HI
*Sac*II	*Eco*RI
	*Hin*dIII
	*Pst*I

[a]*Not*I can also be used. However, although almost all sites for this enzyme occur within islands, only a minority of CpG islands contain a *Not*I site.
[b]Any enzyme that recognizes a 6-base sequence that does not contain the CpG dinucleotide is suitable as the second enzyme. The enzymes listed here are suggested because they cleave DNA efficiently and most commonly used vectors contain cloning sites that are compatible with these enzymes.

PROTOCOL

Preparing the pBluescript II Vector DNA

1. Digest 5 µg of pBluescript II DNA to completion in a reaction volume of 50 µl. For each microgram of DNA, use 4 units of *Eag*I in the buffer specified by the enzyme's manufacturer and incubate at 37°C for 1 hour.

   Notes: This protocol specifies *Eag*I from New England Biolabs. However, *Eag*I from any manufacturer (Boehringer Mannheim, Stratagene) can be used if the units of enzyme are adjusted to match the recommended activity.

   The cloning vector chosen for this protocol is the plasmid pBluescript II (Short et al. 1988; Stratagene), which contains sites for *Bss*HII, *Eag*I, *Not*I, and *Sac*II and allows a blue/white selection in the presence of antibiotic for insert-containing subclones. However, any general plasmid vector with suitable restriction enzyme recognition sequences within its cloning site may be substituted.

   High-quality preparations of the plasmid DNA are required to ensure the integrity of the prepared vector, the absence of contaminating *E. coli* DNA, and the ability to digest the DNA to completion. It is convenient to prepare a large batch of the digested vector for long-term use (e.g., CsCl method of Sambrook et al. 1989; Qiagen Plasmid Midi Kit 12143; Promega Wizard Midipreps A7640).

   Alternatively, set up two separate digestions of 5 µg of plasmid DNA with *Eag*I and *Sac*II. Perform the steps below in parallel for each digested sample to increase by 1.5-fold the chance of finding an island.

2. Add 5 µl of BP/XC/glycerol gel-loading solution to the digestion mixture. Load on a 0.8% agarose minigel along with a 1-kb ladder (Life Technologies 15615-016). Run the gel until the linearized vector DNA is well separated from any undigested DNA.

   Notes: For pBluescript II or a similar sized plasmid (2.9 kb), electrophoresis on a 0.8% gel (10 cm long) in 1x TAE buffer at 5 V/cm for 4 hours is sufficient to give good separation. Voltage in excess of this range is not recommended, since the DNA will smear and be difficult to isolate from the undigested DNA.

   The linearized plasmid should be clearly visible as a single band migrating at 2.9–3 kb.

   If a method such as treatment with agarase will be used to extract the vector DNA from the agarose, the gel should be prepared with low-melting-temperature agarose.

3. Stain the DNA with **ethidium bromide** and view under **UV** illumination.

   ethidium bromide, UV radiation (see Appendix for Caution)

4. Use a sterile razor blade to excise the band of linearized DNA from the gel. Extract the linearized vector DNA from the agarose.

   Note: The GENECLEAN kit (BIO 101) for DNA purification is recommended here, since it is rapid and gives a good yield of pure DNA. Alternatively, other kits (e.g., Prep-A-Gene [Bio-Rad], QIAquick [Qiagen], or BandPrep [Pharmacia]) can also be used.

5. Dissolve the purified, linearized vector DNA (4–5 µg) in TE (pH 7.6) and adjust the DNA concentration to 0.1 µg/µl. Set aside 0.2 µg of the linearized DNA at 4°C for the control transformation in step 9.

   Note: If there is a question regarding the yield, the DNA concentration in a 1-µl aliquot of the linearized plasmid DNA can be determined by measuring the OD_{260} (see Appendix).

6. Repeat step 1 using the remaining DNA, 20 units of a non-RCRE (e.g., *Bam*HI, *Eco*RI, *Hin*dIII, or *Pst*I; see Table 4), and the restriction enzyme buffer specified by the manufacturer.

   Note: If two separate digestions with an RCRE (i.e., *Eag*I and *Sac*II) were performed in step 1, digest each linearized DNA sample with a different non-RCRE in this step.

7. Separate the digested vector DNA (2.9 kb) from the fragment containing the multiple cloning site (which should be 50–100 bp) on a 0.8% gel and repurify the digested vector DNA from the agarose as described in steps 2–4.

8. Dissolve the DNA in TE (pH 7.6) at a concentration of 0.1 µg/µl. Set aside 0.2 µg of DNA from the double digestion at 4°C for the control transformation in step 9. Store the remainder at –20°C until needed for the ligation and transformation on p. 234.

   Note: The DNA can be stored at –20°C indefinitely.

9. Test the vector DNA preparation by performing four control transformations as follows:

   a. Place 0.1 µg of vector DNA from the single and double digestions (from steps 5 and 8) in separate 1.5-ml tubes at room temperature. Keep the remainder of the DNA from each digestion at 4°C.

   b. Add the following to each tube:

10X ligase buffer (supplied with the enzyme)	1 µl
bacteriophage T4 DNA ligase (Boehringer Mannheim)	0.5 Weiss unit
H₂O	to a final volume of 10 µl

   Note: Using bacteriophage T4 DNA ligase from different suppliers can dramatically affect the results. Good results are usually obtained with 0.5 Weiss unit of ligase from Boehringer Mannheim, but equivalent units of ligase from New England Biolabs (400 units/µl) can also be used.

   c. Incubate the ligation mixtures at room temperature for 1 hour.

   d. Use standard procedures (Sambrook et al. 1989, pp. 1.76–1.84) to transform competent *E. coli* (XL1-Blue, DH5α; Stratagene) with 1 µl of each of the two ligation mixtures and with 0.1 µg of unligated vector DNA from the single and double digestions (a total of four transformations). Plate transformation mixtures on LB plates containing ampicillin (50 µg/ml), X-gal (20 µg/ml), and IPTG (20 µg/ml).

Note: Transformation of cells with the unligated DNA from the single digestion should yield very few (or no) colonies, whereas transformation of cells with the ligated DNA from the single digestion should yield several hundred colonies (an average of 300–500), indicating that the vector DNA at this stage was digested almost to completion and that the sticky ends were intact. Transformation of cells with the unligated and the ligated vector DNA from the double digestions should yield approximately the same small number of colonies (0–20), indicating that the second digestion was complete. All of the colonies should be blue, indicating that the preparation is not contaminated with non-vector DNA.

Estimating the Density of RCRE Sites in Individual Clones

1. For each of the genomic clones in the minimal overlapping set (covering each 100-kb segment), set up two separate 20-µl digestions of 0.5 µg of DNA, one with 2 units of *Eco*RI and the other with 2 units of *Eco*RI plus 2 units of the RCRE. Incubate as specified by the enzyme's manufacturer.

   Notes: If the RCRE is incompatible with the buffer recommended for *Eco*RI, digestions must be performed sequentially, according to the manufacturer's instructions or as outlined in Wolff and Gemmill (1997).

   Pools of overlapping cosmid clones with a combined coverage of approximately 100 kb are used as the starting DNA in this protocol. Alternatively, large DNA cloning systems with average insert sizes of at least 100 kb (e.g., YAC, BAC, or bacteriophage P1 vectors) can be used and each clone analyzed individually.

   An excess of *Eco*RI (four times the amount specified by the manufacturer) is used here.

2. Add 0.1 volume of BP/XC/glycerol gel-loading solution to each digestion mixture. Analyze the digestion products along with a 1-kb ladder (Life Technologies 15615-016) on a 1% agarose gel. Run the gel until the linearized vector DNA is well separated from any digested DNA.

   Note: *Eco*RI cuts on average every 5 kb in the genome. If there is an RCRE site within this fragment, products can be anywhere from 0 to 5 kb.

3. Stain the DNA with **ethidium bromide** and view under **UV** illumination. To estimate the number of RCRE sites in each cosmid, subtract the number of bands resulting from the single digestion from the number of bands resulting from the double digestion.

   Note: It is useful to transfer the DNA onto nylon membranes as described for Southern blotting and fix the DNA to the membranes (see Wolff and Gemmill 1997) since these blots can be used later for mapping the subclones (see pp. 243–244).

   ethidium bromide, UV radiation (see Appendix for Caution)

PROTOCOL

Digesting the Cloned Genomic DNA

1. Place 1 μg of cloned genomic DNA from a pool of cosmid clones in a 1.5-ml microcentrifuge tube and digest the DNA to completion in a reaction volume of 50 μl. Use 4 units each of the RCRE (*Eag*I or *Sac*II) and the non-RCRE (*Bam*HI, *Eco*RI, *Hin*dIII, or *Pst*I) in the buffer(s) specified by the enzymes' manufacturers and incubate at 37°C for 1–4 hours. Use the same RCRE that was used to prepare the vector DNA.

   Notes: Pools of overlapping cosmid clones are used as the starting DNA in this protocol. Each pool should cover approximately 100 kb of genomic DNA. The amount of DNA from each clone within a pool should be approximately equal. Alternatively, large DNA cloning systems with average insert sizes of at least 100 kb (e.g., YAC, BAC, or bacteriophage P1 vectors) can be used and each clone analyzed individually. For further discussion, see pp. 225–227.

   Some of these enzyme combinations can cleave effectively in the same buffer. Others must be digested sequentially, according to the manufacturers' instructions or as outlined in Wolff and Gemmill (1997).

   To maximize the chance of recognizing a CpG island, set up one digestion with *Sac*II and one with *Eag*I and use two different non-RCREs. Perform the steps below in parallel for each digested sample.

2. Add 0.5 μl of BP/XC/glycerol gel-loading solution to 5 μl of the digestion mixture. Analyze on a 0.8% agarose gel along with a 1-kb ladder (Life Technologies 15615-016) to determine whether or not digestion is complete. Keep the remainder of the DNA on ice while the gel is running. Confirm that the DNA is completely digested and then proceed with step 3.

   Note: A large number of DNA bands should be visible. These bands should range from approximately 0.2 to 12 kb. If the DNA is not completely digested, most of the bands will still be in the high-molecular-weight range (>12 kb) and the digestion mixture should be incubated for a longer period of time. If the DNA is still just partially digested, the original DNA preparation may not be pure enough and fresh DNA will have to be prepared.

3. Adjust the volume of the digestion mixture to 100 μl with TE (pH 8.0) and extract the DNA sequentially with 1 volume of **phenol**, 1 volume of phenol:**chloroform** (1:1), and then 1 volume of chloroform (see Appendix). Use phenol saturated with TE (pH 8.0).

   phenol, chloroform (see Appendix for Caution)

4. Precipitate the DNA in the aqueous phase with 0.1 volume of 3 M sodium acetate (pH 5.2) and 2 volumes of absolute ethanol and dry the pellet (see Appendix).

5. Dissolve the DNA in 10 μl of TE (pH 7.6) to make a concentration of approximately 0.1 μg/μl. Store the DNA at –20°C until needed for the ligation and transformation on p. 234.

   Note: The DNA can be stored at –20°C indefinitely.

PROTOCOL

Ligating the DNAs and Transforming *E. coli*

1. Pipette 1 μl (0.1 μg) of vector DNA from the double digestion (from p. 230, step 8) into a 1.5-ml microcentrifuge tube on ice. Add 5 μl (~0.5 μg) of digested genomic DNA from p. 233, step 5, to give a 1:5 ratio of vector DNA to insert DNA.

   Note: These volumes should be sufficient to achieve a 1:5 ratio since the average size of the insert will be approximately 3 kb (which is the same as the vector).

2. For each double digestion performed, ligate the DNAs and transform competent *E. coli* as described p. 230, steps 9b–d.

   Note: If the transformation is successful, 0–50 blue colonies that are vector only and several hundred white colonies that are vector plus should be seen.

3. Prepare DNA from the white colonies and analyze the CpG island library as described on pp. 235–244.

DATA ANALYSIS FOR IDENTIFICATION OF GENES ASSOCIATED WITH CpG ISLANDS FROM CLONED GENOMIC DNA

Analyzing the CpG island library prepared from cloned genomic DNA requires the following steps:

- Eliminating cloned vector DNA
- Determining the sizes of inserts and eliminating duplicate subclones
- Analyzing the sequence of candidate subclones containing CpG islands
- Determining where CpG-rich subclones map within a contig
- Demonstrating that mapped subclones with coding potential are transcribed

Two of these steps are presented as protocols; all steps are discussed below.

Eliminating Vector-derived Subclones

Since a number of vectors used to make genomic DNA libraries contain sites for RCREs (e.g., the cosmid vectors sCos and LoristX, the YAC vector pYAC4, and the BAC vector pBeloBAC), vector DNA fragments are subcloned along with the genomic DNA. Once a CpG island library has been generated, it is necessary to remove subclones that appeared because the vector DNA fragments contained target sequences for the RCRE and the non-RCRE being used. Vector-derived subclones can be eliminated at an early stage by hybridization of a radiolabeled probe of the vector DNA to colonies bound onto a membrane/filter and whose DNA is chemically cross-linked to the membrane (Sambrook et al. 1989, pp. 1.90–1.104, or as described by the manufacturer of the nylon membrane filter). Alternatively, Southern blots of digested clones can be probed with labeled vector (Wolff and Gemmill 1997). The subclones that give a positive signal using this analysis should be eliminated from further analysis.

Eliminating Duplicate Subclones

Elimination of duplicate subclones can be achieved by (1) estimating the size of the insert DNA by digesting the plasmid DNA with the same two enzymes used to produce the product from the genomic clone and (2) fingerprinting the purified plasmid DNA by digestion with a restriction enzyme that has a 4-base recognition site (HaeIII or RsaI) to see if a distinctive digestion pattern is revealed. A protocol for this type of analysis is provided on pp. 241–242.

Analyzing the Coding Potential of Sequences Obtained from the RCRE-derived End of Unique CpG-rich Subclones

Subclones that are unique and that do not contain vector DNA are then sequenced from the end adjacent to the RCRE site. At least 250 bp of DNA sequence should be obtained for each of the unique subclones. Two complementary programs, BLAST and GRAIL (for further discussion, see Baxevanis et al. 1997), are then used in tandem to analyze CpG-selected sequences for their gene coding potential. The BLAST programs identify sequences with significant homology to *known* gene sequences in nucleotide and protein databases by performing local similarity searches (Altschul et al. 1990). The GRAIL program assesses the coding potential of a sequence and shows the positions of candidate exons in a linear sequence independently of any homology with a known gene sequence (Uberbacher and Mural 1991). The results of analysis using BLAST and GRAIL place each subclone in one of three categories: (1) subclones with sequences that match those of known genes or show significant homology with known genes, (2) subclones with coding potential but without homology with previously identified genes (these subclones require additional evaluation using a method for the detection of transcripts, such as northern blotting, RT-PCR, or an RNase protection assay), (3) subclones with sequences that do not demonstrate coding potential (these subclones can be assigned a low priority for further analysis).

BLAST

The two BLAST programs used in the analysis of CpG-selected sequences are BLASTN and BLASTX (Altschul et al. 1990). BLASTN compares a nucleotide query sequence with a nucleotide sequence database and BLASTX compares the six-frame conceptual translation products of a nucleotide query sequence (three frames for the forward strand and three for the reverse strand) with a protein sequence database. BLASTN searches are much quicker than BLASTX searches but generally only detect exact matches or matches with very similar sequences such as homologs or paralogs of the query sequence. Searching with BLASTN is still advised since not all genes in public databases are accompanied by the protein sequence. Furthermore, gene-associated regions that do not code for protein (e.g., untranslated exons or promoter regions) can be detected with the BLASTN program. Protein level comparisons using BLASTX are much more sensitive than DNA level comparisons using BLASTN because of the combination of degeneracy in the genetic code and functional constraints on the encoded protein. In addition, TBLASTX can be used to translate conceptually all of the sequences in a database into all six reading frames and compare these to the translated query sequence. Because of the computational resource requirements of TBLASTX, it is currently restricted to searching only the dbest, dbsts, and alu databases and, therefore, has limited use in the analysis of CpG-selected sequences.

GRAIL

The second type of computer program that can be useful in detecting coding sequences within genomic DNA is GRAIL (Uberbacher and Mural 1991). This program is capable of detecting potential coding sequences within genomic DNA independent of homology with known genes. The first version, GRAIL 1, recognizes coding potential within a fixed size window and evaluates coding potential without looking for additional features. This program scores the DNA as having excellent, good, marginal, or no coding potential. The latest version, GRAIL 2, looks at both the potential exon sequence and at surrounding noncoding sequences. This program scores the coding potential of the DNA and predicts the best "edges" (splice junctions, stop/start signals) for the predicted exon.

USING BLAST AND GRAIL

With BLAST, the sequences can be analyzed by sending an email message to the BLAST server at:

    blast@ncbi.nlm.nih.gov

or through the World Wide Web at:

    http://www.ncbi.nlm.nih.gov/blast/

by using Netscape or similar communications software. With GRAIL, the sequences can be analyzed by either sending an email message to the GRAIL server at:

    grailmail@ornl.gov

or through the World Wide Web at:

    http://combio.ornl.gov/grail-bin/emptygrailform/

The easiest method for analyzing sequences with the BLAST and GRAIL programs is through the World Wide Web. With both programs, the preset parameters are adequate for the analysis of CpG-selected sequences.

INTERPRETING SEQUENCE QUERIES FROM BLAST AND GRAIL

Results from a BLAST search are displayed as a histogram of the statistical significance of the matches found, a one-line description of the database sequences that satisfied the statistical significance threshold, and the aligned sequences. The one-line description gives the database accession number and partial name of the matched sequences followed by the score of the match (High Score), the significance of that score (P[N]) and the number of matches between the two sequences (N). This information allows the user to determine the extent of the match with a particular sequence and to compare regions of alignment. Although the program has preventative filters, any sequence that is rich in a particular residue (e.g., an AT-rich or a proline-rich sequence) will have a sig-

nificant match with all of the sequences in the database that are similarly rich in
these residues; this match may or may not turn out to be significant. In addition,
a search of a sublibrary of repetitive elements is performed before a full database
search. The program has been designed so that it does not search the rest of the
database with a repeat sequence, but it will still search the database with
remaining sequence surrounding a repeat and reports both the match with the
repeat sequence and any additional matched sequences.

The results from a GRAIL search are displayed as a table that lists the predicted exon positions in the forward and reverse strand with their GRAIL scores. In
addition, the results include the estimated limits of the coding region, the most
likely strand for the exon (with a probability for the correctness of the strand assignment), the preferred reading frame for the exon, and an assessment of the
quality of the exon (excellent, good, or marginal).

The coding potential of a CpG-rich sequence can be determined by comparing
the results from both types of searches. The BLASTN and BLASTX results (only
the first five one-line descriptions are shown) and GRAIL results of a typical
search of a human sequence cloned with this approach (John et al. 1994) are
shown in Figure 3.

A sequence is scored as having coding potential when there is a significant
score with one or more of the three programs, i.e., a P(N) of less than 0.01 with
BLASTN or BLASTX or an exon quality score of excellent, good, or marginal
with GRAIL.

In this example, the query sequence has significant scores with BLASTX and
with GRAIL:

Program	Result
BLASTN	P(N) 0.015
BLASTX	P(N) 5.3e-06
GRAIL	good

This result would be a strong indication that this sequence contains part of a
gene. Sequences given a significant score with any of the programs are regarded
as having coding potential and are analyzed further. A subclone without coding
potential may still be derived from a CpG island since an exon may not be
detected by either of the programs if insufficient sequence is obtained. However,
such subclones are given a lower priority for further analysis other than mapping.

Mapping CpG Islands to a Contig

To both define a map position and search for additional coding sequence of a
candidate gene, subclones with coding potential but without homology to previously identified genes and subclones with sequences that do not demonstrate
coding potential in the database searches can be used to map the 5'end of the
gene within a contig. In many cases, a well-defined linear restriction map is generated during the process of cloning a genomic region (during preparation of the
original physical map of the region, e.g., see Zuo et al. 1993). By generating

(a) BLASTN search results

Sequences producing high-scoring segment pairs:			High score	Smallest sum probability P(N)	N
gb I T15339 I T15339	ScO5d10-t7 *Zea mays*	cDNA clone 5c05...	147	0.015	1
gb I L13877 I DROG6PDAB	*Drosophila simulans*	DNA sequence, 3...	147	0.018	1
gb I L13891 I DROG6PDAP	*Drosophila simulans*	DNA sequence, 3...	147	0.018	1
gb I L13894 I DROG6PDAS	*Drosophila simulans*	DNA sequence, 3...	147	0.018	1
gb I L13876 I DROG6PDAA	*Drosophila simulans*	DNA sequence, 3...	138	0.099	1

(b) BLASTX search results

Sequences producing high-scoring segment pairs:		Reading frame	High score	Smallest sum probability P(N)	N
pir I A45082 I A45082	probable protein-tyrosine kinas...	+2	58	5.3e-06	3
gp I U04295 I OSU0495_1	DNA-binding factor of bZIP clas...	+2	44	0.00030	3
gp I L43619 I HUMPKD1G08_1	polycystic kidney disease 1 pro...	+3	42	0.0014	4
gp I L33243 I HUMPKD1A_1	polycystic kidney disease 1 pro...	+3	42	0.0014	4
gp I U24497 I HSU24497_1	autosomal dominant polycystic k...	+3	42	0.0014	4

(c) GRAIL search results

Exon predication on forward strand:

start /acceptor	donor/stop	rf	score	orf
28	– 134	1	41	1 – 171

Exon predication on reverse stand:

start / donor	acceptor / stop	rf	score	orf

Final exon predication:

start / donor	acceptor / stop	strand	rf	quality	orf
28	– 134	f	1	good	1 – 171

(rf = reading frame, orf = open reading frame, and f = forward. In this example, there is no predicted exon in the reverse reading frame.)

Figure 3 Results for BLASTN (*a*), BLASTX (*b*), and GRAIL (*c*) searches with the same sequences. The coding potential of a CpG-rich sequence can be determined by comparing the results of both types of search. The BLASTN and BLASTX results (only the first five-line descriptions are shown) and GRAIL results of a typical search of a human sequence cloned with this approach (John et al. 1994) are shown.

Southern blots of a minimal overlapping set of the genomic DNA clones (covering each 100-kb segment), the map position of the CpG-rich subclones can be readily determined. A protocol for mapping is provided on pp. 243–244.

It is important to determine the position of each subclone with respect to the others since this information can help identify CpG island regions. For example, two subclones containing inserts generated with *Eag*I/*Eco*RI that map close to

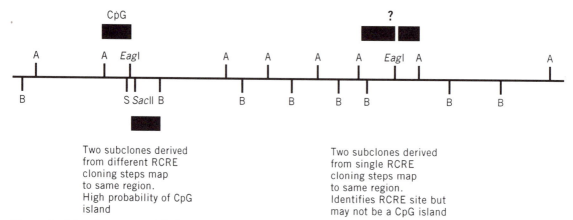

Two subclones derived from different RCRE cloning steps map to same region. High probability of CpG island

Two subclones derived from single RCRE cloning steps map to same region. Identifies RCRE site but may not be a CpG island

Figure 4 Mapping CpG-rich fragments to the contig: Clustering of fragments. A and B denote any enzyme with a 6-base recognition sequence.

each other may represent DNA from either side of the same *Eag*I site. In most cases, this *Eag*I site lies within a CpG island. Two subclones obtained with different pairs of restriction enzymes (e.g., one subclone obtained with *Sac*II/*Hin*dIII and one obtained with *Eag*I/*Eco*RI from the same starting material) that map to the same position are almost certain to define a single CpG island (Figure 4).

Demonstrating That Subclones with Coding Potential Are Transcribed

Once a CpG island has been demonstrated to have coding potential and has been assigned a map position, the next step is to verify that the subclone containing the CpG island actually contains a segment of a gene. If the computer analysis indicates that a subclone contains a known gene sequence, further verification is not needed, as in the case of a BLASTN search of available databases identifying the Huntington's disease gene (BLASTN score of 5.9×10^{-88}) and α-adducin gene (6.8×10^{-116}) in subclones isolated by this method (John et al. 1994). In fact, it is likely that a full-length cDNA clone would be available for immediate analysis from the investigators who originally submitted the sequence to the database. Subclones with coding potential that do not contain a previously identified gene segment can be used to probe northern blots (Sambrook et al. 1989) of multiple tissues, since the gene might be expressed only in a subset of tissues. The disadvantages of northern blotting as a means of verification is that transcripts expressed at a low level or only in a subset of tissues are often missed. This problem can be overcome by using a more sensitive technique such as RT-PCR (using primers designed for the region predicted by either GRAIL or BLAST to encode a gene) or the RNase protection assay (Sambrook et al. 1989), each of which has the advantage that multiple tissues can be screened simultaneously. Once a transcript has been identified, the next stage is to isolate the full-length cDNA by screening (Sambrook et al. 1989, Chapter 8) a library prepared from the tissue that expresses the transcript most abundantly.

PROTOCOL

Determining the Sizes of Inserts and Eliminating Duplicate Subclones

1. Select 20 white colonies from each transformation plate on p. 234. For each colony, place 5 ml of LB medium containing ampicillin (30 μg/μl) plus methicillin (1 mg/ml) in a sterile 50-ml conical tube and inoculate with a single white colony. Incubate the cultures at 37°C with agitation at 300 rpm overnight.

   Notes: The number of subclones to be analyzed is dictated by the density of RCRE sites in the region of interest (determined on p. 232). Analyzing 20 subclones from each 100-kb segment proved to be sufficient to obtain good coverage of the CpG-rich regions with minimum redundancy for human genomic DNA cloned into cosmid vectors (John et al. 1994). If the density of RCRE sites is high (i.e., more than 10 per 100 kb or if the genomic region is much larger than 100 kb), more than 20 subclones should be analyzed to ensure that each subclone is represented at least once. As an approximate guideline, analyze twice as many subclones as the number calculated (i.e., four times the number of RCRE sites since each site generates two subclones).

   If the cloning vector does not encode ampicillin resistance, substitute the appropriate selective medium. Use of cloning vectors that allow a blue/white selection for identifying insert-containing subclones is strongly recommended for this protocol.

2. Prepare plasmid DNA from each culture using one of the many commercially available column-based methods for small-scale DNA preparations (e.g., Wizard Plus Miniprep DNA Purification System [Promega A7100] or QIAprep Spin Plasmid Kit [Qiagen 27104]).

3. Perform the sets of steps below in parallel for each DNA minipreparation.

   To determine the size of the insert:

   a. Digest 0.5 μg of DNA in a reaction volume of 20 μl. Use the same pair of RCRE/non-RCREs that were used to prepare vector in the original subcloning steps.

      Note: This should release the insert from the DNA.

   b. Add 0.1 volume of BP/XC/glycerol gel-loading solution to the digestion mixture. Load on a 1% agarose gel (10 cm long) in 1x TAE buffer along with 1-kb ladder (Life Technologies 15615-016). Run the gel at 5 V/cm for 4 hours. The islands will be approximately 1–2 kb.

   c. Set aside this gel until needed for step 4.

   To identify unique subclones by fingerprinting:

   a. Digest 2 μg of DNA in a reaction volume of 50 μl in a 1.5-ml microcentrifuge tube. Use 10 units of a restriction enzyme with a 4-base recognition sequence (e.g., *Hae*III or *Rsa*I) and digest as specified by the enzyme's manufacturer.

Note: Unique subclones often cannot be identified on the basis of the size of the insert alone, particularly if the density of RCRE sites in a region is high or if the original genomic clones have insert sizes much greater than 100 kb. The average fragment size generated by a restriction enzyme recognizing a 4-base sequence is approximately 250 bp and digestion with any enzyme of this type is likely to generate a unique pattern of restriction fragments, allowing the identification of unique subclones.

b. Add 0.1 volume of BP/XC/glycerol gel-loading solution to the digestion mixture. Load on a 2% agarose gel (10 cm long) in 1x TAE buffer along with a 1-kb ladder (Life Technologies 15615-016). Run the gel at 5 V/cm for 3–4 hours.

c. Stain the DNA with **ethidium bromide** and view under **UV** illumination. Within each set of subclones from a pool, identify those subclones with unique banding patterns.

Note: If all of the subclones are unique at this stage, the number of RCRE sites in the contig may have been underestimated or the cloning strategy may have failed. Redetermine the number of RCRE sites on p. 232 and select another batch of 20 subclones and prepare plasmid DNA from these subclones. If the subclones are still unique, it is possible that either the vector DNA or the insert DNA has been prepared incorrectly. Make sure that the subclone inserts are released only by digestion with both the RCRE and the non-RCRE enzyme and not by digestion with a single enzyme since this would indicate that the vector was not digested to completion by one of the enzymes used. Alternatively, the insert DNA may be contaminated with bacterial or yeast genomic DNA.

ethidium bromide, UV radiation (see Appendix for Caution)

4. Use a sterile razor blade to excise the bands of insert DNA (from the gel used to determine the size of the insert in step 3) that are derived from unique subclones. Extract the DNA from the agarose (see p. 229, step 4).

Note: Unique subclones are selected either by size or by analysis of fingerprinting pattern.

5. Dissolve the DNA in 20 μl of TE (pH 8.0). Store at −20°C until needed for further analysis.

Notes: The concentration of DNA in the insert can be estimated by comparison with the molecular-weight markers on the gel or can be determined from the OD_{260} of a 1-μl aliquot of the gel-purified DNA (see Appendix).

The insert DNAs purified in this manner can be used to map the subclones within the contig (see pp. 243–244). If subsequent analysis (see pp. 236–238) indicates that a subclone has coding potential, the insert DNA can be used as a probe for northern blots and/or screening cDNA libraries (Sambrook et al. 1989, pp. 7.37–7.50 and pp. 8.46–8.49).

The DNA can be stored at −20°C indefinitely.

PROTOCOL

Determining Where Subclones Containing a CpG Island Map within a Contig

Southern blots generated in the estimation of the density of RCRE sites (see p. 232) can be used below for initial screening of CpG islands generated from large-insert clones. If these blots are available, skip steps 1–4 below.

1. For each minimal overlapping set of the genomic DNA clones (the minimal set covering each 100-kb segment), digest 0.5–1 µg of DNA with a non-RCRE as specified by the manufacturer.

   Note: The enzyme can be one of those used in the subcloning step (see p. 233, step 1) or any restriction enzyme with a 6-base recognition sequence (see Table 4).

2. Add 0.1 volume of BP/XC/glycerol gel-loading solution to the digestion mixture. Load on a 0.8% agarose gel (10 cm long) in 1x TAE buffer along with a 1-kb ladder (Life Technologies 15615-016). Run the gel at 5 V/cm for 4 hours.

   Note: The DNA fragments will generally range from 0.2 to 12 kb.

3. Stain the DNA with **ethidium bromide** and view under **UV** illumination.

   ethidium bromide, UV radiation (see Appendix for Caution)

4. Transfer the DNA onto a nylon membrane as described for Southern blotting and fix the DNA to the membrane (see Wolff and Gemmill 1997).

5. Radiolabel an insert DNA fragment (from p. 242, step 5) using standard procedures (for random priming, see Wolff and Gemmill 1997).

   radioactive substances (see Appendix for Caution)

6. In vertebrate genomes that contain many repetitive sequences, prehybridize the probe to block repetitive sequences.

   a. Add the following to the radiolabeled DNA in a 1.5-ml microcentrifuge tube:

0.5 M sodium phosphate (pH 7.2) (see Appendix)	50 µl
sheared genomic DNA (10 mg/ml) from the appropriate species	100 µl
H$_2$O	100 µl

   b. Boil for 10 minutes and then incubate at 65°C for 3 hours.

   Note: Do not boil the probe again after the repetitive sequences are blocked; the unique DNA sequence will still be denatured after the blocking procedure.

7. Prehybridize, hybridize, and wash the membranes at high stringency as described for Southern blotting (see Wolff and Gemmill 1997).

8. Autoradiograph at room temperature for 1–4 hours with an intensifying screen.

Note: The probes should hybridize with a single, unique DNA fragment. If many fragments of different sizes are observed, it is likely that the blocking step failed and the probe is detecting repeat sequences. Occasionally, a unique probe may hybridize with two fragments from different large-insert clone digestions. This is likely to be a region where the large-insert clones overlap.

USING A COLUMN THAT BINDS METHYLATED DNA TO PURIFY CpG ISLANDS FROM UNCLONED GENOMIC DNA

Fragments of DNA can be separated according to their level of methylation by using their affinity for a protein that binds methyl CpG (Cross et al. 1994). Heavily methylated DNA binds tightly to a column containing this protein attached to a solid support, whereas unmethylated DNA binds weakly. CpG island DNA has a relatively high G + C content and contains the dinucleotide CpG at a frequency of one CpG per 10 bp. In contrast, non-CpG-island DNA contains one CpG per 100 bp (see Figure 1). Since CpG islands are 1–2 kb in size, they therefore contain 100–200 CpG dinucleotides. With a few exceptions, CpG islands are unmethylated in their native state and therefore show little affinity for binding to the column. However, when methylated in vitro, they bind tightly and can be isolated from the rest of the genome, which binds only weakly because of the relatively low density of methylated CpG dinucleotides.

The strategy on pp. 266–273 (Figure 5) involves isolating intact CpG islands on a large scale from uncloned genomic DNA by using a column that binds methylated DNA (Cross et al. 1994). This approach for isolation exploits the unusual methylation status and DNA sequence content of CpG islands. Besides preparation and calibration of the column, this strategy involves the following four steps:

- Preparing and digesting genomic DNA
- Removing heavily methylated fragments from the digested genomic DNA to produce the "stripped" fraction
- Methylating CpG dinucleotides on DNA fragments in the stripped fraction
- Binding and then selectively eluting heavily methylated fragments (this fraction will be enriched for CpG islands)

Protocols for each of these steps are provided below. Evaluation is carried out at various points during the purification to monitor the success of the steps. Once the evaluation is complete, a library is prepared and a number of clones are tested to ascertain whether or not their sequence composition and methylation status in genomic DNA are like those of CpG islands.

Preparation of the HMBD Column and Optimization of the Purification of Methylated DNAs

MeCP2, a chromosomal protein that binds to methylated DNA, has been purified, and the gene that encodes it has been cloned (Lewis et al. 1990). MeCP2 contains two DNA-binding domains. The domain for binding methyl CpG (known as MBD) was identified by deletion analysis and shown to be specific for binding to methylated DNA (Meehan et al. 1992; Nan et al. 1993). The 85-amino-acid MBD has strong affinity for methylated DNA and appears to have little or no affinity for unmethylated DNA. The DNA encoding MBD was

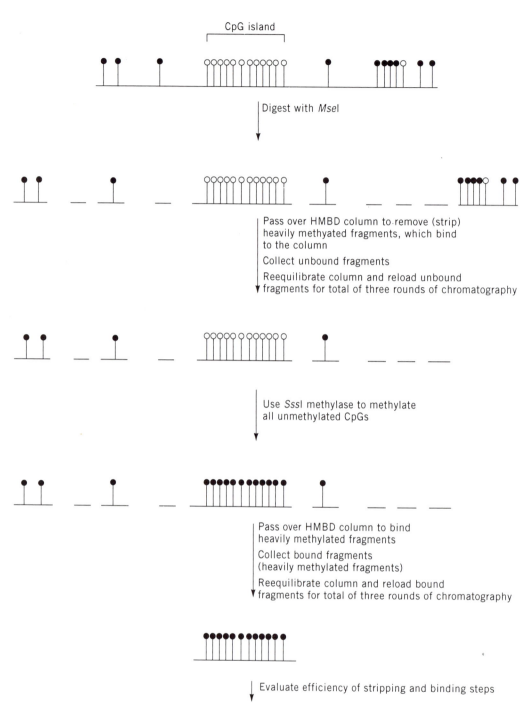

Figure 5 Strategy for purifying *Mse*I-generated fragments containing CpG islands from uncloned genomic DNA. This strategy uses a column that binds methylated DNA to exploit the unusual methylation status and DNA sequence content of CpG islands. The position of CpG dinucleotides is indicated by vertical lines. Open and closed circles denote unmethylated and methylated CpG dinucleotides, respectively. See text and Table 5 for details. (Modified, with permission, from Cross et al. 1994.)

cloned into a bacterial expression vector; the expressed recombinant protein (HMBD) carries a histidine tag at its amino-terminal end fused with MBD (Cross et al. 1994). The plasmid pET6HMBD contains the part of the MeCP2 gene that encodes MBD cloned into the bacterial expression vector pET6H (Cross et al. 1994). To express the protein, the bacterial strain BL21 (DE3) pLysS is transformed with the plasmid. The production of the HMBD protein uses the bacteriophage T7 RNA polymerase expression system (Studier et al. 1990). The HMBD protein is coupled to a nickel-agarose matrix via the histidine tag (Hochuli et al. 1987) to form a column that can be used to fractionate DNA according to its degree of methylation (see Figure 1 in Cross et al. 1994). Protocols for preparing the HMBD protein and for coupling the protein to the column matrix are provided on pp. 251–258.

Before an HMBD column can be used to separate fragments of genomic DNA, it must be calibrated using plasmid DNA (pp. 259–265). The amount of HMBD coupled to the column determines the salt concentration at which DNAs with different amounts of methylation will elute. Two types of tests are used to optimize purification of methylated DNAs on the column. The most convenient way to perform the first test and calibrate the HMBD column is to determine the elution profile of a mixture of plasmid DNAs that contain different numbers of symmetrically methylated CpG dinucleotides. A mixture of one unmethylated plasmid and one completely methylated plasmid is used in the protocol on pp. 262–264. This type of analysis has been described for the plasmid pCG11 (Cross et al. 1994), but any plasmid with a known sequence, and therefore a known number of CpG dinucleotides, can be used. A cloning vector such as pUC19, which contains 173 CpG dinucleotides (EMBL databank accession number M77789), is an ideal choice and is used in the protocol below.

Typically, heavily methylated DNA fragments (those containing more than 100 methyl CpG dinucleotides) elute from the HMBD column at a NaCl concentration of 0.7–0.9 M. The salt concentration at which a fragment elutes from the HMBD column is determined principally by the total number of methylated CpG dinucleotides it contains, not the number of CpG dinucleotides per unit length (Cross et al. 1994). Fragments containing fewer methylated CpG dinucleotides elute at lower NaCl concentrations. Partially methylated plasmid DNA (i.e., plasmid DNA in which only some of the CpG dinucleotides are methylated; a typical test plasmid containing approximately 33 methyl CpG dinucleotides could be used) will elute from the HMBD column at an NaCl concentration of approximately 0.1–0.2 M less than that required to elute heavily methylated DNA fragments. (Heavily methylated DNA contains 100 plus methyl CpG dinucleotides, partially methylated DNA contains 30–50 methyl CpG dinucleotides, and lightly methylated DNA contains 0–10 methyl CpG dinucleotides.)

The second test for optimizing purification on the column is often useful for the construction of CpG island libraries. It involves loading DNA on the HMBD column such that the unmethylated DNA, which includes the CpG island DNA, remains in the flow-through. Unmethylated DNA binds to the HMBD column under conditions of low salt concentration (0.1 M NaCl), presumably because the HMBD protein is very basic (Cross et al. 1994). If the salt concentration of the loading buffer is increased to approximately 0.5 M (the exact molarity will vary according to the amount of HMBD on the column), unmethylated DNA no longer binds, but methylated DNA still does (S.H. Cross, unpubl.).

To determine the appropriate salt concentration to keep the unmethylated DNA in the flow-through, end-labeled unmethylated and partially methylated plasmid DNAs are prepared as described on pp. 259–261. Treatment of the test plasmid pUC19 with a combination of *Hpa*II and *Hha*I methylases results in the methylation of 33 CpG dinucleotides. The unmethylated plasmid serves as a model for unmethylated human DNA *Mse*I-generated fragments (the vast majority of the fragments), and the partially methylated plasmid serves as a model for human DNA *Mse*I-generated fragments that contain 33 or more methylated CpG dinucleotides (a minority of the fragments). The plasmids are loaded on the HMBD column individually at various NaCl concentrations and the eluate is examined to identify the salt concentration at which the unmethylated plasmid does not bind to the column, but the partially methylated one does (pp. 264–265).

Preparation and Digestion of Uncloned Genomic DNA

In the first step of the isolation strategy (pp. 266–267), genomic DNA is fragmented such that CpG islands are left intact, whereas bulk DNA is highly fragmented. This is accomplished by using a restriction enzyme that cuts frequently (every 100–200 bp on average) in bulk DNA but rarely in CpG island DNA. A suitable restriction enzyme is *Mse*I because its recognition site, TTAA, is predicted to occur rarely within CpG island DNA (once per 1000 bp) but frequently in genomic DNA (once per 140 bp). Digestion with *Mse*I results in predominantly intact CpG island fragments (~0.5–2 kb in size) and small fragments (<500 bp containing, on average, one to five methylated CpG dinucleotides) from the rest of the genome. Another reason that *Mse*I is a convenient enzyme to use is that fragments produced by the enzyme can be cloned into the *Nde*I site of the pGEM-5Zf(–) cloning vector (Promega). Other restriction enzymes with a 4-base recognition site containing only Ts and As can also be used.

The starting genomic DNA for the preparation of CpG island DNA does not need to be of very high molecular weight since the DNA will be digested so that the maximum fragment size is approximately 2 kb. Therefore, the DNA can be prepared using standard procedures (see Wolff and Gemmill 1997). The DNA can be prepared from blood or tissue samples but not from placenta or sperm, because satellite sequences, which contain many CpG dinucleotides that are *not* associated with genes, are often undermethylated in these tissues (Sanford et al. 1985).

Cross et al. (1994) demonstrated that plasmid DNAs methylated to different degrees could be separated on HMBD columns. They also described the purification of CpG island DNA from human genomic DNA. The resulting library, human CGI-1 (human CpG island 1), is available from the UK HGMP Resource Centre (Hinxton Hall, Hinxton, Cambridgeshire CB10 1RQ, UK; http://www.hgmp.mrc.ac.uk/). HMBD columns have proved to be useful for the production of CpG island libraries from other vertebrates; mouse (Cross et al. 1997), chicken (McQueen et al. 1996), and pig (McQueen et al. 1997) CpG island libraries are also available from the UK HGMP Resource Centre. Other potential sources of starting material from which CpG island libraries could be made are somatic cell hybrids, which are useful when interest is focused on a particular region of the human genome, or sorted human chromosomes (S.H. Cross, unpubl.).

Removal of Heavily Methylated Fragments from the Genomic DNA

In the second step (pp. 267–269), the digested genomic DNA is passed through the calibrated HMBD column under conditions that bind both fully and partially methylated DNA. The flow-through contains both CpG island fragments and bulk genomic DNA fragments that are unmethylated or contain only a few methyl CpG dinucleotides (typically one to five) and is designated the stripped fraction (i.e., stripped of methylated fragments). Fragments containing a cluster of methyl CpG dinucleotides are retained on the column.

Methylation of the Stripped Fraction

The third step (pp. 269–270) uses a CpG methylase (in this case, *Sss*I) to methylate all of the nonmethylated CpG dinucleotides isolated in the stripped fraction. Nearly all non-CpG-island CpG dinucleotides in the majority of DNA fragments in the stripped fraction are already methylated; therefore, the affinity of the HMBD column for these fragments does not change after the methylation step and the fragments still pass through. Because of the high density of CpG dinucleotides in CpG islands, the methylation step converts CpG islands in the stripped fraction into heavily methylated stretches of DNA that anneal very tightly to the column.

Selection of Fractions Enriched for CpG Islands

In the fourth step (pp. 270–272), the heavily methylated CpG islands are bound to the column and then selectively eluted. Selecting the fragments that elute at a high salt concentration yields a fraction that is highly enriched for CpG islands. At least three rounds of binding are usually required to obtain a highly purified CpG island fraction. This purified CpG island fraction can then be cloned into a plasmid vector to generate a CpG island library (pp. 274–275) after evaluation of the success of the binding and stripping steps.

Evaluation of the Isolation and Purification of CpG Islands

To determine whether or not heavily methylated DNA has been removed during the stripping steps and whether or not CpG island DNA has been successfully

Table 5 PCR Analysis of *Mse*I-generated Fragments from the Stripping and Binding Steps in Purifying CpG Islands

*Mse*I-generated fragment tested	L	Stripping steps						Binding steps					
		S1		S2		S3		B1		B2		B3	
		U	B	U	B	U	B	U	B	U	B	U	B
CpG island DNA	+	+	−	+	−	+	−	−	+	−	+	−	+
Non-CpG-island genomic DNA	+	+	−	+	−	+	−	+	−	+	−	+	−
Heavily methylated GC-rich DNA	+	−	+	−	+	−	+	−	−	−	−	−	−

The expected presence (+) or absence (−) of PCR products is shown for the three different *Mse*I-generated fragments in the three stripping steps (S1, S2, and S3) and the three binding steps (B1, B2, and B3) used in purifying CpG islands. L denotes the genomic DNA originally loaded on the column, U denotes the unbound fraction, and B denotes the bound fraction from each round of chromatography.

purified during the binding steps, PCR assays can be performed at the different stages of the purification procedure to check for the presence of representative *Mse*I-generated fragments from known sequences (pp. 272–273). Each of the *Mse*I-generated fragments chosen for the assay should be derived from one of three groups of eluted DNA: (1) CpG island DNA, (2) non-CpG-island genomic DNA, and (3) heavily methylated GC-rich DNA. Cross et al. (1994) used the following test sequences during the construction of a human CpG island library: a CpG-island fragment from the monoamine oxidase A gene, a fragment from the interferon-α gene and a fragment from the aldolase B gene that are representative of non-CpG-island DNA containing only a few CpG dinucleotides, and a fragment containing a cluster of methylated CpG dinucleotides from the 3′ end of the apolipoprotein A-IV gene.

Table 5 shows which fractions should contain PCR products for the three types of fragments. The CpG island fragment is found in the unbound fractions of the stripping steps (S1-U, S2-U, and S3-U) and the bound fractions of the binding steps (B1-B, B2-B, and B3-B). The lightly methylated non-CpG-island genomic DNA fragment is found in only the unbound fractions of both the stripping and binding steps. The heavily methylated fragment is found in the bound fractions (S1-B, S2-B, and S3-B) of the stripping steps and is not present in any of the fractions from the binding steps. In B3-B, the only fragment detectable by PCR is the one from the CpG island.

For the first two stripping (S1 and S2) and binding (B1 and B2) steps, it is likely that there will be residual amounts of the three representative *Mse*I-generated fragments in the "wrong" fractions. For example, there may be trace amounts of the heavily methylated fragment present in the unbound fractions from the first two stripping steps (S1-U and S2-U). It is also possible that if the first two stripping steps have been particularly efficient, the heavily methylated fragment will be *absent* from the bound fraction in the third stripping step (S3-B). The presence of these residual amounts does not present a problem as long as the only fragment detectable by PCR in the final bound fraction from the binding steps (B3-B) is the one from the CpG island.

PROTOCOL

Preparing the HMBD Column

These protocols should yield enough HMBD protein for a 1-ml column and can be adjusted to obtain different yields.

PREPARATION OF THE HMBD PROTEIN

1. Streak an LB agar plate containing ampicillin (50 μg/ml) plus chloramphenicol (30 μg/ml) with BL21 (DE3) pLysS (pET6HMBD) from a frozen glycerol stock. Incubate at 37°C overnight.

   Note: The bacterial strain BL21 (DE3) pLysS (pET6HMBD) (*E. coli* B F⁻ *dcm ompT hsdS* (r_{B-} m_{B-}) *gal* λ(DE3) [pLysS Cam^r]) and the plasmid pET6HMBD in the bacterial strain XL1-Blue (*recA1 endA1 gyrA96 thi-1 hsdR17 supE44 relA1 lac* [F' *proAB lacI^{q}ZΔM15* Tn*10* (Tet^r)]) are available from S. Cross.

2. Place 100 ml of LB medium containing ampicillin (50 μg/ml) plus chloramphenicol (30 μg/ml) in a 0.5-liter flask and inoculate with a single bacterial colony from the plate in step 1. Incubate at 37°C with agitation at approximately 300 rpm overnight.

3. Place 1.5 liters of LB medium containing ampicillin (50 μg/ml) plus chloramphenicol (30 μg/ml) in a 2-liter flask and inoculate with 45 ml of the overnight culture. Measure the OD_{600}.

   Notes: The OD_{600} should be approximately 0.1. If the OD_{600} is considerably higher than this (>0.2), dilute an aliquot of the culture to a final volume of 1.5 liters with LB medium containing ampicillin (50 μg/ml) plus chloramphenicol (30 μg/ml) in another 2-liter flask and incubate until the culture reaches an OD_{600} of 0.1.
     The strain should be freshly grown from step 2.

4. Divide the 1.5-liter culture (from step 3) into two 2-liter flasks and incubate at 37°C with vigorous agitation (300 rpm) until the cultures reach an OD_{600} of 0.3–0.5. Transfer a 0.5-ml aliquot of these uninduced cells from one of the cultures into a 1.5-ml microcentrifuge tube (sample A) and place on ice until needed in step 6.

   Note: The cultures with an OD_{600} of 0.1 should reach an OD_{600} of 0.3–0.5 in 2–3 hours.

5. To each flask from step 4, add IPTG to a final concentration of 0.4 mM (i.e., add 3 ml of 25 mg/ml IPTG to 750 ml of culture and continue incubating the cultures at 37°C with vigorous agitation (300 rpm) for an additional 3 hours. Transfer a 0.5-ml aliquot of these induced cells from one of the cultures into a 1.5-ml microcentrifuge tube (sample B) and place on ice until needed in step 6.

6. Prepare the aliquots of uninduced and induced cells for analysis on p. 255, step 6, as follows:

   a. Centrifuge in a microcentrifuge at 12,000g at room temperature for 5 minutes. Discard the supernatants.

b. Resuspend each cell pellet in 100 µl of H_2O plus 100 µl of 2x SMASH buffer.

2x SMASH buffer

Component and final concentration	Amount to add per 100 ml
125 mM Tris-Cl	25 ml of 0.5 M (pH 6.8 at 25°C)
20% glycerol	20 ml of 100%
4% **SDS**	4 g
10 µg/ml bromophenol blue	1 mg
286 mM β-**mercaptoethanol**	2 ml of 14.3 M
H_2O	to make 100 ml

Divide into 10-ml aliquots and store at −20°C for up to 2 years. Keep one aliquot at room temperature for immediate use.

SDS, β-**mercaptoethanol** (see Appendix for Caution)

c. Store at −20°C until needed for analysis.

Note: These cells are probably stable indefinitely but are usually analyzed in 1–2 days.

7. Transfer each induced culture (750 ml) from step 5 into a 1-liter centrifuge bottle. Centrifuge in a Beckman J6-B centrifuge (or equivalent) at 3600 rpm (2000*g*) at 4°C for 20 minutes. Discard the supernatants.

Note: Perform this step and all remaining steps on ice or in the coldroom with ice-cold solutions.

8. Resuspend each cell pellet in 12.5 ml of HMBD extraction buffer containing protease inhibitors, each at a final concentration of 5 µg/ml.

HMBD extraction buffer

Component and final concentration	Amount to add per 100 ml
5 M urea	30 g
50 mM NaCl	1 ml of 5 M
20 mM HEPES	2 ml of 1 M (pH 7.9)
1 mM EDTA	200 µl of 0.5 M (pH 8.0)
10% glycerol	10 ml of 100%
0.5 mM **PMSF**	0.5 ml of 100 mM
H_2O	to make 100 ml

Prepare just before use and place on ice.

Protease inhibitors

The protease inhibitors **leupeptin** HCl, **antipain** HCl, chymostatin, **pepstatin A**, and **aprotinin** are available from Sigma. Prepare stock solutions of each as specified by the manufacturer and store at −20°C indefinitely. Just before use, add each protease inhibitor to 100 ml of HMBD extraction buffer to make a final concentration of 5 µg/ml each.

PMSF, leupeptin, antipain, pepstatin A, aprotinin (see Appendix for Caution)

9. Pool the two resuspended cell pellets in one 50-ml plastic conical tube. Add Triton X-100 (from a 20% stock solution) to a final concentration of 0.1% and mix by swirling gently.

10. Sonicate the cells with the thin probe of a Sonifer model 250/450 (Branson Ultrasonics) or equivalent at a power setting of 6, cycle 4, for a total of 50 1-second pulses.

11. Transfer a 100-µl aliquot of the disrupted cells into a 1.5-ml microcentrifuge tube. Add 100 µl of 2x SMASH buffer to the aliquot, mix by flicking the tube, and store at –20°C (sonicated sample C) until needed for analysis on p. 255, step 6. Pour the remainder of the disrupted cells into a 30- or 50-ml plastic tube capable of withstanding centrifugation at speeds up to 35,000g.

12. Centrifuge the disrupted cells in a JA-20 rotor in a Beckman J2-21 centrifuge (or equivalent) at 16,000 rpm (31,000g) at 4°C for 30 minutes.

13. Carefully decant the supernatant into a 50-ml plastic tube and discard the pellet. Transfer a 100-µl aliquot of supernatant into a 1.5-ml microcentrifuge tube. Add 100 µl of 2x SMASH buffer to the aliquot, mix by flicking the tube, and store at –20°C (centrifuged sample D) until needed for analysis on p. 255, step 6.

14. If desired, store the remaining supernatant (~25 ml of HMBD protein) at –80°C overnight at this point. Otherwise, proceed with step 2 of the partial purification of HMBD protein.

PARTIAL PURIFICATION OF THE HMBD PROTEIN

Perform all steps on ice or in the coldroom with ice-cold solutions. Run the column in the coldroom.

1. *If the supernatant containing the HMBD protein has been stored at –80°C:*

   a. Thaw in cold H_2O or on ice.

   b. Add protease inhibitors (**leupeptin, antipain**, chymostatin, **pepstatin A**, and **aprotinin**), each at a final concentration of 5 µg/ml, and mix by swirling.

      Note: Since protease inhibitors are unstable after a freeze/thaw cycle, it is advisable to add them again at this point.

      leupeptin, antipain, pepstatin A, aprotinin (see Appendix for Caution)

   c. Centrifuge in a JA-20 rotor in a Beckman J2-21 centrifuge (or equivalent) at 16,000 rpm (31,000g) at 4°C for 30 minutes to remove the insoluble material.

   d. Decant the supernatant into a 50-ml plastic tube and discard the pellet.

   e. Proceed with step 2.

If the supernatant containing the HMBD protein has not been stored at −80°C:

Proceed with step 2.

2. Prepare 12 ml of Fractogel EMD SO3^e-650(M) (Merck) resin as specified by the manufacturer. Pipette 5 ml of the resin into each of two 5-ml disposable plastic chromatography columns (e.g., Econo-Pac column; Bio-Rad 732-1010).

   Notes: Two 5-ml columns are used instead of one 10-ml column to increase the speed of the elutions below.

   The basic HMBD protein (predicted pI 9.75) will bind tightly to this cation-exchange resin, allowing partial purification of the HMBD protein from the crude protein extract (Cross et al. 1994).

3. Equilibrate each column by washing it sequentially with 20 ml of Fractogel washing buffer #1, 20 ml of Fractogel washing buffer #2, and then 20 ml of Fractogel buffer #1.

   Note: Run the column under gravity flow in all steps in this protocol.

   Fractogel washing buffer #1

Component and final concentration	Amount to add per 200 ml
5 M urea	60 g
50 mM NaCl	2 ml of 5 M
10% glycerol	20 ml of 100%
20 mM HEPES	4 ml of 1 M (pH 7.9)
0.1% Triton X-100	1 ml of 20%
0.5 mM **PMSF**	1 ml of 100 mM
10 mM **β-mercaptoethanol**	140 µl of 14.3 M
H$_2$O	to make 200 ml

   Prepare just before use and place on ice.

   Fractogel washing buffer #2

Component and final concentration	Amount to add per 100 ml
2 M urea	12 g
1 M NaCl	20 ml of 5 M
10% glycerol	10 ml of 100%
20 mM HEPES	2 ml of 1 M (pH 7.9)
0.1% Triton X-100	0.5 ml of 20%
0.5 mM PMSF	0.5 ml of 100 mM
10 mM β-mercaptoethanol	70 µl of 14.3 M
H$_2$O	to make 100 ml

   Prepare just before use and place on ice.

   PMSF, β-mercaptoethanol (see Appendix for Caution)

4. Arrange the two columns side-by-side so that they drip into the same 50-ml plastic Falcon tube. Divide the supernatant containing the HMBD protein into two aliquots (~12.5 ml each) and simultaneously load one aliquot on

each column. Collect the flow-through (~25 ml) in the 50-ml tube and place on ice until needed for analysis in step 6.

5. Perform the following 12 washes simultaneously for the two columns. For each wash, collect the two 5-ml fractions of eluate from the pair of columns in a single numbered 15-ml snapcap or screwcap tube.

   a. Wash each column four times with 5 ml of Fractogel washing buffer #1, collecting the eluate in tubes 1–4.

   b. Wash each column four times with 5 ml of Fractogel buffer #1/#2, collecting the eluate in tubes 5–8.

   Note: Fractogel buffer #1/#2 is a mixture of 27.5 ml of Fractogel washing buffer #1 and 12.5 ml of Fractogel washing buffer #2.

   c. Wash each column four times with 5 ml of Fractogel washing buffer #2, collecting the eluate in tubes 9–12.

   d. Place the tubes on ice until needed for analysis in step 6.

6. Determine which fractions contain the HMBD protein as follows:

   a. Transfer 10-µl aliquots of the flow-through (set aside in step 4) and each of the 12 fractions collected in step 5 into separate, correspondingly labeled 1.5-ml microcentrifuge tubes.

   b. Add 10 µl of 2x SMASH buffer (for preparation, see p. 252) to each tube.

   c. Heat these samples and the four samples set aside on pp. 251–253 (the uninduced, induced, sonicated, and centrifuged samples A–D set aside in steps 6, 11, and 13) at 90°C for 90 seconds.

   Note: Although the aliquots of collected eluate are only 20 µl, the entire volume (200 µl) of samples A–D is analyzed during this step.

   d. Load each sample (along with the appropriate molecular-weight markers) on a 15% SDS-polyacrylamide gel and separate the proteins. Detect the proteins by staining the gel with Coomassie Brilliant Blue and then destaining.

   Notes: Protocols for preparing, running, and staining and destaining **SDS-polyacrylamide** gels can be found in Sambrook et al. (1989, pp. 18.47–18.59). A mini-PROTEAN II gel system (Bio-Rad) can be used, with the gel run at 200 V for 40 minutes (or until the blue dye reaches the bottom of the gel).
   The HMBD protein has a molecular mass of 11.4 kD. Use a commercially available broad-range marker as specified by the manufacturer (e.g., Protein Marker, Broad Range [2–212 kD]; New England Biolabs 7701S). Alternatively, use 2 µg of cytochrome C (Sigma C 3256), which has a molecular mass of 12.4 kD. Prepare a cytochrome C stock solution (200 µg/ml) in HMBD extraction buffer without urea (for preparation, see p. 252); store at –20°C indefinitely. To prepare the marker for loading, mix 100 µl of the cytochrome C stock solution with 100 µl of 2x SMASH buffer and heat at 90°C for 90 seconds; store at –20°C indefinitely.
   The band of HMBD protein will migrate just below the cytochrome C band. It should be present in samples B, C, and D and in fractions 9–12. It should be *absent* from samples A, 1–8, and the flow-through. It may be difficult to see the band of HMBD protein in samples B, C, and D because of the vast excess of bacterial proteins. This should not be a

problem for the fractions eluted from the Fractogel EMD SO3^e-650(M) column since most bacterial proteins do not bind to the resin and will therefore be found in the flow-through fraction.

SDS, acrylamide (see Appendix for Caution)

7. Pool all eluted fractions from step 5 that are enriched for the HMBD protein in a single 50-ml plastic tube.

   Notes: Fractions enriched for HMBD are assessed by detecting the expression of the protein in the gel.
      The pooled fractions can be stored at –80°C indefinitely at this point.

8. Couple the HMBD protein to the nickel-agarose resin as described below.

COUPLING THE HMBD PROTEIN TO THE NICKEL-AGAROSE RESIN

Perform all steps on ice or in the coldroom with ice-cold solutions. Run the column in the coldroom.

1. If the pooled HMBD protein has been stored at –80°C, thaw on ice or in cold H$_2$O. Use the Bradford assay (Bradford 1976) to measure the protein concentration in a 10-μl aliquot.

   Notes: The protein assay can also be performed using the Protein Assay Kit (Bio-Rad 500-0002) or equivalent.
      Expect a total of approximately 20–50 mg of protein (~1 mg/ml).

2. Pipette 1 ml of nickel-agarose matrix (Ni-NTA-agarose; Qiagen 30210) into a 5-ml disposable chromatography column (Poly-Prep chromatography column; Bio-Rad 731-1550).

   Notes: Alternative matrices are available, but some of the nickel-agarose resins must be charged before use. Consult the manufacturer's instructions for charging the resin. A matrix called Ni-NTA Superflow (Qiagen 30410) is reported to have handling properties superior to those of the Ni-NTA-agarose matrix.
      Approximately 20–50 mg of HMBD protein is enough to saturate a 1-ml column.

3. Equilibrate the column by washing it with 4 ml of nickel-agarose column washing buffer.

   Note: Run the column under gravity flow in all steps in this protocol.

   Nickel-agarose column washing buffer

Component and final concentration	Amount to add per 100 ml
50 mM NaCl	1 ml of 5 M
20 mM HEPES	2 ml of 1 M (pH 7.9)
10% glycerol	10 ml of 100%
0.1% Triton X-100	0.5 ml of 20%
0.5 mM **PMSF**	0.5 ml of 100 mM
10 mM β-mercaptoethanol	70 μl of 14.3 M
H$_2$O	85.9 ml

   Prepare just before use and place on ice.

   PMSF, β-mercaptoethanol (see Appendix for Caution)

4. Transfer a 50-µl aliquot of the pooled HMBD protein into a 1.5-ml micro-centrifuge tube. Place on ice (load sample E) until needed in step 7.

5. Load the remainder of the pooled HMBD protein on the column and collect the flow-through in a 50-ml plastic tube. Remove a 50-µl aliquot (sample F) and place on ice until needed in step 7. Set aside the rest of the flow-through for step 8.

6. Perform the following three washes. Collect 12 1-ml fractions in numbered 1.5-ml microcentrifuge tubes.

   a. Wash the column with 4 ml of nickel-agarose column washing buffer, collecting the fractions in tubes 1–4.

   b. Wash the column with 4 ml of imidazole/nickel-agarose column washing buffer, collecting the fractions in tubes 5–8.

   Imidazole/nickel-agarose column washing buffer

Component and final concentration	Amount to add per 100 ml
50 mM NaCl	1 ml of 5 M
20 mM HEPES	2 ml of 1 M (pH 7.9)
10% glycerol	10 ml of 100%
0.1% Triton X-100	0.5 ml of 20%
0.5 mM PMSF	0.5 ml of 100 mM
10 mM β-mercaptoethanol	70 µl of 14.3 M
8 mM imidazole	0.8 ml of 1 M
H_2O	85.1 ml

   Prepare just before use and place on ice.

   1 M Imidazole
   Add 34.04 g of imidazole (m.w. = 68.08) to 500 ml of H_2O. Sterilize the solution by passing it through a 0.22 µm filter. Store at room temperature for up to 1 year.

   c. Wash the column with 4 ml of nickel-agarose column washing buffer, collecting the fractions in tubes 9–12. Place the tubes on ice until needed for analysis in steps 7–8.

7. Determine whether or not the coupling of the HMBD protein to the nickel-agarose resin was successful as follows:

   a. Transfer a 10-µl aliquot of each of the 12 fractions collected in step 6, the load sample E (set aside in step 4), and sample F from the flow-through (set aside in step 5) into separate, correspondingly labeled 1.5-ml micro-centrifuge tubes.

   b. Add 10 µl of 2x SMASH buffer to each aliquot.

   c. Heat the samples at 90°C for 90 seconds.

   d. Load each 20-µl sample (along with the appropriate molecular-weight markers) on a 15% SDS-polyacrylamide gel. Separate and detect the proteins as described on p. 255, step 6d.

Note: If the HMBD protein has been successfully coupled to the nickel-agarose resin, there should be very little HMBD protein in the flow-through or in the fractions from the washing step. The 11.4-kD HMBD protein should be *present* in the load sample E but *absent* from the other samples on the gel. If a small amount of the HMBD protein is present in sample F from the flow-through and in the fractions from the washing step, it is likely that the capacity of the nickel-agarose matrix has been exceeded. If there is little difference between load sample E and sample F, this implies that the coupling has been unsuccessful. The most likely explanation for this failure is that the nickel-agarose resin is not charged. This can be remedied by recharging the resin according to manufacturer's instructions and repeating the procedure.

8. Estimate the amount of the HMBD protein coupled to the nickel-agarose resin as follows:

   a. Pool the flow-through from step 5 and the 12 fractions eluted in the washing step in a single 50-ml tube and measure the protein concentration as described in step 1.

   b. Estimate the amount of HMBD protein coupled to the column by subtracting the total amount of protein eluted from the amount of protein loaded.

   Note: The capacity of the nickel-agarose matrix should be approximately 30 mg of HMBD per milliliter.

9. Store the column at 4°C until needed for calibration on pp. 262–264.

   Note: The HMBD protein coupled to the nickel-agarose matrix is stable at 4°C for at least 6 months.

PROTOCOL

Optimizing the Purification of Methylated DNAs on the Column

PREPARATION OF DIFFERENTIALLY METHYLATED STANDARD PLASMID DNAs

1. Linearize 5 μg of pUC19 plasmid DNA in a reaction volume of 50 μl. Use 20 units of *Eco*RI (e.g., New England Biolabs 101S) in the buffer specified by the enzyme's manufacturer and incubate at 37°C for 1 hour.

   Note: Other plasmids (e.g., pBluescript or pGEM) can be used as long as the plasmid has a known sequence and therefore a known number of CpG dinucleotides. A different restriction enzyme (e.g., *Bam*HI) can also be used, but it must have only one site in the plasmid and must create a convenient overhang for end-labeling.

2. Analzye 2 μl (200 ng) of the digested plasmid on a 1% agarose gel in 1x TAE buffer to determine whether or not digestion is complete. Also include a control lane with 200 ng of undigested plasmid for comparison and a lane with the appropriate molecular-weight markers (e.g., 1 μg of a 1-kb ladder [Life Technologies 15615-016]). Run the gel at 80 V for approximately 1 hour. Keep the remainder of the DNA on ice while the gel is running.

   Note: **Ethidium bromide** can be included in the gel at a final concentration of 0.2 μg/ml.

   ethidium bromide (see Appendix for Caution)

3. If ethidium bromide was not included in the gel, stain the DNA with ethidium bromide. View the DNA using a **UV** transilluminator. Confirm that the DNA is linearized and then proceed with step 4.

   Note: A completely digested plasmid will appear as one discrete band (for pUC19, the size of this band is 2.7 kb).

   UV radiation (see Appendix for Caution)

4. Extract the remaining digested plasmid with 1 volume of **phenol:chloroform** (1:1) and then with 1 volume of chloroform (see Appendix). Use phenol saturated with TE (pH 8.0).

   phenol, chloroform (see Appendix for Caution)

5. Precipitate the DNA in the aqueous phase with 0.1 volume of 3 M sodium acetate (pH 5.2) and 2.5 volumes of absolute ethanol and dry the pellet under vacuum (see Appendix).

6. Dissolve the DNA in 100 μl of TE (pH 7.5).

   Note: If desired, the DNA can be stored at –20°C indefinitely at this point.

7. Prepare completely methylated and partially methylated DNA samples as well as mock-methylated (i.e., unmethylated) samples.

For completely methylated plasmid DNA:

Place 40 μl (~2 μg) of the linearized plasmid DNA in each of two 1.5-ml microcentrifuge tubes labeled A and B. In tube B, set up a 100-μl reaction containing 20 units of the CpG methylase *Sss*I (New England Biolabs 226S), which methylates all CpG dinucleotides, and the buffer specified by the enzyme's manufacturer. In tube A, set up a 100-μl reaction containing the buffer but no methylase; this will be the mock-methylated sample. Incubate both reactions at 37°C for 2 hours.

For partially methylated plasmid DNA:

Place 40 μl (~2 μg) of the linearized plasmid DNA in each of two 1.5-ml microcentrifuge tubes labeled A and B. In tube B, set up a 100-μl reaction containing 20 units of *Hha*I methylase and/or 20 units of *Hpa*II methylase (New England Biolabs 217S and 214S, respectively) and the buffer specified by the manufacturer. (This combination of enzymes yields 33 methyl CpG dinucleotides.) In tube A, set up a 100-μl reaction containing the buffer but no methylase; this will be the mock-methylated sample. Incubate both reactions as specified by the manufacturer.

8. Determine whether or not the methylation in step 7 has been successful by digesting aliquots from reaction tubes A and B with a methylation-sensitive restriction enzyme. Perform the steps below separately for the pair of tubes from the partial methylation and the pair of tubes from the complete methylation in step 7.

   a. Transfer three 3-μl (~60-ng) aliquots from methylation tube A into separate 1.5-ml microcentrifuge tubes labeled A1, A2, and A3; also transfer three 3-μl aliquots from methylation tube B into tubes B1, B2, and B3. Place the remainder of each reaction mixture on ice until needed in step 9.

   b. Set up 10-μl digestions using the buffer supplied by the restriction enzyme's manufacturer. In tubes A1 and B1, set up 10-μl reactions containing the buffer but no restriction enzyme; these will be the mock-digested samples. In tubes A2 and B2, set up 10-μl reactions containing 10 units of *Hha*I (e.g., New England Biolabs 139S) and the buffer. In tubes A3 and B3, set up 10-μl reactions containing 10 units of *Hpa*II (e.g., New England Biolabs 171S) and the buffer.

      Note: Use a buffer that is suitable for maximal activity of both *Hha*I and *Hpa*II in all of the reactions.

   c. Incubate all reactions at 37°C for 1 hour.

   d. Analyze the digestion products on a 1% agarose gel as described in steps 2–3.

      Notes: The mock-methylated samples should be digested to completion with both *Hha*I and *Hpa*II in tubes A2 and A3, producing many small bands with pUC19.

The completely methylated samples should be resistant to digestion by both enzymes (tubes B2 and B3).

If *Hha*I methylase was used to partially methylate the DNA, the methylated sample should be resistant to digestion with *Hha*I (tube B2) and should be digested to completion with *Hpa*II (tube B3). If *Hpa*II methylase was used, the methylated sample should be digested to completion with *Hha*I (tube B2) and should be resistant to digestion with *Hpa*II (B3). If both methylases were used simultaneously, the methylated sample should be resistant to digestion by both enzymes (tubes B2 and B3). The mock-digested samples (A1 and B1) will be the original undigested plasmid.

9. Extract each of the remaining mock-methylated (i.e., unmethylated) and methylated samples sequentially with 1 volume of phenol, 1 volume of phenol:chloroform (1:1), and then 1 volume of chloroform. Use phenol saturated with TE (pH 8.0).

10. For each sample, precipitate the DNA in the aqueous phase with 0.1 volume of 3 M sodium acetate (pH 5.2) and 2.5 volumes of absolute ethanol and dry the pellet under vacuum (see Appendix).

11. Dissolve each DNA pellet in 100 μl of TE (pH 8.0). Determine the DNA concentration in each of the samples by measuring the OD_{260} (see Appendix) of a 1-μl aliquot.

    Note: The concentration should be 50 ng/μl.

12. For the unmethylated, methylated, and partially methylated linearized plasmid DNAs, end-label 12 μl (600 ng) of DNA using the Klenow fragment of *E. coli* DNA polymerase I and [α-^{32}P]dATP and dTTP (see Sambrook et al. 1989).

    Note: These labeled dNTPs are used for plasmids linearized with *Eco*RI because digestion with *Eco*RI creates a 4-bp 5' overhang containing only As and Ts. If a different restriction enzyme was used, choose the appropriate labeled and unlabeled dNTPs for the end-labeling reaction on the basis of the composition of the overhang.

    radioactive substances (see Appendix for Caution)

13. Precipitate the DNAs with 0.1 volume of 3 M sodium acetate (pH 5.2) and 2–2.5 volumes of absolute ethanl and dry the DNA. In this case, wash the pellets twice with 70% ethanol.

14. Dissolve each DNA pellet in 0.6 ml of HMBD column buffer containing the appropriate NaCl concentration (see below). (The DNA concentration will be approximately 1 ng/μl.) Store at 4°C until needed for calibrating the column and for determining the appropriate NaCl concentration for binding only methylated DNA to the column.

    Note: This amount of DNA is sufficient to run six columns (100 μl/column). The efficiency of the labeling reaction can be estimated by detecting the cpm in the labeled DNA with a handheld Geiger counter.

CALIBRATION OF THE HMBD COLUMN USING THE END-LABELED PLASMID DNAs

Always use the HMBD column in a coldroom with ice-cold solutions.

1. Prepare HMDB column buffer containing no salt and 1 M NaCl. If an FPLC system is available, program the machine to mix 0.1 M NaCl/HMBD column buffer with 1 M NaCl/HMBD column buffer to obtain the desired NaCl concentration during the elution of the column in the steps below. For gravity flow, prepare a series of buffers (50 ml of each) by mixing 0.1 M NaCl/HMBD column buffer and 1 M NaCl/HMBD column buffer in the correct proportions to obtain buffers with NaCl concentrations varying by 0.1 M NaCl.

   Notes: To fractionate differentially methylated genomic DNAs, DNA is loaded on the HMBD column at a low salt concentration and then eluted using a salt gradient. The salt concentration at which DNAs with different degrees of methylation will elute varies with the amount of HMBD protein on the column. Therefore, each column should be calibrated using artificially methylated plasmid DNAs that contain known numbers of methyl CpG dinucleotides (Cross et al. 1994).

   Ideally, an apparatus such as the FPLC system (Pharmacia 18-1035-00) or GradiFrac system (Pharmacia 8-1993-01) should be used to run the column. Use of an FPLC or GradiFrac system allows easy control of flow rate and the salt gradients.

   If an FPLC or GradiFrac system is not available, the column can be run under gravity flow, and the protocols should be modified such that step salt gradients are used instead of the continuous (linear) gradients specified. Good separation can be obtained by increasing the NaCl concentration in steps of 0.1 M during the elution (S.H. Cross, unpubl.).

   HMBD column buffer

Component and final concentration	Amount to add per 0.5 liter
0 M or 1 M NaCl	0 ml or 100 ml (respectively) of 5 M
20 mM HEPES	10 ml of 1 M (pH 7.9)
10% glycerol	50 ml of 100%
0.1% Triton X-100	2.5 ml of 20%
0.5 mM **PMSF**	2.5 ml of 100 mM
H$_2$O	to make 0.5 liter

   Just before use, prepare HMBD column buffer with the appropriate NaCl concentration and place at 4°C. (Buffers are denoted 0.1 M NaCl/HMBD column buffer, 1 M NaCl/HMBD column buffer, and so on throughout the chapter.)

   PMSF (see Appendix for Caution)

2. Pipette 1 ml of the HMBD matrix into a 1-ml column.

   Note: For an FPLC system, use an HR 5/5 column (Pharmacia 18-0382-01) as specified by the manufacturer. For gravity flow, use a 1-ml plastic disposable column or a column from the protocol on pp. 256–258.

3. Equilibrate the column by washing it sequentially with 5 column volumes (5 ml) of 0.1 M NaCl/HMBD column buffer, 5 column volumes of 1 M NaCl/HMBD column buffer, and then 5 column volumes of 0.1 M NaCl/HMBD column buffer.

Notes: If an FPLC or GradiFrac system is being used, elute the column at 1 ml/minute throughout this protocol. Use gravity flow if an FPLC system is not available.

Do not allow the HMBD column to dry out! If an FPLC system is being used, be careful not to pump air over the column. For gravity flow, be careful to leave buffer on top of the column.

4. Mix 100 μl (100 ng) each of the **end-labeled, unmethylated plasmid DNA** with 100 μl (100 ng) of the **end-labeled, completely methylated plasmid DNA** (from p. 261, step 14) in a 1.5-ml microcentrifuge tube.

   radioactive substances (see Appendix for Caution)

5. Load the entire DNA mixture on the HMBD column. Wash the column with 4.8 ml of 0.1 M NaCl/HMBD column buffer and collect five 1-ml fractions in 5-ml tubes or 1.5-ml microcentrifuge tubes.

6. To elute the DNA, wash the column sequentially with 5 ml of 0.4 M NaCl/HMBD column buffer, 40 ml of a linear salt gradient in which the NaCL concentration in HMBD column buffer increases from 0.4 M to 1 M, and then 5 ml of 1 M NaCl/HMBD column buffer. Collect 50 1-ml fractions in 5-ml tubes or 1.5-ml microcentrifuge tubes.

   Note: Typically, heavily methylated DNA fragments elute between 0.7 and 0.9 M NaCl.

7. After use, reequilibrate the HMBD column with low-salt buffer by washing it with 5 column volumes (5 ml) of 0.1 M NaCl/HMBD column buffer. Store at 4°C or in a coldroom.

   Note: Reequilibrated HMBD columns are stable for at least 6 months.

8. Determine the radioactivity in each of the 55 1-ml fractions in a scintillation counter.

   Notes: The radioactivity should elute from the column in two peaks during the linear gradient part of the elution. Typically, the first peak (unmethylated DNA) elutes at approximately 0.5–0.6 M NaCl and the second peak elutes at 0.8 M NaCl.

   If the fractions were collected in 5-ml tubes, they may have to be transferred into 1.5-ml microcentrifuge tubes so that they fit into the vials used in the scintillation counter. No scintillation fluid is needed.

9. Transfer a 400-μl aliquot of DNA from each of the two peak fractions into 1.5-ml microcentrifuge tubes. Precipitate with 0.1 volume of 3 M sodium acetate (pH 5.2) and 2–2.5 volumes of absolute ethanol and dry the DNA pellets under vacuum (see Appendix).

   Note: It may be necessary to pool fractions from each peak before precipitation of the 400-μl aliquot.

10. Dissolve each DNA pellet in 10 μl of TE (pH 8.0). Store at −20°C until needed in step 11.

11. Assess the methylation status of the two end-labeled DNA aliquots as described on pp. 260–261, step 8. After running the analytical gel, transfer it onto Whatman DE81 paper, place on two pieces of Whatman 3MM paper, cover with plastic wrap, and dry. Autoradiograph at −80°C overnight using an intensifying screen to view the end-labeled DNA fragments.

Note: The DNA in the first peak should be digested with both *Hpa*II and *Hha*I, showing that it is unmethylated; the DNA in the second peak should be resistant to digestion with both enzymes, showing that it is completely methylated.

DETERMINING THE HIGHEST NaCl CONCENTRATION AT WHICH METHYLATED DNA BINDS TO THE HMBD COLUMN AND ALL UNMETHYLATED DNA ELUTES

Always use the HMBD column in a coldroom with ice-cold solutions.

1. Equilibrate the column by washing it sequentially with 5 column volumes (5 ml) of 0.5 M NaCl/HMBD column buffer, 5 column volumes of 1 M NaCl/HMBD column buffer, and then 5 column volumes of 0.5 M NaCl/HMBD column buffer (for buffer preparation, see p. 262).

   Notes: If an FPLC system is being used, elute the column at 1 ml/minute throughout this protocol. Use gravity flow if an FPLC system is not available.
   Do not allow the HMBD column to dry out! If an FPLC system is being used, be careful not to pump air over the column. For gravity flow, be careful to leave buffer on top of the column.

2. Mix 100 µl of 0.5 M NaCl/HMBD column buffer with 100 µl (100 ng) of the **end-labeled, unmethylated plasmid DNA** (from p. 261, step 14).

   radioactive substances (see Appendix for Caution)

3. Load the entire mixture on the column. Wash the column with 9.8 ml of 0.5 M NaCl/HMBD column buffer and then with 10 ml of 1 M NaCl/HMBD column buffer. Collect 20 1-ml fractions in 5-ml tubes or 1.5-ml microcentrifuge tubes.

   Note: The flow-through will be collected as part of the first fraction.

4. Determine the radioactivity in each of the 20 1-ml fractions in a scintillation counter.

   Notes: All of the radioactivity should elute from the column in the first ten fractions.
   If the fractions were collected in 5-ml tubes, they may have to be transferred into 1.5-ml microcentrifuge tubes so that they fit into the vials used in the scintillation counter. No scintillation fluid is needed.

5. Reequilibrate the column by washing it with 10 ml of 0.5 M NaCl/HMBD column buffer.

6. Repeat steps 1–5 using end-labeled, partially methylated plasmid DNA (e.g., pUC19 methylated with a combination of *Hpa*II and *Hha*I methylases so that 33 CpGs are methylated; see pp. 259–261) instead of unmethylated plasmid DNA.

   Note: With the partially methylated plasmid DNA, the radioactivity should elute from the column in fractions 11–20.

7. If the unmethylated plasmid DNA elutes from the column in fractions 11–20, repeat steps 1–6 until no unmethylated DNA is found in fractions 11–20, by increasing the NaCl concentration of the starting buffer in incre-

ments of 0.05 M (i.e., substitute 0.55 M NaCl and then 0.6 M NaCl for the 0.5 M NaCl/HMBD column buffer in steps 1–3) each time the steps are repeated. In step 5, reequilibrate the column for each round of steps by substituting HMBD column buffer containing the NaCl concentration that will be used to load the DNA in step 2.

Note: This procedure will determine the highest NaCl concentration at which the unmethylated DNA does not bind to the column, but the partially methylated DNA does (i.e., the NaCl concentration at which the unmethylated DNA elutes in the loading buffer without binding to the column but the partially methylated DNA binds and then elutes only with the buffer containing 1 M NaCl). HMBD column buffer (see p. 262) containing NaCl at the highest appropriate concentration for loading is subsequently designated UM (unmethylated) buffer. It may be necessary to use up to three different NaCl concentrations in steps 1–5 to determine the appropriate concentration empirically.

8. After use, reequilibrate the HMBD column with low-salt buffer by washing it with 5 column volumes (5 ml) of 0.1 M NaCl/HMBD column buffer. Store at 4°C or in a coldroom.

Note: Reequilibrated HMBD columns are stable for at least 6 months.

PROTOCOL

Preparing the CpG Island Fraction from Genomic DNA Using the HMBD Column

PURIFICATION AND DIGESTION OF GENOMIC DNA

1. Prepare genomic DNA from **blood** or **tissue** samples using standard procedures (see Wolff and Gemmill 1997). Measure the DNA concentration (see Appendix). Adjust the concentration to 1–10 mg/ml with TE 9 (pH 8.0).

   human blood, blood products, and tissues (see Appendix for Caution)

2. In a 1.5-ml microcentrifuge tube, digest 100–200 μg of genomic DNA in a reaction volume of 400 μl. For 10 μg of DNA, use 2 units of *Mse*I in the buffer specified by the enzyme's manufacturer. Set up a test reaction for monitoring the digestion by transferring 10 μl of the digestion mixture into another 1.5-ml microcentrifuge tube and adding 200 ng of linearized plasmid DNA (e.g., *Eco*RI-digested pUC19 DNA [2.7 kb]) dissolved in 1 μl of TE (pH 8.0). Incubate both reactions at 37°C overnight.

   Notes: The test reaction for monitoring the digestion contains 2.5–5 μg of genomic DNA and 200 ng of plasmid DNA.
   For the construction of a genomic CpG island library, 100–200 μg of DNA is sufficient.

3. Analyze 10 μl of the digested genomic DNA and the entire sample from the test reaction on a 1.5% agarose gel in 1x TAE buffer to determine whether or not digestion is complete. Also include a control lane with 200 ng of undigested plasmid for comparison and a lane with the appropriate molecular-weight markers (e.g., 1 μg of a 1-kb ladder [Life Technologies 15615-016]). Run the gel at 80 V for approximately 1 hour. Keep the remainder of the digested genomic DNA on ice while the gel is running.

4. If **ethidium bromide** was not included in the gel, stain the DNA with ethidium bromide. View the DNA using a **UV** transilluminator. Confirm that the DNA is linearized and then proceed with step 5.

   Notes: If the digestion is complete, none of the intact linearized plasmid DNA in the test reaction will be visible. This is because pUC19 contains 14 sites for *Mse*I and is therefore highly fragmented after digestion with *Mse*I. The sample of digested genomic DNA should appear as a smear.
   If intact linearized plasmid DNA from the test reaction is visible, digestion is not complete. In this case, add more enzyme (1 unit per 10 μg of DNA) to the genomic DNA digestion mixture and mix thoroughly by gently flicking the tube. Set up a test reaction for monitoring the digestion as described in step 2. Incubate the two reactions at 37°C for an additional 3 hours and reassess for the completeness of digestion on a 1.5% agarose gel.

   ethidium bromide, UV radiation (see Appendix for Caution)

5. Extract the *Mse*I-digested genomic DNA with 1 volume of **phenol:chloroform** (1:1) and then with 1 volume of chloroform (see Appendix). Use phenol saturated with TE (pH 8.0).

   phenol, chloroform (see Appendix for Caution)

6. Precipitate the DNA in the aqueous phase with 0.1 volume of 3 M sodium acetate (pH 5.2) and 2.5 volumes of absolute ethanol and dry the pellet under vacuum (see Appendix).

7. Dissolve the DNA in 250 µl of UM buffer (for preparation, see p. 265, step 7, Note). Store until needed for the stripping steps below.

   Note: The DNA can be stored at –20°C indefinitely.

REMOVAL OF HEAVILY METHYLATED SEQUENCES (STRIPPING STEPS)

Always use the HMBD column in a coldroom with ice-cold solutions.

1. Equilibrate the column by washing it sequentially with 5 column volumes (5 ml) of UM buffer, 5 column volumes of 1 M NaCl/HMBD column buffer, and then 5 column volumes of UM buffer (for buffer preparations, see p. 262 and p. 265, step 7, Note).

   Notes: If an FPLC system is being used, elute the column at 0.5 ml/minute throughout this protocol unless otherwise stated. Use gravity flow if an FPLC system is not available.
      Do not allow the HMBD column to dry out! If an FPLC system is being used, be careful not to pump air over the column. For gravity flow, be careful to leave buffer on top of the column.

2. Load 200 µl of the *Mse*I-digested genomic DNA dissolved in UM buffer (from the protocol above) on the HMBD column. Wash the column with 3.8 ml of UM buffer and collect the eluate in a 15-ml plastic tube labeled S1-U. Set aside the remaining genomic DNA for evaluation on pp. 272–273.

   Notes: The *Mse*I-digested DNA in UM buffer is passed over the HMBD column to remove (strip) any *Mse*I-generated fragments that contain a cluster of methylated CpG dinucleotides and would thus contaminate the final library. For efficient removal of such heavily methylated fragments, three rounds of column chromatography are necessary (see Table 5 and Figure 5). For each round of stripping (S1, S2, and S3), aliquots of both the unbound (U) and bound (B) fractions are set aside for evaluation on pp. 272–273. All fractions and aliquots set aside during this procedure should be kept at 4°C until needed or at –20°C if they must be stored overnight.
      This S1-U fraction should be enriched for unmethylated DNA.

3. Wash the HMBD column with 5 ml of 1 M NaCl/HMBD column buffer and collect the eluate in a 15-ml plastic tube labeled S1-B.

   Note: The S1-B fraction consists primarily of methylated DNA.

4. Set aside 200-µl aliquots of S1-U and S1-B in separate labeled 1.5-ml microcentrifuge tubes until needed for evaluation on pp. 272–273.

5. Reequilibrate the HMBD column by washing it with 5 ml of 1 M NaCl/ HMBD column buffer and then with 10 ml of UM buffer. If an FPLC system is being used, elute the column at 1 ml/minute.

6. Load the 3.8-ml S1-U fraction on the HMBD column. Wash the HMBD column with 3.2 ml of UM buffer and collect the eluate in a 15-ml plastic tube labeled S2-U.

   Note: The S2-U fraction should be highly enriched for unmethylated DNA.

7. Wash the HMBD column with 1 M NaCl/HMBD column buffer and collect the eluate in a 15-ml plastic tube labeled S2-B.

   Note: The S2-B fraction consists almost exclusively of methylated DNA.

8. Set aside 200-μl aliquots of S2-U and S2-B in separate labeled 1.5-ml micro-centrifuge tubes until needed for evaluation on pp. 272–273.

9. Reequilibrate the HMBD column as described in step 5.

10. Load the 6.8-ml S2-U fraction on the HMBD column. Wash the HMBD column with 3.2 ml of UM buffer and collect the eluate in a 30-ml Corex tube labeled S3-U.

    Note: The S3-U fraction should contain highly purified unmethylated DNA.

11. Wash the HMBD column with 5 ml of 0.1 M NaCl/HMBD column buffer and collect the eluate in a 15-ml plastic tube labeled S3-B.

    Note: The S3-B fraction consists of residual methylated DNA.

12. Set aside 200-μl aliquots of S3-U and S3-B in separate labeled 1.5-ml micro-centrifuge tubes until needed for evaluation on pp. 272–273.

13. Precipitate the DNA in the S3-U fraction as follows:

    a. Add 1 μl of glycogen solution (20 mg/ml) as a carrier and 2.5 volumes of ice-cold absolute ethanol to the S3-U fraction and mix by inverting the tube five or six times. Place at –20°C overnight.

    b. Centrifuge in a JS-13.1 rotor in a Beckman J2-21 centrifuge (or equivalent) at 8000 rpm (10,000*g*) at 4°C for 30 minutes. Discard the supernatant.

    c. Wash the DNA pellet by adding 1 ml of 70% ethanol, recentrifuging for 15 minutes, and discarding the supernatant.

    d. Air dry the DNA pellet.

14. Dissolve the DNA in 100 μl of TE (pH 8.0) and transfer it into a 1.5-ml microcentrifuge tube. Store the DNA at –20°C until needed for the methylation procedure below.

15. After use, reequilibrate the HMBD column with low-salt buffer by washing it with 5 column volumes (5 ml) of 0.1 M NaCl/HMBD column buffer. Store at 4°C or in a coldroom.

    Note: Reequilibrated HMBD columns are stable for at least 6 months.

METHYLATION OF CpG DINUCLEOTIDES IN THE FINAL STRIPPED FRACTION

1. Set up a 400-μl methylation for all of the S3-U fraction (from p. 268, step 14) using 10–20 units of CpG methylase per microgram of DNA (*SssI*; New England Biolabs 226S) as specified by the manufacturer. Mix thoroughly by gently flicking the tube. Set up a test reaction for monitoring the methylation by transferring 10 μl of the methylation mixture into 1 μl of TE (pH 8.0) containing 200 ng of linearized plasmid DNA (e.g., *Eco*RI-digested pUC19 DNA [2.7 kb]) in a 1.5-ml microcentrifuge tube. Incubate both reactions at 37°C overnight.

    Notes: This procedure methylates all of the nonmethylated CpG dinucleotides in the DNA of the S3-U fraction.
    Keep the methylated S3-U fraction on ice during step 2.

2. Determine whether or not the plasmid DNA has been methylated to completion (which also indicates that the genomic DNA has been methylated to completion) as follows:

    a. Transfer three 3-μl aliquots from the test reaction mixture containing plasmid DNA into separate 1.5-ml microcentrifuge tubes.

    b. Set up the following 10-μl digestions as specified by the restriction enzyme's manufacturer: (1) a mock digestion containing plasmid DNA and restriction enzyme buffer but no restriction enzyme; (2) a digestion containing DNA, buffer, and *Hpa*II; and (3) a digestion containing DNA, buffer, and *Msp*I. Use 10 units of restriction enzyme per microgram of DNA.

       Note: Use a buffer that is suitable for both *Msp*I and *Hpa*II in all three reactions.

    c. Incubate all reactions at 37°C for 1 hour.

    d. Analyze the digestion products on a 1% agarose gel as described on p. 259, steps 2–3. If the methylation is complete, proceed with step 4; otherwise, proceed with step 3.

       Notes: *Hpa*II and *Msp*I recognize the same sequence (CCGG). However, digestion with *Hpa*II is blocked if the recognition site is methylated, but digestion with *Msp*I is not. Methylation is therefore complete if there is no cleavage of the DNA in the digestion with *Hpa*II (i.e., the DNA band is the same size and intensity as that in the mock digestion). The methylation is not complete if any cleavage is noted in that lane.
       Large amounts of genomic DNA often require at least two rounds of methylation.

3. If methylation is not complete, perform the following steps with the remainder of the DNA from step 1:

a. Extract sequentially with 1 volume of **phenol**, 1 volume of phenol:**chloroform** (1:1), and then 1 volume of chloroform (see Appendix). Use phenol saturated with TE (pH 8.0).

   phenol, chloroform (see Appendix for Caution)

b. Precipitate the DNA in the aqueous phase with 0.1 volume of 3 M sodium acetate (pH 5.2) and 2.5 volumes of absolute ethanol and dry the pellet under vacuum (see Appendix).

c. Dissolve the DNA in 100 μl of TE (pH 8.0).

d. Repeat steps 1 and 2.

4. When the methylation is complete, extract and precipitate the DNA as described in steps 3a–b and dissolve the DNA in 0.5 ml of UM buffer (for preparation, see p. 265, step 7, Note). Store the sample at –20°C until needed for the binding steps below.

   Note: The DNA can be stored at –20°C indefinitely.

PURIFICATION OF CpG ISLANDS FROM THE METHYLATED DNA OF THE FINAL STRIPPED FRACTION (BINDING STEPS)

Always use the HMBD column in a coldroom with ice-cold solutions.

1. Equilibrate the column by washing it sequentially with 5 column volumes (5 ml) of UM buffer, 5 column volumes of 1 M NaCl/HMBD column buffer, and then 5 column volumes of UM buffer (for buffer preparations, see p. 262 and p. 265, step 7, Note).

   Notes: If an FPLC system is being used, elute the column at 0.5 ml/minute throughout this protocol unless otherwise stated. Use gravity flow if an FPLC system is not available.
   Do not allow the HMBD column to dry out! If an FPLC system is being used, be careful not to pump air over the column. For gravity flow, be careful to leave buffer on top of the column.

2. Load the 0.5-sample of methylated S3-U DNA (from the protocol above) on the HMBD column. Collect the flow-through in a 1.5-ml tube.

   Note: To obtain a highly purified CpG island fraction, three rounds (in addition to the equilibration step) of column chromatography are necessary (see Table 5 and Figure 5). For each round of binding (B1, B2, and B3), aliquots of both the unbound (U) and bound (B) fractions are set aside for evaluation on pp. 272–273. All fractions and aliquots set aside during the procedure should be kept at 4°C until needed or at –20°C if they must be stored overnight. Samples stored at –20°C are stable indefinitely.

3. Reload the flow-through from step 2 and wash the column sequentially with 9.5 ml of UM buffer, 40 ml of a linear salt gradient in which the NaCl concentration in HMBD column buffer increases from its empirically determined concentration in UM buffer to 1 M, and then 6 ml of 1 M NaCl/HMBD column buffer. Collect the first 10 ml in a single 15-ml plastic tube labeled B1-U. Set aside a 200-μl aliquot in a 1.5-ml microcentrifuge

tube (labeled B1-U) until needed for evaluation on pp. 272–273 and then collect 23 2-ml fractions in 5-ml plastic tubes labeled B1-1–23.

Notes: Either plot a graph of fraction number versus expected NaCl concentration or use a chart recorder attached to the FPLC system to monitor the salt concentration. Determining the appropriate NaCl concentration for eluting heavily methylated DNA is highly dependent on the accuracy of the calibration.

If an FPLC system is not available, the column can be run under gravity flow, and the protocol should be modified such that step salt gradients are used instead of the continuous (linear) gradients specified. Good separation can be obtained by increasing the NaCl concentration in steps of 0.1 M during the elution (S.H. Cross, unpubl.).

4. Pool the fractions that elute at the approximate NaCl concentration at which heavily methylated DNA elutes from the HMBD column in a single 50-ml plastic tube labeled B1-B. Set aside a 200-µl aliquot in a 1.5-ml microcentrifuge tube (labeled B1-B) until needed for evaluation on pp. 272–273.

Notes: The appropriate NaCl concentration for eluting heavily methylated DNA (see step 6, p. 263) is usually 0.7–0.9 M (to determine the appropriate concentration, see pp. 262–264).
Typically, 3–5 fractions are pooled (6–10 ml).

5. Dilute the remainder of the pooled B1-B fractions with a volume of HMBD column buffer without NaCl that reduces the final NaCl concentration to approximately that of UM buffer (~0.5 M).

Note: For example, if 8 ml of DNA eluted at an NaCl concentration of 0.8 M, add 6 ml of HMBD column buffer.

6. Reequilibrate the HMBD column by washing it with 5 ml of 1 M NaCl/ HMBD column buffer and then with 10 ml of UM buffer. If an FPLC system is being used, elute the column at 1 ml/minute.

7. Load the diluted B1-B fractions from step 5 on the HMBD column. Wash the column with 10 ml of UM buffer and collect both the flow-through and the wash (~22–30-ml) in a single 50-ml plastic tube labeled B2-U. Set aside a 200-µl aliquot in a 1.5-ml microcentrifuge tube (labeled B2-U) until needed for evaluation on pp. 272–273.

8. Wash the HMBD column with 40 ml of a linear salt gradient in which the NaCl concentration in HMBD column buffer increases from its empirically determined concentration in UM buffer to 1 M (i.e., 1 M NaCl/HMBD column buffer) and then with 6 ml of 1 M NaCl/HMBD column buffer. Collect 23 2-ml fractions in 5-ml plastic tubes labeled B2-1–23.

9. Pool the heavily methylated DNA fractions in a single 50-ml plastic tube labeled B2-B as described in step 4. Set aside a 200-µl aliquot in a 1.5-ml microcentrifuge tube (labeled B2-B) until needed for evaluation on pp. 272–273.

Note: These remaining fractions are set aside to evaluate in case the results of the evaluation on pp. 272–273 are not conclusive.

10. Dilute the remainder of the pooled B2-B fractions as described in step 5.

11. Reequilibrate the HMBD column as described in step 6.

12. Load the diluted B2-B fractions from step 10 on the HMBD column and repeat steps 7–8. Label the tube containing the flow-through and wash B3-U. Label the 23 2-ml fractions B3-1–23. Set aside a 200-µl aliquot in a 1.5-ml microcentrifuge tube (labeled B3-U) until needed for evaluation on pp. 272–273.

13. Pool the heavily methylated DNA fractions in a single 50-ml plastic tube labeled B3-B as described in step 4. Set aside a 200-µl aliquot in a 1.5-ml microcentrifuge tube (labeled B3-B) until needed for evaluation on pp. 272–273.

    Notes: The B3-B fractions should contain CpG island DNA.
       These remaining fractions are set aside to evaluate in case the results of the evaluation on pp. 272–273 are not conclusive.

14. After use, reequilibrate the HMBD column with low-salt buffer by washing it with 5 column volumes (5 ml) of 0.1 M NaCl/HMBD column buffer. Store at 4°C or in a coldroom.

    Note: Reequilibrated HMBD columns are stable for at least 6 months.

EVALUATION OF THE STRIPPING AND BINDING STEPS

1. Thaw the 200-µl aliquots of unbound (U) and bound (B) DNA set aside from the S1, S2, S3, B1, B2, and B3 fractions on pp. 267–279 and pp. 270–272 and place on ice.

2. Precipitate the DNA in each sample with 0.1 volume of sodium acetate (pH 5.2) and 2–2.5 volumes of absolute ethanol and dry the pellets (see Appendix).

3. Dissolve each DNA pellet in 20 µl of TE (pH 8.0).

    Note: These DNAs and the genomic DNA originally loaded on the column (set aside in step 2, p. 267) provide the PCR templates for the different test sequences.

4. Consult the sequence database and select suitable PCR primer pairs within *Mse*I-generated fragments from a CpG island, a non-CpG-island genomic DNA fragment, and, if possible, a fragment that contains a cluster of methylated CpG dinucleotides.

    Notes: The best way to design primer pairs is to use a primer design program (e.g., PRIMER, which is available at http://www-genome.wi.mit.edu/cgi-bin/primer2.2/primer_front).
       The primers below were used by Cross et al. (1994) for human DNA.
    Monoamine oxidase A CpG island (monoamine oxidase A gene):
       5′CGGGTATCAGATTGAAACAT3′
       5′CTCTAAGCATGGCTACACTACA3′

Fragment containing a cluster of methylated CpG dinucleotides (3′end apolipoprotein A-IV gene):

    5′GGAGAAGTGAACACTTACGC3′

    5′TTTGAATTCGTCAGCGTAG3′

Non-CpG-island (aldolase B gene):

    5′TCATTGCTTGCTTTCTCAAGCAGGG3′

    5′CAATGCTTCTCCGTGTTGGAAAGTC3′

5. Perform PCR assays on each of the templates for each of the three pairs of primers. Use 1 μl of each of the DNA samples in step 3 as the template for a 50-μl reaction. Set up blank reactions.

   Note: For general guidelines and standard PCR protocols, see Innis et al. (1990); see also Fanning and Gibbs (1997).

6. Analyze 10% of each PCR product on a 1.5% agarose gel as described on p. 259, steps 2–3.

   Note: There is no need to run the control lane with 200 ng of undigested plasmid in this step.

7. Evaluate the efficiency of the stripping and binding steps using the expected results in Table 5 (p. 250).

   Note: If the results are not as expected, evaluate all additional fractions set aside during the earlier steps of the protocol.

PROTOCOL

Cloning of CpG Island Fragments

1. Transfer the B3-B fraction (~6–10 ml from p. 272, step 13), which contains the CpG island DNA, into a 30-ml Corex tube and precipitate the DNA as follows:

   a. Add 1 µl of glycogen solution (20 mg/ml) as a carrier and 2.5 volumes of ice-cold absolute ethanol and mix by inverting the tube five or six times. Place at –20°C overnight.

   b. Centrifuge in a JS-13.1 rotor in a Beckman J2-21 centrifuge (or equivalent) at 8000 rpm (10,000g) at 4°C for 30 minutes to collect the DNA. Discard the supernatant.

   c. Wash the DNA pellet by adding 0.8 ml of 70% ethanol, recentrifuging for 15 minutes, and discarding the supernatant.

   d. Air dry the pellet.

2. Dissolve the CpG island DNA in 100 µl of TE (pH 8.0) and transfer it into a 1.5-ml microcentrifuge tube.

   Note: The DNA can be stored at –20°C until needed below.

3. Digest 20 µg of pGEM-5Zf(–) or pGEM-5Zf(+) (Promega) vector to completion in a reaction volume of 100 µl. For each microgram of DNA, use 10 units of *Nde*I in the buffer specified by the enzyme's manufacturer and incubate at 37°C for 90 minutes.

   Note: An easy way to clone the *Mse*I-generated fragments in the B3-B CpG island fraction is to clone them into the *Nde*I site of plasmid vector pGEM-5Zf(–) or pGEM-5Zf(+) (Promega). *Mse*I and *Nde*I each leaves a 5′TA overhang and are thus compatible for ligation.

4. Purify the digested vector as described on p. 269, steps 2–4. Dissolve the purified vector DNA in TE (pH 7.6) and adjust the DNA concentration to 0.1 µg/µl.

5. Reduce the possibility of background plasmid self-ligation as follows:

   a. Dephosphorylate the linearized vector with CIP according to the manufacturer's instructions.

   b. Extract sequentially with 1 volume of **phenol**, 1 volume of phenol:**chloroform** (1:1), and then 1 volume of chloroform. Use phenol saturated with TE (pH 8.0).

   phenol, chloroform (see Appendix for Caution)

   c. Precipitate the vector DNA with 0.1 volume of 3 M sodium acetate (pH 5.2) and 2–2.5 volumes of absolute ethanol and dry the pellet (see Appendix).

6. Dissolve the vector DNA in TE (pH 8.0) to make a final concentration of 0.1 μg/μl.

7. Ligate one fifth of the CpG island DNA to 0.1 μg of the vector DNA and transform SURE cells (Stratagene) with the DNA as described on p. 230, step 9.

   Note: The bacterial strain chosen for transformation should be one that does not restrict (digest) methylated DNA. A strain such as SURE (Stratagene) is a good choice because, in addition to being methylation-tolerant, it has been shown to produce high yields of plasmid DNA of good quality for sequencing (Taylor et al. 1993).

8. Generate inserts by PCR and analyze the clones in the CpG island library as described on pp. 276–283.

DATA ANALYSIS FOR IDENTIFICATION OF GENES ASSOCIATED WITH CpG ISLANDS FROM UNCLONED GENOMIC DNA

To confirm that the library consists of DNA from bone fide CpG islands, a number of clones should be tested to ascertain whether or not their sequence composition and methylation status in genomic DNA are like those of CpG islands. The sequence of the clones should have a G + C content of greater than 50% and should contain approximately the expected number of CpG dinucleotides (see p. 218). Such clones should also be derived from unmethylated genomic sequences. Analysis of 20 clones is usually sufficient to give a good idea of the validity/accuracy of the designation as a CpG island library. Analyzing the CpG island prepared from uncloned genomic DNA requires the following steps:

- Generating inserts by PCR and estimating the sizes of inserts
- Digesting the inserts with an enzyme that cuts in most CpG islands
- Performing hybridization analysis to map to native genomic DNA and to check the methylation status in the genome
- Performing hybridization analysis to check for the presence of repeat sequences and ribosomal DNA
- Sequencing to test for the presence of a CpG-island-associated transcript or other known sequence (e.g., ribosomal DNA sequence)

Each of these steps is discussed below.

Generation of Inserts by PCR and Estimation of Insert Size

If blue/white selection is being used, 20 white colonies should be selected from the transformation plates for analysis. Since the inserts cannot be excised from the vector because the cloning site is destroyed during cloning, PCR is used to amplify the inserts directly from the bacterial colonies using primers that flank the cloning site for *Nde*I (Cross et al. 1994). Purified DNA from minipreparations could be used for PCR, but it is generally not necessary. A protocol for generating the inserts and estimating their sizes is provided on p. 279.

Analysis by Digestion with Restriction Enzymes

The amplified inserts are next digested with *Bst*UI, which has the recognition sequence CGCG. This sequence occurs approximately once per 100 bp in CpG island DNA and approximately once per 10 kb in non-CpG-island DNA. If a clone contains a *Bst*UI site, this is a good indication that it is derived from a CpG island; more than 75% of inserts from a CpG island library are expected to contain *Bst*UI sites. Therefore, testing clones for *Bst*UI sites is an easy and reliable way of quickly judging if a library consists of clones containing CpG islands. A protocol for this type of analysis is provided on p. 280.

Hybridization Analysis to Test the Methylation Status in the Genome and to Map to Native Genomic DNA

Hybridization analyses should be performed to confirm that clones are derived from the correct species and from unmethylated parts of the genome. Up to five clones should be selected and the inserts used to prepare probes. The probes are then hybridized to Southern blots of genomic DNA that has been digested with *Mse*I alone and with *Mse*I plus a methylation-sensitive restriction enzyme such as *Bst*UI or *Hap*II (for Southern blotting, see Wolff and Gemmill 1997). If, by chance, a clone contains repeated DNA, it may detect a smear and the results will be difficult to interpret; if this happens, another clone should be selected. If the clone is derived from a CpG island, the *Mse*I-generated fragments should be cleaved by the methylation-sensitive enzymes. A protocol for this type of analysis is provided on p. 281.

Hybridization Analysis to Test for the Presence of Repeat Sequences and Ribosomal DNA

One characteristic of CpG island libraries is that a small proportion of the clones (~10%) will contain highly repeated sequences (Cross et al. 1994). It has been estimated that *Alu* repeats account for up to 10% of the human genome. The majority of the clones with similar insert sizes in libraries made from human genomic DNA would therefore be expected to contain repeated sequences (Tashima et al. 1981).

If the starting DNA contained ribosomal sequences, the CpG island library would be expected to contain these sequences because part of the ribosomal DNA repeat has the same sequence characteristics as a CpG island and is unmethylated in genomic DNA (Bird and Taggart 1980; Brock and Bird 1997). Therefore, ribosomal DNA will copurify with CpG island DNA and will contaminate the final CpG island fraction (therefore, all good CpG island libraries will have measurable ribosomal DNA content). Approximately 10% of the clones in the human CpG island library of Cross et al. (1994) contain ribosomal DNA sequences.

To ascertain the proportion of the library that contains these repeat sequences or ribosomal DNA, use genomic DNA probes, *Alu* repeat probes (e.g., Blur8 [Deininger et al. 1981]), and ribosomal DNA probes (as used by Cross et al. 1994) for hybridization. This type of analysis is best done by performing hybridization analysis on gridded clones instead of just using the original 20 clones (using one 96-well plate is recommended). Nylon membranes of gridded clones can be produced as described in Cross et al. (1994). A protocol for this type of analysis is provided on p. 282.

Sequence Analysis

In the protocol on p. 283, the 20 PCR-amplified cloned inserts should be sequenced completely to determine whether or not they have the characteristics of CpG islands. One easy way to visualize these data is to plot a graph with base composition on the X axis and CpG observed/expected on the Y axis. If the fractionation of the CpG island DNA from the rest of the genome has been successful, 80% or more of the clones will have the sequence characteristics of CpG islands (see, e.g., Cross et al. 1994).

Sequence databases should be searched as outlined on pp. 236–238 to ascertain whether or not known CpG islands are present and whether or not the clones are derived from the species used to make the library (i.e., that there is no major contamination from bacterial sources, for example). Analysis of this kind was performed for 18 clones from a human CpG island library (Cross et al. 1994). In this case, five strong matches were found: three matches to the 5′ ends of genes, one to human mitochondrial DNA, and one to the *Alu* repeat. Mitochondrial clones were present in the library because it was made from whole blood DNA, which contains mitochondrial DNA (which is unmethylated and GC-rich like CpG island DNA). It is likely that if such a search were performed today, a much higher proportion of clones would have matches in the databases because of the explosion in the amount of EST information available over the past few years.

PROTOCOL

Amplifying Cloned Inserts by PCR and Estimating Insert Size from PCR-amplified Inserts

1. For each bacterial colony from pp. 274–275 to be analyzed, use a sterile toothpick to streak an appropriate selective plate with the colony and then mix the remainder of the colony with 100 µl of sterile H_2O in a 1.5-ml microcentrifuge tube. Incubate the selective plates at 37°C overnight.

   Notes: If blue/white selection is being used, select 20 white colonies from the transformation plates for analysis.

   LB agar plates containing ampicillin (50 µg/ml) are appropriate to preserve the clone if either pGEM-5Zf(–) or pGEM-5Zf(+) was used as the cloning vector.

   These plates will provide colonies for sequence analysis, mapping, or any other analysis that is required.

2. Boil the H_2O/colony mixtures for 5 minutes.

3. Centrifuge in a microcentrifuge at 12,000*g* at room temperature for 5 minutes to remove the cellular debris.

4. For each colony being analyzed, pipette the supernatant into a new 1.5-ml tube and discard the pellet. Use 10 µl of this supernatant as template DNA for a 100-µl PCR assay.

   Notes: Use custom-made PCR primers that directly flank the cloning site (see Cross et al. 1994) or the pUC/bacteriophage M13 forward and reverse sequencing primers for amplifying inserts cloned into pGEM-5Zf(–).

   Since GC-rich DNA can be difficult to amplify, a variety of conditions may have to be tried before a suitable one is found. For general guidelines and standard PCR protocols, see Innes et al. (1990); see also Fanning and Gibbs (1997).

   Purified DNA from minipreparations could be used for PCR, but it is generally not necessary.

5. Analyze 10% (10 µl) of each PCR product on a 1.5% agarose gel as described on p. 259, steps 2–3, to estimate the sizes of the inserts.

   Notes: The inserts in these PCR products should range from 200 bp to 2 kb. The average size of the inserts in the human CpG island library of Cross et al. (1994) is 760 bp.

   Store the rest of each PCR product at –20°C until needed in protocols below.

PROTOCOL

Testing Cloned Inserts for *Bst*UI Sites

Carry out steps 1–3 for all the amplified inserts.

1. Transfer duplicate 10-μl aliquots of each PCR product from p. 279 into either 1.5-ml microcentrifuge tubes or the wells of a 96-well plate.

2. For one aliquot of each pair of duplicates, set up a 15-μl digestion using 5 units of *Bst*UI in the buffer as specified by the manufacturer. For the other aliquot, set up a 15-μl mock digestion including all reagents except the enzyme. Incubate both reactions at 60°C for 1 hour.

   Note: If a 96-well plate is being used, seal the plate with either a piece of tape or a Titertek Plate Sealer (ICN 77-400-05) to prevent evaporation.

3. Load each pair of duplicates in adjacent lanes of a 1.5% agarose gel and analyze as described on p. 259, steps 2–3.

   Note: If the insert contains one or more *Bst*UI sites, the size of the fragment(s) in the sample that was digested with *Bst*UI will be smaller than the full-length insert in the mock-digested (undigested) sample. If a clone contains a *Bst*UI site, this is a good indication that it is derived from a CpG island; more than 75% of inserts from a CpG island library are expected to contain *Bst*UI sites.

PROTOCOL

Testing the Methylation Status in the Genome by Hybridization

1. Prepare Southern blots using standard procedures (see Wolff and Gemmill 1997; see also Sambrook et al. 1989). For this analysis, digest 10 μg of genomic DNA (from p. 266, step 1) with *Mse*I alone and with *Mse*I plus a methylation-sensitive restriction enzyme such as *Bst*UI or *Hpa*II. Size fractionate the digestion products on a 1.5% agarose gel (1-kb ladder) to obtain good resolution and then transfer the DNA to nylon membranes.

   Note: The fragments of interest will be less than 3 kb in size.

2. For up to five of the clones, purify the PCR-amplified inserts (from p. 279) by using the QIAquick PCR Purification Kit (Qiagen 28104) or equivalent.

   Note: Choose the clones randomly so that there is no bias in the analysis (e.g., use the first five).

3. Prepare hybridization **probes** by using the random-priming method (Feinberg and Vogelstein 1983; Wolff and Gemmill 1997) to label the DNA purified from the PCR products.

   radioactive substances (see Appendix for Caution)

4. Prehybridize and hybridize the probes to the DNA on the filters, and then wash the filters at high stringency (0.2x SSC/0.1% **SDS** at 68°C) using standard protocols (Sambrook et al. 1989; Wolff and Gemmill 1997).

   Note: If the clone is derived from a CpG island, the *Mse*I-generated fragments should be cleaved by the methylation-sensitive enzyme. If the methylation-sensitive enzyme does not cleave the fragment, either the *Mse*I-generated fragment contains methylated sites for these enzymes or there are no sites present in the fragment. Sequencing (see p. 283) will distinguish between these two possibilities.

   SDS (see Appendix for Caution)

PROTOCOL
***

Testing for the Presence of Repeats and Ribosomal DNA by Hybridization

1. Pipette 100 μl of LB medium containing ampicillin (50 μg/ml) into each of the wells of a sterile 96-well plate (with a lid). Inoculate the medium in each well with a single white bacterial colony and incubate at 37°C overnight.

2. Produce nylon membranes of the gridded clones as follows:

   a. Place a circular nylon membrane (82-mm diameter) on an LB plate containing ampicillin.

   b. Place the prongs of the sterile 96-pronged tool into the cultures from step 1 and then stamp the prongs onto the membrane.

   Note: This step is performed with a 96-pronged tool. The prongs should match the wells of the 96-well plate in step 1 and should also fit inside the agar plate in step 2a.

   c. Incubate at 37°C overnight.

3. Remove the membrane from the agar plate and prepare it for hybridization by processing as described for colony/plaque blotting in the Hybond N⁺, positively charged nylon membrane handbook (p. 9, steps 7–9). Fix the DNA to the membrane using one of the protocols in the same handbook (see p. 9).

4. Prepare hybridization **probes** by using the random-priming method (Feinberg and Vogelstein 1983; Wolff and Gemmill 1997). Use genomic DNA probe, *Alu* repeat probe (e.g., Blur8 [Deininger et al. 1981]), and, if the starting DNA contained ribosomal sequences, ribosomal DNA probe (as used by Cross et al. 1994) to ascertain the proportion of the library that contains these sequences.

   radioactive substances (see Appendix for Caution)

5. Prehybridize and hybridize the probes to the membranes and then wash the membranes at high stringency (0.2x SSC/0.1% **SDS** at 68°C) using standard procedures (Sambrook et al. 1989; Wolff and Gemmill 1997).

   SDS (see Appendix for Caution)

PROTOCOL

Analyzing the Sequence of Clones for the Presence of CpG Islands

1. For each of 20 clones, inoculate 10 ml of LB medium containing ampicillin (50 µg/ml) with a single colony from one of the plates on p. 279, step 1, and incubate at 37°C with agitation at 250–300 rpm overnight.

2. Prepare minipreparations of DNA using the QIAprep Spin Plasmid Miniprep Kit (Qiagen 27104) or equivalent.

3. Sequence the inserts using standard procedures (e.g., Wilson and Mardis 1997).

   Note: Be sure to sequence the 20 clones completely to determine whether or not they have the characteristics of CpG islands.

4. Plot a graph with base composition on the *X* axis and CpG observed/expected on the *Y* axis (for an example of this, see Cross et al. 1994).

   Note: If the fractionation of the CpG island DNA from the rest of the genome has been successful, 80% or more of the clones will have the sequence characteristics of CpG islands.

5. Search sequence databases as described on pp. 236–238 to ascertain whether or not known CpG islands are present and whether or not the clones are derived from the species used to make the library (i.e., that there is no major contamination from bacterial sources, for example).

REFERENCES

Altschul, S.F., W. Gish, W. Miller, E.W. Myers, and D.J. Lipman. 1990. Basic local alignment search tool. *J. Mol. Biol.* **215:** 403–410.

Antequera, F. and A. Bird. 1993. Number of CpG islands and genes in human and mouse. *Proc. Natl. Acad. Sci.* **90:** 11995–11999.

Antequera, F., J. Boyes, and A. Bird. 1990. High levels of *de novo* methylation and altered chromatin structure at CpG islands in cell lines. *Cell* **62:** 503–514.

Baxevanis, A.D., M.S. Boguski, and B.F. Ouellette. 1997. Computational analysis of DNA and protein sequences. In *Genome analysis: A laboratory manual.* Vol. 1 *Analyzing DNA* (ed. B. Birren et al.), pp. 533–586. Cold Spring Harbor Laboratory Press, Cold Spring Harbor, New York.

Bickmore, W.A. and A.P. Bird. 1992. Use of restriction endonucleases to detect and isolate genes from mammalian cells. *Methods Enzymol.* **216:** 224–245.

Bird, A.P. 1986. CpG-rich islands and the function of DNA methylation. *Nature* **321:** 209–213.

———. 1987. CpG islands as gene markers in the vertebrate nucleus. *Trends Genet.* **3:** 342–347.

Bird, A.P. and M.H. Taggart. 1980. Variable patterns of total DNA and rDNA methylation in animals. *Nucleic Acids Res.* **8:** 1485–1497.

Bird, A., M. Taggart, M. Frommer, O.J. Miller, and D. Macleod. 1985. A fraction of the mouse genome that is derived from islands of nonmethylated, CpG-rich DNA. *Cell* **40:** 91–99.

Bradford, M. 1976. A rapid and sensitive method for the quantitation of microgram quantities of protein utilizing the principle of protein dye binding. *Anal. Biochem.* **72:** 248–254.

Brock, G.J.R. and A. Bird. 1997. Mosaic methylation of the repeat unit of the human ribosomal RNA genes. *Hum. Mol. Genet.* **6:** 451–456.

Buckler, A.J., D.D. Chang, S.L. Graw, J.D. Brook, D.A. Haber, P.A. Sharp, and D.E. Housman. 1991. Exon amplification: A strategy to isolate mammalian genes based on RNA splicing. *Proc. Natl. Acad. Sci.* **88:** 4005–4009.

Cross, S.H. and A.P. Bird. 1995. CpG Islands and genes. *Curr. Opin. Genet. Dev.* **5:** 309–314.

Cross, S.H., J.A. Charlton, X. Nan, and A.P. Bird. 1994. Purification of CpG islands using a methylated DNA binding column. *Nat. Genet.* **6:** 236–244.

Cross, S.H., M. Lee, V.H. Clark, J.M. Craig, A.P. Bird, and W.A. Bickmore. 1997. The chromosomal distribution of CpG islands in the mouse: Evidence for genome scrambling in the rodent lineage. *Genomics* **40:** 454–461.

Deininger, P.L., D.J. Jolly, C.M. Rubin, T. Friedmann, and C.W. Schmid. 1981. Base sequence studies of 300 nucleotide renatured repeated human DNA clones. *J. Mol. Biol.* **151:** 17–33.

Duyk, G.M., S.W. Kim, R.M. Myers, and D.R. Cox. 1990. Exon trapping: A genetic screen to identify candidate transcribed sequences in cloned mammalian genomic DNA. *Proc. Natl. Acad. Sci.* **87:** 8995–8999.

Fanning, S. and R.A. Gibbs. 1997. PCR in genome analysis. In *Genome analysis: A laboratory manual.* Vol. 1 *Analyzing DNA* (ed. B. Birren et al.), pp. 249–299. Cold Spring Harbor Laboratory Press, Cold Spring Harbor, New York.

Feinberg, A.P. and B. Vogelstein. 1983. A technique for radio-labeling DNA restriction endonuclease fragments to high specific activity. *Anal. Biochem.* **132:** 6–13.

Gao, J., P. Erickson, D. Patterson, C. Jones, and H. Drabkin. 1991. Isolation and regional mapping of *Not*I and *Eag*I clones from human chromosome 21. *Genomics* **10:** 166–172.

Gardiner-Garden, M. and M. Frommer. 1987. CpG islands in vertebrate genomes. *J. Mol. Biol.* **196:** 261–282.

Green, E.D., R.M. Mohr, J.R. Idol, M. Jones, J.M. Buckingham, L.L. Deaven, R.K. Moyzis, and M.V. Olson. 1991. Systematic generation of sequence-tagged sites for physical mapping of human chromosomes: Application to the mapping of human chromosome 7 using yeast artificial chromosomes. *Genomics* **11:** 548–564.

Hochuli, E., H. Dîbeli, and A. Schacher. 1987. New metal chelate adsorbents selective for proteins and peptides containing neighbouring histidine residues. *J. Chromatogr.* **411:** 177–184.

Innis, M A., D.H. Gelfand, J.J. Sninsky, and T.J. White. 1990. *PCR protocols.* Academic Press, London.

John, R.M., C.A. Robbins, and R.M. Myers. 1994. Identification of genes within CpG-enriched DNA from human chromosome 4p16.3. *Hum. Mol. Gen.* **3:** 1611–1616.

Larsen, F., G. Gunderson, R. Lopez, and H. Prydz. 1992. CpG islands as gene markers in the human genome. *Genomics* **13:** 1095–1107.

Lavia, P., D. Macleod, and A. Bird. 1987. Coincident start sites for divergent transcripts at a randomly selected CpG-rich island of mouse. *EMBO J.* **6:** 2773–2779.

Lewis, J.D., R.R. Meehan, W.J. Henzel, I. Maurer-Fogy, P. Jeppesen, F. Klein, and A. Bird. 1992. Purification, sequence and cellular localisation of a novel chromosomal protein that binds to methylated DNA. *Cell* **69:** 905–914.

Lindsay, S. and A.P. Bird. 1987. Use of restriction enzymes to detect potential gene sequences in mammalian DNA. *Nature* **327:** 336–338.

Lovett, M., J. Kere, and L. M. Hinton. 1991. Direct selection: A method for the isolation of cDNAs encoded by large genomic regions. *Proc. Natl. Acad. Sci.* **88:** 9628–9632.

McQueen, H.A., V.H. Clark, A.P. Bird, M. Yerle, and A.L.

Archibald. 1997. CpG islands of the pig. *Genome Res.* 7: 924–931.

McQueen, H.A., J. Fantes, S.H. Cross, V.H. Clark, A.L. Archibald, and A.P. Bird. 1996. CpG islands of chicken are concentrated on microchromosomes. *Nat. Genet.* 12: 321–324.

Meehan, R.R., J.D. Lewis, and A.P. Bird. 1992. Characterization of MeCP2, a vertebrate DNA binding protein that binds methylated DNA. *Nucleic Acids Res.* 20: 5085–5092.

Nan, X., R.R. Meehan, and A. Bird. 1993. Dissection of the methyl-CpG binding domain from the chromosomal protein MeCP2. *Nucleic Acids Res.* 21: 4886–4892.

Parimoo, S., S.R. Patanjali, H. Shukla, D.D. Chaplin, and S.M. Weissman. 1991. cDNA selection: Efficient PCR approach for the selection of cDNAs encoded in large chromosomal DNA fragments. *Proc. Natl. Acad. Sci.* 88: 9623–9627.

Razin, A. and H. Cedar. 1994. DNA methylation and genomic imprinting. *Cell* 77: 473–476.

Riethman, H., B. Birren, and A. Gnirke. 1997. Preparation, manipulation, and mapping of HMW DNA. In *Genome analysis: A laboratory manual* Vol. 1 *Analyzing DNA* (ed. B. Birren et al.), pp. 83–248. Cold Spring Harbor Laboratory Press, Cold Spring Harbor, New York.

Sambrook, J., E.F. Fritsch, and T. Maniatis. 1989. *Molecular cloning: A laboratory manual*, 2nd edition. Cold Spring Harbor Laboratory Press, Cold Spring Harbor, New York.

Sanford, J., V.M. Chapman, and J. Rossant. 1985. DNA methylation in extra-embryonic lineages of mammals. *Trends Genet.* 1: 89–93.

Shiraishi, M., L.S. Lerman, and T. Sekiya. 1995. Preferential isolation of DNA fragments associated with CpG islands. *Proc. Natl. Acad. Sci.* 92: 4229–4233.

Short, J.M., J.M. Fernandez, J.A. Sorge, and W.D. Huse. 1988. Lambda ZAP: A bacteriophage lambda expression vector with in vivo excision properties. *Nucleic Acids Res.* 16: 7583–7600.

Stoger, R., P. Kubicka, C.G. Liu, T. Kafri, A. Razin, H. Cedar, and D.P. Barlow. 1993. Maternal-specific methylation of the imprinted mouse Igf2r locus identifies the expressed locus as carrying the imprinting signal. *Cell* 73: 61–71.

Studier, F.W., A.H. Rosenberg, J.J. Dunn, and J.W. Dubendorff. 1990. Use of T7 RNA polymerase to direct expression of cloned genes. *Methods Enzymol.* 185: 60–89.

Tashima, M., B. Calabretta, G. Torelli, M. Scofield, A. Maizel, and G.F. Saunders. 1981. Presence of a highly repetitive and widely dispersed DNA sequence in the human genome. *Proc. Natl. Acad. Sci.* 78: 1508–1512.

Taylor, R.G., D.C. Walker, and R.R. McInnes. 1993. *E. coli* host strains significantly affect the quality of small

scale plasmid DNA preparations used for sequencing. *Nucleic Acids Res.* 21: 1677–1678.

Toniolo, D., M. D'Urso, G. Martini, M. Persico, V. Tufano, G. Battistuzzi, and L. Luzzatto. 1984. Specific methylation pattern at the 3'end of the human housekeeping gene for glucose 6-phosphate dehydrogenase. *EMBO J.* 3: 1987–1995.

Triboli, C., F. Tamanini, C. Patrosso, L. Milanesi, A. Villa, R. Pergolizza, E. Maestrini, S. Rivella, S. Bione, M. Mancini, P. Vezzoni, and D. Toniolo. 1992. Methylation and sequence analysis around *Eag*I sites: Identification of 28 new CpG islands in XQ24–XQ28. *Nucleic Acids Res.* 20: 727–733.

Tykocinski, M.L. and E.E. Max. 1984. CG dinucleotide clusters in MHC genes and in 5' demethylated genes. *Nucleic Acids Res.* 12: 4385–4396.

Uberbacher, E.C. and R.J. Mural. 1991. Locating protein-coding regions in human DNA sequences by a multiple sensor-neural network approach. *Proc. Natl. Acad. Sci.* 88: 11261–11265.

Valdes, J.M., D.A. Tagle, and F.S. Collins. 1994. Island rescue PCR: A rapid and efficient method for isolating transcribed sequences from yeast artificial chromosomes and cosmids. *Proc Natl. Acad. Sci.* 91: 5377–5381.

Weber, B., C. Collins, D. Kowbel, O. Riess, and M.R. Hayden. 1991. Identification of multiple CpG islands and associated conserved sequences in a candidate region for the Huntington disease gene. *Genomics* 11: 1113–1124.

Wilson, R.K. and E.R. Mardis. 1997. Fluorescence-based DNA sequencing. In *Genome analysis: A laboratory manual.* Vol. 1 *Analyzing DNA* (ed. B. Birren et al.), pp. 301–395. Cold Spring Harbor Laboratory Press, Cold Spring Harbor, New York.

Wolf, S.F., S. Dintzis, D. Toniolo, G. Persico, K.D. Lunnen, J. Axelman, and B.R. Migeon. 1984. Complete concordance between glucose-6-phosphate dehydrogenase activity and hypomethylation of 3'CpG clusters: Implications for X chromosome dosage compensation. *Nucleic Acids Res.* 12: 9333–9348.

Wolff, R. and R. Gemmill. 1997. Purifying and analyzing genomic DNA. In *Genome analysis: A laboratory manual.* Vol. 1 *Analyzing DNA* (ed. B. Birren et al.), pp. 1–81. Cold Spring Harbor Laboratory Press, Cold Spring Harbor, New York.

Yen, P.H., P. Patel, A.C. Chinault, T. Mohandas, and L.J. Shapiro. 1984. Differential methylation of hypoxanthine phosphoribosyl transferase genes on active and inactive human X chromosomes. *Proc. Natl. Acad. Sci.* 81: 1759–1763.

Zuo, J., C. Robbins, S. Baharloo, D.R. Cox, and R.M. Myers. 1993. Construction of cosmid contigs and high-resolution restriction mapping of the Huntington disease region of human chromosome 4. *Hum. Mol. Genet.* 2: 889–899.

6

Detection of DNA Variation

RICHARD M. MYERS, LORA HEDRICK ELLENSON,
AND KENSHI HAYASHI

The discipline of genetics is based on the presence of differences in nucleic acid sequences between individuals of species. This variation is generated through mutations, which in some cases have devastating consequences for an individual. On the population level, mutations are the events that lead to evolution and speciation. Geneticists use naturally occurring and experimentally induced mutations to study the principles of genetics and their relationship to diverse biological processes in a wide variety of organisms. Most often, the investigator is interested in the particular change in the DNA that is responsible for a phenotype. In cases of simple Mendelian traits, particularly in inbred strains of experimental organisms, it is often straightforward to assign causation of the phenotype to a specific mutation. Even in human beings, a strong correlation of a disease with a mutation (or better yet, several different mutations in the same gene in different families) is generally accepted as evidence that the mutant gene causes the disease. However, many phenotypes result from the contributions of DNA variation in several genes, which greatly complicates the ability to assign causation of such complex traits to particular mutations. One of the major challenges facing geneticists today is sorting out the roles of multiple genes in determining complex traits.

This chapter provides experimental details of several methods for initial detection and, if desired, large-scale screening of single-base changes and other small DNA sequence alterations.

Overview of DNA Variation

Outside the laboratory, individual members of a particular species differ in DNA sequence at a significant number of positions in their chromosomes. In human beings, the frequency of variation between two homologs is approximately one every 1–2 kb, although it deviates widely at different regions of the chromosome (Cooper et al. 1985; Kwok et al. 1994, 1996). A fraction of this polymorphism is responsible for the differences between individuals, although most of it is probably neutral. Regardless of the effects of these DNA variants, the ability to detect them provides an immensely powerful approach for tracking the inheritance of chromosomes and for mapping and cloning genes responsible for diseases and other phenotypes on the basis of the chromosomal locations of the genes (Collins 1992; Ballabio 1993; Chakravarti 1998; see also Chapter 1).

Before the tools of molecular biology were available, DNA variation could be detected only by observing and following the transmission of phenotypes or, in some cases, by measuring alterations in the behaviors of proteins in electrophoretic systems. As the ability to manipulate, clone, and dissect nucleic acids by recombinant DNA technology was developed, many mutations responsible for genetic diseases in human beings and for phenotypes in experimental organisms were identified by DNA sequencing. However, this work was laborious because the mutant genes had to be cloned from affected individuals to provide templates for DNA sequencing, and in general, there was no clue about where in the gene a mutation might reside before the DNA sequencing was performed. Beginning in the late 1970s and early 1980s, these limitations led to the development of a number of techniques that make it possible to screen a portion of the nucleotides in a gene or chromosomal region for genetic variation directly from genomic DNA or mRNA, thus obviating the need to clone each allele for sequencing. The development of PCR (Saiki et al. 1985; Mullis et al. 1986) further simplified most of these methods.

Methods for Detecting DNA Variation

The RFLP method was the first screening technique to be used for detection of DNA variation, successfully identifying both disease-causing muta-

tions and apparently neutral variants (Botstein et al. 1980; White et al. 1985; Donis-Keller et al. 1986). This method detects base changes among individuals by digestion of the DNA with restriction enzymes, Southern blotting, and hybridization using labeled probes. Occasionally, single-base changes, insertions, and deletions in the DNA segment detected by a probe lead to a difference in the sizes of the fragments on the blot. DNA changes in a restriction enzyme recognition site itself are detected only when that enzyme is tested, whereas in the cases where an insertion or deletion of detectable size occurs, the variation is often detected when different restriction enzymes are used. In cases where changes are in a restriction enzyme recognition site, only a small fraction of the DNA variants are identified with a single test. Although this fraction can be increased by testing for RFLPs with a large number of restriction enzymes in separate digestions or, in some cases, by increasing the number of individuals tested, this substantially increases the cost and labor and still leaves a large fraction of base changes in a fragment undetected.

These limitations motivated the development of new methods that screen a larger fraction of the bases in a DNA segment. Many of these methods have been applied directly to genomic DNA samples, although all can be easily adapted for, or were initially developed for, use with PCR. These methods, many of which are listed in Table 1, can be broadly defined in the classes discussed below.

ELECTROPHORETIC MOBILITY ALTERATION METHODS

This class includes the SSCP method (Orita et al. 1989; Hayashi 1991), DGGE (Fischer and Lerman 1983; Myers et al. 1985c,d, 1987; Lerman et al. 1986; Sheffield et al. 1990), and CDGE (Børresen et al. 1991). The implementation of each of these methods varies, but all three are based on the principle that differences in the nucleotide sequence of a DNA fragment, even as small as a single-base substitution, can alter the shape of the fragment to an extent sufficient to cause its migration to vary in an electrophoretic gel matrix.

Although DGGE and CDGE can be effective for comparing a homoduplex of suspected mutant DNA with a homoduplex of wild type, the efficacy of these two methods is enhanced by using a heteroduplex generated by denaturing and rean-

Table 1 Methods for Detection of DNA Variation
_____

Electrophoretic mobility alteration methods
 single-strand conformational polymorphism (SSCP; Orita et al. 1989; Hayashi 1991)
 restriction enzyme fingerprinting (REF; Liu and Sommers 1995)
 ddNTP fingerprinting (ddF; Sarkar et al. 1992b)
 denaturing gradient gel electrophoresis (DGGE; Fischer and Lerman 1983; Myers et al.
  1985c,d,e, 1987; Lerman et al. 1986)
 constant denaturant gel electrophoresis (CDGE; Børresen et al. 1991)
 denaturing high-performance liquid chromatography (dHPLC; P. Oefner, unpubl.)
 nondenaturing gel mismatch detection (White et al. 1992)
 carbodiimide mismatch detection (Novack et al. 1986; Ganguly and Prockop 1990)

Mismatch cleavage methods
 RNase cleavage (Myers et al. 1985a; Winter et al. 1985)
 chemical cleavage at mismatches (CCM; Cotton et al. 1988)
 bacteriophage T4 endonuclease VII cleavage at mismatches (ECM; Youil et al. 1995)
 mismatch repair enzyme cleavage with _E. coli_ MutY protein (MREC; Lu and Hsu 1991)

Mismatch recognition methods
 mismatch repair detection (MRD; Faham and Cox 1996)
 oligonucleotide microarray hybridization ("DNA chips"; Chee et al. 1996)

Sequencing methods
 direct sequencing of uncloned PCR products
 sequencing of cloned PCR products

Protein truncation test (PTT; Roest et al. 1993; Hogervorst et al. 1995)
_____

nealing wild-type and mutant DNA strands, which results in a mismatch at the position of the single-base change. In many cases, the effect can be very large, such that the two mismatched heteroduplexes have a 25–50% decrease in mobility relative to the homoduplexes.

Various other nondenaturing gel mobility alteration methods also rely on mismatches between double-stranded wild-type and mutant DNA fragments (White et al. 1992). In these methods, mutations are detected by electrophoresis of DNA fragments containing mismatches through standard or specialized polyacrylamide matrices under conditions that sometimes alter the migration rate, presumably because of changes in the shape of the DNA fragments conferred by the mismatches.

Another version of this general class enhances the alterations in electrophoretic behaviors by attaching a chemical moiety, such as carbodiimide, to the mismatched bases under conditions where the reactive chemical distinguishes between double- and single-stranded DNA (Novack et al. 1986). A variation of this method was developed in which mismatches are reacted with carbodiimide, but the reacted bases are detected and mapped by measuring the position of a block to primer extension instead of by measuring altered electrophoretic mobility (Ganguly and Prockop 1990).

Finally, a recently developed method that is largely based on the same principles as DGGE and CDGE is denaturing HPLC (P. Oefner, unpubl.). This method allows separation of heteroduplexes from homoduplexes, and because running times are very short, it has the potential to analyze many samples rapidly.

MISMATCH CLEAVAGE METHODS

These approaches involve forming heteroduplexes between the complementary strands of a DNA fragment from a reference sample (usually considered a wild-type sample) and a test (mutant) sample, cleaving the mismatched bases with a reagent that reacts preferentially with single-stranded bases, and analyzing the samples by electrophoresis for the presence and location of the mismatch.

In the RNase cleavage method, RNase A is used to cleave mismatches in RNA:DNA and RNA:RNA heteroduplexes (Myers et al. 1985a; Winter et al.

1985). Another technique, the CCM method (Cotton et al. 1988), uses various chemicals that cleave mismatches in DNA:DNA duplexes. These chemicals were previously used in developing the chemical degradation DNA sequencing method of Maxam and Gilbert (1980). A similar technique, enzyme cleavage at mismatches, uses bacteriophage T4 endonuclease VII, which is involved in resolving branched DNA molecules during recombination, to cleave at mismatches in DNA:DNA duplexes (Youil et al. 1995). A fourth cleavage technique, mismatch repair enzyme cleavage, uses *E. coli* MutY protein to recognize and cleave at mismatches in DNA (Lu and Hsu 1991).

MISMATCH RECOGNITION METHODS

In this class of methods, mismatches present in heteroduplexes formed between wild-type and mutant DNA fragments are detected by means other than cleavage at the mismatch or an alteration in electrophoretic mobility. One of these methods, called MRD, for mismatch repair detection (Faham and Cox 1996), identifies mutations by means of a bacterial colony assay that is based on the mismatch repair system of *E. coli*. Perhaps the most promising developments in technologies for identifying DNA variation are various approaches for detecting mismatches by DNA hybridization. The most developed of these approaches use DNA ''chips,'' which are high-density arrays of oligonucleotides to which labeled test DNA is hybridized (see, e.g., Chee et al. 1996; Hacia et al. 1996). In some configurations of these DNA chips, multiple oligonucleotides for testing the sequence at any given base are designed into the chip, which allows multiple sampling of each base to be performed. These approaches have been applied to identifying mutations in known disease genes as well as to identifying DNA polymorphisms on a large scale (see, e.g., D. Wang and E. Lander et al., in prep.).

SEQUENCING METHODS

Many investigators have used direct sequencing of uncloned PCR products to search for naturally occurring mutations in candidate disease genes or in experimentally induced gene systems and to identify new biallelic polymorphisms (Nickerson et al. 1992; Kwok et al. 1994). Although it is also possible to clone PCR products and perform DNA sequencing with single cloned templates to identify DNA variation, the relatively high rates of mutations introduced by the DNA polymerases during PCR make such an approach risky unless multiple clones are sequenced from each individual and/or more accurate polymerases are used.

PROTEIN TRUNCATION TEST

The PTT method (Roest et al. 1993; Hogervorst et al. 1995) can detect those mutations that result in an alteration in the size of the protein encoded by a gene. Coupled in vitro transcription and translation is used to produce labeled protein fragments that are screened for abnormalities in electrophoretic migration. These mutations include nonsense, frameshift, some insertion and deletion, and splice-site mutations.

LARGE-SCALE SCREENING METHODS FOR DETECTING DNA VARIATION

The above methods were developed mainly as initial screening approaches to scan segments of DNA for the presence of sequence differences (primarily single-base substitutions and other small alterations in the DNA sequence) in a small number of different individuals. In some cases, once the presence of a sequence difference is found, a different method is used to screen large numbers of individuals for the presence of the particular base change. These methods generally rely on a set of oligonucleotides designed to detect a specific base change and include the following:

- the oligonucleotide ligation assay (OLA; Landegren et al. 1988; Dietrich et al. (1998)
- allele-specific amplification by PCR (Gibbs et al. 1989; Wu et al. 1989)
- minisequencing, or single-base extension methods (Syvänen et al., 1990, 1992; Chen and Kwok 1997)
- fluorescence quenching during PCR (the TaqMan assay; Holland et al. 1991; Lee et al. 1993; Livak et al. 1995)
- the ligase chain reaction (Wu and Wallace 1989)
- allele-specific oligonucleotide hybridization (Conner et al. 1983)

Another class of DNA variation that has been enormously useful for meiotic linkage analysis is length variation of tandem repeats, including VNTRs and simple-sequence repeats (see Dietrich et al. 1998).

Choosing a Suitable Approach for Mutation Detection

This chapter provides experimental details of several of the methods described above for initial detection and, if desired, large-scale screening of single-base changes and other small DNA sequence alterations. In choosing among the large number of methods for mutation detection, the expertise of the particular investigator and the needs of the project should be considered. For example, investigators that perform large amounts of DNA sequencing for other purposes may find it easier to establish a direct sequencing strategy for mutation detection than would investigators that do little sequencing. Factors such as the level of technical expertise in the laboratory, the size of the target region, the total number of base pairs that need to be screened, the need for radioisotopes, and the cost of equipment are all considerations in choosing a technique.

A limitation for all mutation detection techniques is the length of the target fragment that can easily be tested for the presence of sequence variation. For most methods, a signal-to-noise ratio sufficient for unambiguous mutation detection and the resolution limits of various gel matrices generally result in a practical limit of DNA fragments up to only a few hundred base pairs in length. This limitation can be partly alleviated in some of the methods by multiplexing (i.e., performing the procedure with multiple DNA fragments in the several hundred base pair size range in a single unit of analysis, such as a lane on a gel for the electrophoretic techniques). Most versions of the DNA chip hybridization methods can be performed with a very high degree of multiplexing. However, multiplexing is not easily applied to some of the approaches, such as direct DNA sequencing of uncloned PCR products and the PTT method.

A related issue that should be considered is the fraction of base pairs in a target DNA fragment that is actually sampled for variation by a particular technique. This parameter is often somewhat inappropriately referred to as the *sensitivity* of the mutation detection method. For example, the SSCP, DGGE, CCM, and RNase cleavage methods work best with DNA fragments 200–500 bp in length, but under some experimental conditions, a single assay may screen only about half of the test DNA fragment. Similarly, direct DNA sequencing

of uncloned PCR products screens all of the base pairs in the part of the DNA fragment that is sequenced, which is usually approximately 300–400 bp, but does not screen the remainder of the PCR fragment unless additional sequencing is performed.

Most of the techniques in this chapter can be used to screen a large fraction of the total base pairs in a DNA fragment. For some DNA fragments, close to 100% coverage of the base pairs can be achieved if multiple experimental conditions are used (e.g., if a primer walking step is used in direct sequencing, if several different mismatch cleavage conditions are used in RNase cleavage and CCM, if GC clamps and heteroduplexes are used in DGGE, and if multiple gel conditions are used in SSCP). The choice of whether or not to spend additional effort to test a target fragment or set of fragments depends on how important it is to identify every possible base change in the target region. For example, in cases where a single-base change at one or a very few positions in a gene is responsible for a disease or phenotype, a thorough screening of all base pairs is likely to be required. However, in cases where many alleles of mutations in a gene are available, screening only a portion of the base pairs in the gene is likely to be sufficient. Unfortunately, in many situations, it may not be possible to predict accurately whether or not multiple alleles are present in the mutant population. For this reason, the most efficient route may be to use a simpler, less thorough version of a technique to perform an initial screen for mutations in a gene or coding region and to go to the additional trouble of screening every base pair in the target region if the initial tests are insufficient.

In addition to using the techniques described in this chapter to identify mutations in genes that result in phenotypes, all of the approaches can be applied to identifying new neutral polymorphic DNA segments for genetic analysis. For example, although meiotic mapping in human beings and mice has been greatly aided by the development of simple-sequence-repeat DNA markers (see Dietrich et al. 1998), additional biallelic polymorphisms need to be identified in particular target regions and in particular individuals in many genetic studies involving standard linkage, linkage disequilibrium, and identity-by-descent approaches. Given the estimated frequency of polymorphic variation in human beings, at least several kilobase pairs generally need to be screened at a locus when a new polymorphism is desired. Fortunately, the DNA that is screened does not need to be completely contiguous. Therefore, an efficient way of identifying polymorphisms is to apply one of the simpler, less thorough versions of the various mutation detection techniques (e.g., the SSCP or DGGE method) to multiple DNA fragments from the target locus.

The SSCP Method

In the SSCP (sometimes called PCR SSCP) method, the target DNA sequence is amplified by PCR (and typically labeled concomitantly with a radioisotope), denatured, and separated in a single-stranded form by electrophoresis in a nondenaturing polyacrylamide gel (Orita et al. 1989). DNA sequence variants often result in a shift in electrophoretic mobility compared with wild type, such that mutations are detected by the appearance of new bands in the resulting autoradiograph. The SSCP method is very simple and requires perhaps the minimum of technical proficiency and instrumentation relative to many other mutation scanning techniques. The mobility shift is believed to be caused by sequence-dependent alteration in the tertiary structure of single-stranded DNA.

The tertiary structure of single-stranded DNA varies under different physical conditions (e.g., temperature and the ionic environment), and SSCP, not surprisingly, is sensitive to changes in such parameters. Empirically derived rules have been proposed for electrophoretic conditions that efficiently separate mutant alleles in particular sequence contexts (Glavac and Dean 1993), but whether or not any particular mutation can be detected under a particular experimental condition is not predictable, especially when the mutation is in a new sequence context.

ADVANTAGES AND DISADVANTAGES OF THE SSCP METHOD

The sensitivity of the SSCP method is generally high, such that more than 80% of mutations in most DNA fragments 300 bp and shorter can be detected in a single electrophoretic run (Hayashi and Yandell 1993), although the estimate is lower in some publications (Sarkar et al. 1992a). As with most mutation detection techniques, the inability to detect 100% of all possible mutations with the SSCP method means that the absence of a new band in the autoradiograph does not provide evidence that no mutation is present. In contrast, false-positive results are not likely with the SSCP method because mutations lead to the appearance of new bands and alternative mobilities do not appear to occur in the absence of mutations. This feature of SSCP analysis, which is shared by DGGE and CDGE, provides some advantage over mutation detection methods that depend on the absence

of signal (e.g., such as allele-specific hybridization [Conner et al. 1983], allele-specific amplification by PCR [Gibbs et al. 1989; Wu et al. 1989], and RFLP methods [White et al. 1985]), where accidental failure of amplification or restriction enzyme reactions can lead to misdiagnosis.

As with most other mutation detection techniques, the SSCP method can detect mutations that appear at any position within the PCR-amplified target DNA fragments. The technique can also detect mutations that constitute a less than haploid equivalent of the total target DNA fragment in a sample (e.g., in tumor tissues containing nontumor contaminants). Indeed, the technique has been used to detect mutated DNA segments present in as low as a few percent of the corresponding wild-type segments in a sample; in these cases, the mutations can be confirmed by extracting the DNA from the shifted band in a dried gel, reamplifying by PCR, and directly sequencing the uncloned PCR product (Suzuki et al. 1990, 1991).

VARIATIONS OF THE SSCP METHOD

In some cases, the use of a gel matrix other than that described in this chapter can enhance the separation of wild-type and mutant DNA fragments in the SSCP method. Some alternative matrices are higher-concentration polyacrylamide gels (Savov et al. 1992) and a new gel matrix called MDE gel (available from AT Biochem; Keen et al. 1991). In both cases, the time required for electrophoresis is much longer than with the standard conditions described in the protocols in this chapter. Therefore, whether or not to use an alternative matrix is a compromise between efficiency (the number of mutations found for the amount of effort and length of time spent) and sensitivity (the number of mutations found compared to the number of mutations present).

Several variants of the SSCP method do not rely on radioisotopes. These include an approach that uses silver staining to detect DNA bands (Ainsworth et al. 1991) and a method that uses fluorescently labeled oligonucleotide primers during PCR and then analysis on an automated DNA sequencer (Makino et al. 1992). Advantages and disadvantages of these alternatives are discussed in Hayashi (1992).

Two other variants of the SSCP method are restriction enzyme fingerprinting (Liu and Sommer 1995) and ddNTP fingerprinting (Sarkar et al.

1992b). In restriction enzyme fingerprinting, PCR products are screened for variations that affect restriction enzyme cleavage sites and the conformation of single-stranded DNA fragments. In ddNTP fingerprinting, SSCP is combined with ddNTP-mediated chain-termination sequencing.

EXPERIMENTAL STRATEGIES FOR SSCP ANALYSIS

The sensitivity (i.e., the fraction of base pairs screened in a DNA fragment) of the SSCP method gradually decreases with increasing fragment length. Therefore, it is preferable to design oligonucleotide primers such that the length of the PCR products will be less than 300 bp. In cases where the target segments are longer than this length (e.g., in full-length cDNA segments), multiple pairs of oligonucleotide primers should be designed that allow the region to be covered by overlapping segments of 300 bp or less. Alternatively, long PCR products (e.g., 2–3 kb) can be synthesized and digested with restriction enzymes before electrophoresis, although this choice does not allow reamplification of each DNA fragment using the initial primer pair after elution of the DNA from the gel.

As with many other PCR-based techniques, specific amplification of the target sequence is critical for unambiguous results. In the protocols provided in this chapter, the oligonucleotide primer concentrations or dNTP concentrations used in the amplification step are lower than those generally recommended by suppliers of the enzymes or kits. This feature allows a reduction in consumption and possible hazards of radioisotopes and helps ensure a high specific radioactivity in the PCR product. PCR products labeled to low specific radioactivities require large amounts of each sample to be loaded on the gel, which often results in reduced resolution and ambiguous interpretation (Cai and Touitou 1993). A simple way to check for efficiency and specificity of the PCR assay is to analyze a small aliquot (as little as 1% of the total sample) of the labeled amplification products by electrophoresis in a small (e.g., 0.03 x 20 x 20 cm) 5% polyacrylamide gel containing 7 M urea. In this test, electrophoresis can be performed at 700 V for

approximately 15 minutes (until the bromophenol blue dye migrates approximately one third of the gel length). The short period of electrophoresis not only makes the experiment more efficient, but also retains all of the radioactivity, including radiochemical impurities, unincorporated dNTPs, labeled primers (when end-labeling is used), and all of the PCR products on the gel. The gel is then dried on Whatman 3MM paper and placed on film for several hours of exposure.

The efficacy of the SSCP method depends on the formation of different conformations of the single strands of wild-type and mutant DNA molecules. Because the conformations are determined by the balance between intramolecular interactions and thermal fluctuations, it is important to maintain the gel at a constant temperature during electrophoresis. Because ohmic heat production is unavoidable in electrophoresis in a high electrical field, efficient heat dissipation is essential to maintain constant temperature. Different approaches for achieving this purpose are discussed on p. 295 and pp. 301–304. Recently, two important technical improvements of SSCP analysis have been made, both of which significantly extended the length range (up to 800 bp) for which mutations can be detected at a high rate. These improvements are the use of low-pH gels for separation (Kukita et al. 1997) and the use of an automated capillary electrophoresis system.

Electrophoresis in the SSCP method is often carried out under two or more conditions (e.g., using gels with and without glycerol) to increase the fraction of mutations detected (Hayashi 1991). Effects on mobility similar to that seen with the inclusion of glycerol (which slows mobility) are observed when electrophoresis is performed at lower temperature (e.g., 10–15°C without glycerol is similar to 25°C with glycerol; Hayashi 1991). The glycerol effect on mobility is mainly due to a reduction of pH caused by the formation of complexes of glycerol and borate ion, which is present in TBE buffer. No glycerol effects are observed with electrophoresis buffer that lacks borate. One hypothesis for the glycerol effect is that the lower pH suppresses the charge of phosphate in the nucleic acid backbone, allowing the formation of tertiary structures in the single-stranded DNA.

PROTOCOLS FOR DETECTING DNA VARIATION USING SSCP ANALYSIS

Amplification of Products for SSCP Analysis

Two alternative protocols for amplification of target samples from multiple individual genomic DNA samples are provided on pp. 297–300. The protocol in which PCR is carried out with 5'-end-labeled oligonucleotide primers is more economical than that in which internal base pairs are labeled with radioisotopes and is particularly recommended when 20 or more individuals are being tested for mutations. In step 1 of the protocol on pp. 297–299, oligonucleotide primers are labeled at a very high specific radioactivity (7000 Ci/ mmole) with [γ-^{32}P]ATP. The use of high concentrations (>20 μM) of [γ-^{32}P]ATP is essential. Under the conditions described in the protocol, virtually every molecule of each oligonucleotide primer is labeled with ^{32}P, producing enough 5'-end-labeled primers to amplify 80 genomic DNA samples for SSCP (Mashiyama et al. 1990).

Although more costly, the protocol that generates internally labeled PCR products is more convenient because amplification and radiolabeling are carried out simultaneously. In this protocol (see p. 300), target sequence from ten genomic DNA samples is amplified by PCR and directly labeled by including [α-^{32}P]dCTP during the amplification reaction. This results in a specific radioactivity of approximately 0.3 Ci/mmole in the C residues in the amplified products.

Gel Electrophoresis of PCR-amplified DNA Fragments for SSCP Analysis

It is important to maintain the gel at a constant temperature during electrophoresis for SSCP analysis (see p. 294). To achieve reproducible results with SSCP analysis, the gel plates should generally not feel warm to the touch during electrophoresis. To avoid the possibility of electrocution, turn off the power supply before testing to see if the plates are warm to the touch. *Do not touch the gel plates while the gel is running.*

In the protocol on pp. 301–304, a water-jacketed electrophoresis apparatus is used for a gel approximately 0.3 mm thick, 30 cm wide, and 40 cm high that is fitted with a 50-lane shark's tooth comb to create lanes 5 mm in width. Alternatively, the gel can be efficiently kept at room temperature by blowing air with strong fans (e.g., two cross-flow fans of approximately 30 W, one set on each side of the gel plates). Otherwise, electrophoresis should be performed at low wattage (~10 W), but the gels must run approximately twice as long as those that are cooled by one of the methods above.

Reamplification and Sequencing of DNA from Single Gel Bands Resolved During SSCP Analysis

In many cases, it is desirable to isolate each allele resolved during SSCP analysis and separately determine its nucleotide sequence. The protocol on pp. 305–306 provides methods for gel purification and reamplification of SSCP fragments and preparation for sequencing. This technique is critically important when a faint band is found in the autoradiograph from the SSCP analysis and only a minor fraction of the amplified DNA fragment is suspected of carrying a mutation (Suzuki et al. 1990). Such cases frequently occur during searches for oncogene mutations in surgically obtained cancer tissues or during searches for mutations in cDNA when levels of transcripts between two alleles differ. For example, the level of mRNA of an allele carrying a nonsense mutation is often much lower than that of normal allele.

PROTOCOL

Amplification Using 5′-End-labeled Oligonucleotide Primers

1. Prepare the labeled primer mixture.

   a. Prepare a 5-μl labeling mixture by combining the components below (in the order listed) in a 0.5-ml microcentrifuge tube. Mix by gently pipetting up and down.

H_2O	1 μl
10x bacteriophage T4 polynucleotide kinase buffer	0.5 μl
left primer (10 μM in H_2O)	1 μl
right primer (10 μM in H_2O)	1 μl
[γ-^{32}P]ATP (160 mCi/ml, 7000 Ci/mmole; ICN or NEN Life Science Products)	1 μl
bacteriophage T4 polynucleotide kinase (10 units/μl)	0.5 μl

   Note: Any commercially available bacteriophage T4 polynucleotide kinase can be used. The appropriate buffer is usually supplied with the enzyme.

   10x Bacteriophage T4 polynucleotide kinase buffer

Component and final concentration	Amount to add per 1 ml
0.5 M Tris-Cl	0.5 ml of 1 M (pH 8.3 at room temperature)
100 mM $MgCl_2$	100 μl of 1 M
50 mM DTT	50 μl of 1 M
H_2O	350 μl

   Store at −20°C for up to 1 year.

   radioactive substances (see Appendix for Caution)

   b. Incubate at 37°C for 30 minutes.

   c. Add the components below (in the order listed) to the 5-μl labeling mixture to make 320 μl of labeled primer mixture.

H_2O	223 μl
10x amplification buffer	40 μl
1.25 mM dNTP mixture (see Appendix)	20 μl
25 mM $MgCl_2$	32 μl

   Notes: The resulting labeled primer mixture can be stored at −20°C for a few weeks. Thaw just before use in the PCR assays.

   Aliquots of this mixture are used to perform the PCR assays below. Ten genomic DNA samples are amplified here. If necessary, adjust the volume of each component to assay additional samples.

10x Amplification buffer

Component and final concentration	Amount to add per 1 ml
0.5 M KCl	0.5 ml of 1 M
100 mM Tris-Cl	100 µl of 1 M (pH 8.3 at room temperature)
H$_2$O	400 µl

Store at –20°C indefinitely.

2. Transfer 40 µl of labeled primer mixture into a 0.5-ml microcentrifuge tube and add 0.5 µl of *Taq*/anti-*Taq*.

Taq/*anti*-Taq
Prepare and store a 1:1 (v/v) mixture of *Taq* DNA polymerase (5 units/µl) and anti-*Taq* antibody (1.1 mg/ml; TaqStart antibody, Clontech Laboratories) as recommended by the supplier.

The use of anti-*Taq* antibody is recommended to suppress unwanted *Taq* DNA polymerase action, such as the polymerization of primers that sometimes occurs (particularly at room temperature) before template DNA is added during preparation of reaction mixtures.

3. Transfer 4 µl of the mixture prepared in step 2 into each of ten labeled, thin-walled PCR tubes.

4. Add 1 µl of genomic DNA to each tube and mix by gentle vortexing. Use test DNA prepared from a different source for each tube, and include at least one control wild-type DNA sample (e.g., genomic DNA from a normal individual) among the samples so that the electrophoretic mobility of its band can be compared with those of the DNAs suspected of carrying mutation(s).

Notes: Genomic DNA can be prepared as described in Wolff and Gemmill (1997). DNA from a complex genome (e.g., human genomic DNA) should then be diluted to 50 ng/µl with 1 mM Tris-Cl (pH 7.8 at 25°C)/100 µM EDTA (pH 8.0) for this step.
For thermal cyclers that do not have a top heating block, overlay each reaction mixture with 15 µl of light mineral oil, briefly centrifuge, and then cover the tube.

5. Perform thermal cycling under the appropriate conditions. For a typical amplification, perform an initial denaturation cycle of 94°C for 1 minute and then perform thermal cycling as follows for 30 cycles:

94°C for 30 seconds
60–65°C for 2 minutes

At the end of the last cycle, hold the samples at 4°C.

Note: Although the conditions described here typically work well, it may be necessary to adjust the cycling parameters with some primer-template combinations or with some thermal cyclers.

6. Add 45 µl of formamide/dye gel-loading solution to each tube, mix by vortexing, centrifuge briefly to collect the liquid at the bottom of the tubes, mix by vortexing, and centrifuge briefly again.

Formamide/dye gel-loading solution

Component and final concentration	Amount to add per 10 ml
95% deionized **formamide**	9.5 ml
2 mM EDTA	40 µl of 0.5 M (pH 8.0)
0.05% bromophenol blue	5 mg
0.05% xylene cyanole FF	5 mg
H_2O	460 µl

Store at 4°C for up to 1 year.

formamide (see Appendix for Caution)

7. Store the PCR products at room temperature until needed for loading the gel on p. 302, step 5. *Do not place on ice;* cooling PCR products in formamide/dye gel-loading solution to 0–4°C may allow strand reassociation and sometimes strand–primer complex formation.

Amplification and Internal Labeling of the PCR Products

1. Prepare a 40.5-μl reaction mixture by combining the components below (in the order listed) in a 0.5-ml microcentrifuge tube.

H_2O	26 μl
10x amplification buffer (for preparation, see p. 298)	5 μl
25 mM $MgCl_2$	4 μl
left primer (10 μM in H_2O)	1 μl
right primer (10 μM in H_2O)	1 μl
1.25 mM dNTP mixture (see Appendix)	1 μl
[α-^{32}P]dCTP (10 mCi/ml, 3000 Ci/mmole; Amersham or NEN Life Science Products)	2 μl
Taq DNA polymerase/anti-*Taq* antibody mixture (for preparation, see p. 298)	0.5 μl

radioactive substances (see Appendix for Caution)

2. Use the mixture from step 1 above and follow steps 3–7 on pp. 298–299.

PROTOCOL

Gel Electrophoresis for SSCP Analysis

1. Assemble the gel plates for a water-jacketed electrophoresis apparatus.

   Note: The electrophoresis apparatus (e.g., Atto Corp.) used here has a temperature-controlled circulating-water jacket. It holds a gel approximately 0.3 mm thick, 30 cm wide, and 40 cm high. The 50-lane shark's tooth comb creates lanes 5 mm in width.

2. For a gel with dimensions of 0.03 x 30 x 40 cm, prepare 45 ml of TBE-based gel mixture for SSCP analysis (with or without glycerol) or prepare 45 ml of low-pH TME-based gel mixture just before use. Adjust the volumes proportionately for a gel of different dimensions.

   To prepare TBE-based gel mixture for SSCP analysis:
   Combine the components below in a 50-ml Falcon tube and mix by inverting.

50% glycerol (*optional*)	4.5 ml
5x TBE buffer	4.5 ml
acrylamide:methylenebisacrylamide (99:1) solution	4.5 ml
1.6% ammonium persulfate	1.5 ml
H_2O	to a final volume of 45 ml

   To prepare low-pH TBE-based gel mixture for SSCP analysis:
   Combine the components below in a 50-ml Falcon tube and mix.

10x TME buffer	4.5 ml
acrylamide:methylenebisacrylamide (99:1) solution	4.5 ml
1.6% ammonium persulfate	1.5 ml
H_2O	34.5 ml

   Notes: Although not understood, the use of polyacrylamide gels with a low ratio of cross-linking reagent (*N,N'*-methylenebisacrylamide) to acrylamide is important for efficient detection of mutations by SSCP analysis.

   A recommended initial condition is electrophoresis at 25°C in a polyacrylamide gel containing 5% glycerol, since a large fraction of mutations in most DNA fragments are detected under such conditions. Additional mutations may be detected by using other electrophoresis conditions (e.g., running a gel without glycerol at 25°C).

   A higher detection rate of mutations in long PCR products (up to 800 bp) can be achieved by using low-pH buffer (Kukita et al. 1997). A practical design for efficient screening of mutations is to start from approximately 600-bp PCR fragments and to substitute low-pH TME-based gel mixture for the TBE-based gel mixture in the SSCP analysis. Broadening of bands is occasionally seen in the low-pH gels for SSCP analysis, depending on sequence contexts that are difficult to predict. Performing electrophoresis at higher temperature (e.g., 35°C) often, but not always, alleviates this problem but costs a little sensitivity. A possible explanation of the enhanced sensitivity under the acidic condition is that fewer phosphate charges in the nucleic acid backbone at lower pH reduces intramolecular repulsion and encourages involvement of more nucleotide residues in the stabilization of the tertiary structure.

Acrylamide:methylenebisacrylamide (99:1) solution

Component and final concentration	Amount to add per 100 ml
49.5% acrylamide	49.5 g
0.5% *N,N′*-methylenebisacrylamide	0.5 g
H_2O	to make 100 ml

Store at 4°C for up to 1 year. Before use, thaw any crystals in the solution by swirling at room temperature.

10x TME buffer

Component and final concentration	Amount to add per 1 liter
300 mM Tris base	36.4 g
350 mM MES (free acid, monohydrate; Sigma)	74.6 g
10 mM EDTA	20 ml of 0.5 M (pH 8.0)
H_2O	to make 1 liter

Store at room temperature indefinitely.

acrylamide, methylenebisacrylamide (see Appendix for Caution)

3. Add 45 μl of TEMED to the gel mixture from step 2 and mix. Immediately pour into the assembled gel plates. Keep the plates in a horizontal position and insert the flat side of a shark's tooth comb in the top of the gel mixture to a depth of approximately 5 mm. Allow the gel to polymerize at room temperature for a sufficient length of time (e.g., 2 hours).

4. Remove the comb and fit the gel plates into the electrophoresis apparatus. Fill the reservoir with 0.5x TBE buffer. Use a pasteur pipette filled with the buffer to thoroughly rinse the top surface of gel, and then insert the teeth of the comb.

   Notes: If the low-pH TME-based gel mixture was used to prepare the gel, substitute 0.5x TME buffer for the TBE buffer.

   If an air-cooling system is being used, attach an aluminum plate after fitting the gel plates into the apparatus.

5. Heat the PCR products in the formamide/dye gel-loading solution (from pp. 297–299 or p. 300) at 80°C for 5 minutes. Load 1 μl of each in a separate lane of the gel.

   Notes: Once PCR products in formamide/dye gel-loading solution have been heated, chilling the samples on ice before loading them on the gel is not recommended. Such treatment may allow remaining oligonucleotide primers to associate with the single-stranded PCR products, thus complicating the analysis of the gel (Cai and Touitou 1994). For similar reasons, diluting the PCR products with formamide/dye gel-loading solution before heating them is

recommended to reduce the concentration of primers and facilitate strand separation.
If a mineral oil overlay was used, it does not need to be removed.

formamide (see Appendix for Caution)

6. Begin electrophoresis at 40 W. Keep the gel cool by using the temperature-controlled water jacket. *Do not touch the gel plates while the gel is running.*

   Notes: The time required for electrophoresis depends on the length and nucleotide sequence of the target fragment. For a gel containing glycerol run at 25°C, suggested times for the first trial are 2 hours for 150-nucleotide DNA fragments (the bromophenol blue dye should reach 5 cm from the bottom) and 4 hours for 400-bp fragments (the xylene cyanole FF should reach 5 cm from the bottom). The mobility in gels without glycerol is faster by 1.5- to 2-fold because the reaction of glycerol with borate ions lowers the pH. Electrophoresis at lower temperature requires longer times. Better resolution is achieved by longer migration distances. The duration of electrophoresis should be set so that the bands of interest migrate into the bottom quarter of the gel.

   Alternatives for maintaining a constant temperature during electrophoresis are discussed on p. 294 and p. 295.

7. When electrophoresis is completed, transfer the gel onto a sheet of filter paper, cover with plastic wrap, and dry on a gel dryer.

8. Place X-ray film on top of the covered, dried gel. Mark the position and orientation of the gel on the film by stapling at three corners of the gel.

9. Expose the film at room temperature for a few hours to overnight, depending on the level of radioactivity in the PCR product. Remove the staples and develop the film.

   Note: Expose first for 2–3 hours. If the signal is not strong enough, continue the exposure overnight. If necessary, use an intensifying screen and expose at –80°C.

10. Interpret the bands on the autoradiograph to determine whether or not other alleles are present in each of the test samples used for SSCP analysis.

    Note: In general, comparison of autoradiographic bands present in test samples to a control wild-type sample provides sufficient information to determine whether or not other alleles are present in the test sample used for SSCP analysis (Figure 1). Autoradiographs usually contain two bands for each DNA fragment, one for each of the separated strands. For some DNA fragments under some gel conditions, the two strands may not be resolved, and a single band is seen for each DNA fragment. In addition, three or more bands often appear for a single allele of a DNA fragment. In each of these cases, it is thought that the nucleotide sequence of the DNA fragment can adopt more than one stable conformation or that the 3′ ends of the PCR product are heterogeneous because of the addition of an extra nucleotide (usually A) at the 3′ ends of DNA fragments by *Taq* DNA polymerase. DNA strands with the extra nucleotide have a conformation different from that of DNA strands without the addition. Such extra bands also appear reproducibly, both in their positions and in their relative abundances, in all lanes with control DNA fragments, allowing a definitive interpretation of the results.

11. If desired, isolate each allele resolved during SSCP analysis and separately determine its nucleotide sequence as described on pp. 305–306.

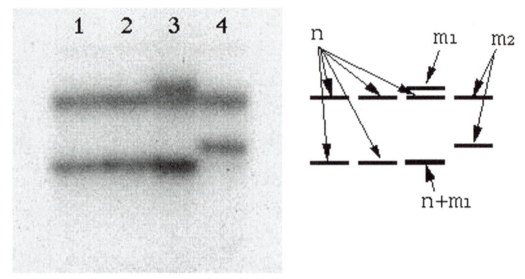

Figure 1 Autoradiograph of a gel for detection of mutations by SSCP analysis. Exon 8 of the *p53* gene (a tumor suppressor gene) in four tumor-derived cell lines was amplified by using primers annealing in nearby introns (amplification unit 330 bp) and examined by the SSCP method. (*Left*) An appropriate area of the autoradiograph; (*right*) its interpretation. Bands denoted n correspond to strands of normal allele; bands denoted m1 and m2 correspond to strands of two different mutant alleles. The faster-migrating strand (lower band) of the m1 mutant (lane 3) comigrated with that of normal allele, and slower-migrating strand (upper band) of the m2 mutant allele had mobility indistinguishable from that of the normal allele (lane 4). Two of the cell lines (lanes 1 and 2) had no mutation in this exon. The third cell line (lane 3) was a heterozygote—one allele carrying a mutation in this exon, the other allele without a mutation in this exon. The fourth cell line (lane 4) also carried a mutation in this exon, but the mutation was different from that found in the third cell line. The normal allele was absent in this cell line.

PROTOCOL

Gel Purification, Reamplification, and Sequencing of SSCP Fragments

1. Place the filter paper carrying the dried gel, with the gel side up, on top of the autoradiograph from pp. 301–304. Align the gel and the film exactly at the position used for autoradiography by referring to the holes made by the staples, and fix the position again by stapling.

2. Place the film/filter paper/gel on a light box. Make the filter paper on top of the band of interest partially transparent by using a pasteur pipette filled with ethanol to wet the appropriate area of the paper. Use a razor blade to cut a small rectangular area of approximately 1 mm x 2 mm in the area of the gel containing the band of interest.

3. Remove the plastic wrap with fine forceps, and peel the rectangle of dried gel off the filter paper. Place the piece of dried gel in a 0.5-ml microcentrifuge tube and immerse in 20 μl of H_2O.

4. Heat the tube at 80°C for 3 minutes, and then allow to cool to room temperature. Centrifuge briefly to collect the liquid at the bottom of the tube.

5. Prepare a 40-μl PCR mixture by combining the components below (in the order listed) in a thin-walled PCR tube. Perform thermal cycling for 20–25 cycles as described on p. 298, step 5.

PCR template (material from step 4)	1 μl
H_2O	27.5 μl
10x amplification buffer (for preparation, see p. 298)	4 μl
1.25 mM dNTP mixture (see Appendix)	1 μl
25 mM $MgCl_2$	4 μl
left primer (10 μM in H_2O)	1 μl
right primer (10 μM in H_2O)	1 μl
Taq/anti-*Taq* (for preparation, see p. 298)	0.5 μl

   Note: This mixture should include the same two primers used in the initial PCR, but the primers should not be radiolabeled.

6. Add 200 μl of H_2O to the PCR-amplified sample and apply it to a Microcon-100 microconcentrator (Amicon). Centrifuge in a Sorvall SS34 rotor (or equivalent) at 2500 rpm (r_{center} = 7 cm; $500g$) at 4°C for 10 minutes.

7. Wash the sample twice by adding 200 μl of H_2O and recentrifuging.

   Note: The primers and unincorporated dNTPs from the PCR step will go through the filter and be collected in the bottom tube of the microconcentrator. These can be discarded. PCR products larger than 100 bp will be retained in the upper chamber.

8. Invert the upper chamber of the microconcentrator and fit it to a new bottom tube (supplied by manufacturer). Centrifuge in a Sorvall SS34 rotor (or equivalent) at 3500 rpm (a rotor with r_{center} = 7 cm; 1000g) at 4°C for 4 minutes to recover the concentrated PCR-amplified sample.

9. Adjust the final volume to 20 μl with H_2O.

   Note: This concentrated PCR-amplified sample can be stored at –20°C indefinitely.

10. Mix 1 μl of the concentrated PCR product with 3 μl of 5% glycerol/0.05% bromophenol blue/0.05% xylene cyanole FF. Analyze on a 2% agarose gel along with a known amount of marker DNA (e.g., *Hae*III-digested φX174 DNA). Estimate the DNA concentration of the test sample by comparing the intensities after staining with **ethidium bromide** (see Appendix).

5% Glycerol/0.05% bromophenol blue/0.05% xylene cyanole FF

Component and final concentration	Amount to add per 10 ml
5% glycerol	0.5 ml
0.05% bromophenol blue	5 mg
0.05% xylene cyanole FF	5 mg
H_2O	to make 10 ml

Store at –20°C indefinitely.

ethidium bromide (see Appendix for Caution)

11. Use 20–50 pmoles of the concentrated, reamplified DNA to perform cycle sequencing as described on pp. 367–370 (see also Wilson and Mardis 1997) or as specified by the manufacturer of one of the cycle sequencing kits (e.g., AmpliCycle [Perkin-Elmer]; SequiTherm [Epicentre Technologies]).

The DGGE Method

DGGE, first described by Fischer and Lerman (1983), uses a polyacrylamide gel containing a linear gradient of DNA denaturants to separate DNA fragments that differ by as little as a single base. The basis of the separation is that DNA fragments with nucleotide sequence differences, even differences as small as a single base substitution, have different melting behaviors. The method works with DNA fragments 100 bp to approximately 600 bp in size, although in some cases, larger fragments can be used. One of the advantages of DGGE is that a large fraction (up to 100% if particular care is taken) of the possible DNA changes in most DNA fragments can be detected. In addition, it is possible to examine as many as 10–30 different DNA fragments in a single gel lane, which allows several kilobase pairs of DNA to be screened for variation at once. A disadvantage is that the technique requires special electrophoresis equipment and is more complicated to perform than some other techniques, such as the SSCP method.

An understanding of the theoretical basis of DGGE is helpful for designing experiments appropriate for particular needs.

THEORETICAL BASIS OF DGGE

When the temperature or concentration of chemical denaturant is gradually raised in a solution containing a duplex DNA molecule, the molecule melts (i.e., it undergoes the transition from double-stranded to single-stranded form) in a discrete pattern. Stretches of contiguous DNA sequence, called melting domains, melt cooperatively at a precise temperature or denaturant concentration that is referred to as the T_m of the domain. Melting domains can vary from 25 bp to several hundred base pairs in length, and sharp boundaries appear to exist between the domains in a molecule. The T_m of each melting domain is highly dependent on its nucleotide sequence, not just its composition. This sequence dependence is due to the fact that stacking interactions between adjacent bases on the same strand make a significant contribution to stabilizing the double helix. The amount of stacking in a DNA strand varies with the nearest neighbors in the strand, so that the stability of the helix is highly dependent on the nucleotide sequence. For this reason and the fact that DNA denaturation is highly cooperative, a change in the DNA sequence as small as a single base substitution causes a change in the T_m of a melting domain.

Fischer and Lerman (1983) used this knowledge of the melting behavior of duplex DNA in solution and knowledge of the electrophoretic properties of DNA in a gel matrix to develop the DGGE method. In the most common version of this method, a polyacrylamide gel is prepared so that it contains a linearly increasing gradient of the DNA denaturants formamide and urea from top to bottom. The gel is placed in a special apparatus in an aquarium filled with heated (usually to 60ºC) electrophoresis buffer, and DNA samples are loaded for electrophoresis into the gel.

Initially, the DNA fragments migrate in the gel at a rate determined by their molecular weight. When a DNA molecule reaches the concentration of denaturant in the gel corresponding (in combination with the heated temperature of the gel) to the T_m of the first melting domain, that domain melts to form a Y-shaped molecule or a "bubbled" molecule, depending on whether the domain is at one end or in the middle of the fragment. The partial melting causes the mobility of the DNA fragment to decrease in the gel, resulting in a sharp focused band. Because the position in the gel at which the branching occurs is dependent on the T_m, DNA fragments of even slightly different sequence begin slowing down in the gel at different positions corresponding to their T_m values and thus are separated from one another. Lerman and colleagues have been able to use this theory of melting behavior, combined with experimental measurements of the stabilities of dinucleotides in duplex form determined by Gotoh and colleagues (Gotoh and Tagashira 1981; Gotoh 1983), to devise a computer program called MELTMAP that predicts with remarkable accuracy the melting behaviors of DNA fragments on the basis of their nucleotide sequences. This program is helpful for optimizing gel conditions for particular DNA fragments in applications of DGGE where it is desirable to detect a very high fraction of all possible base changes in a DNA fragment.

Although it is not well understood, there is empirical evidence that different DNA fragments vary widely in their degrees of mobility retardation on melting in denaturing gradient gels. Some fragments slow down dramatically (although they do not stop completely) when their first domain melts, whereas other fragments slow down only a very little. In either case, it appears that any degree of mobility reduction is sufficient to allow the separation of DNA fragments differing by a single base or larger changes.

DNA fragments up to 500 bp in length generally contain one to three melting domains. Although the mobility of a multiple-domain fragment is reduced on melting of the domain with the lowest T_m, it is possible to continue electrophoresis of most partially melted fragments through the gel until the domain with the second lowest T_m melts, allowing separation of DNA fragments with base changes in that domain to be observed (Myers et al. 1985c,d). However, because a branched DNA molecule is necessary for DGGE to separate fragments differing by one base change, it is not possible to detect base changes in the domain with the highest T_m. In general, base changes in all but the domain with the highest T_m can be detected by DGGE. Thus, a DNA fragment with three melting domains can be screened effectively for base changes in the first two domains but not in the last domain. However, only a small fraction of base changes in a DNA fragment that melts as a single domain can typically be detected by DGGE, since the melting separates the strands completely and the sequence dependence of their migration on the gel is lost.

VARIATIONS THAT EXTEND THE USEFULNESS OF THE DGGE METHOD

The GC Clamp

To extend the usefulness of DGGE so that all or most melting domains can be screened for base changes, the concept of the "GC clamp" was developed (Myers et al. 1985c,d). A GC clamp is a GC-rich DNA segment with a very high T_m. It can be attached to a DNA fragment to be tested for base changes, so that the melting domain with the highest T_m will be the GC clamp. This allows all of the other melting domains in the fragment to be screened for base changes, except those unusual cases where the test sequence also has a very high T_m. Although the first GC clamps developed were large 300-bp segments that were attached to target DNA fragments by cloning the targets into a GC-clamp vector (pGC1 or pGC2; see Myers et al. 1985b), theoretical melting calculations based on the MELTMAP program suggested that the melting domain with the highest T_m needs to be only 30–40 bp in length. On the basis of this consideration, short GC clamps of 40 bp were attached to genomic DNA target fragments ranging in length from 100 bp to 500 bp by PCR by synthesizing one of the PCR primers with a 40-nucleotide GC-rich segment at its 5' end (Myers et al. 1989b; Sheffield et al. 1989).

Because the 40 nucleotides of the GC clamp must be synthesized de novo for each DNA fragment, which creates additional expense, two methods were developed that allow a single batch of GC-clamp oligonucleotides to be attached to one of the target-specific oligonucleotides used for PCR. One of these methods (Sheffield et al. 1992) involves carrying out two PCR steps, the first of which uses one standard primer and one primer that contains an 8-nucleotide "clamp attachment sequence" at its 5' end. A third primer, a universal GC-clamp primer that recognizes this 8-nucleotide tail and that contains an additional 40 nucleotides of GC-rich sequence on its 5' end, is then used in the second round of PCR (Figure 2a). The second method, called the "chemi-clamp," uses the reactive agent, psoralen, and UV light to cross-link the DNA strands at one end of the PCR-amplified fragment, the product of which behaves like a GC

Figure 2 GC clamps formed with a clamp attachment sequence (*a*) or with a chemi-clamp (*b*). (*a*) Use of a clamp attachment sequence to produce PCR-amplified DNA fragments with a universal GC clamp. The test DNA fragment is amplified with primer 2 in 20-fold excess over primer 1, which contains an 8-nucleotide addition called a clamp attachment sequence at its 5' end. After this first round of amplification, a second round is performed with primer 2 and primer 3, which is a 40-nucleotide universal GC-clamp primer in which the last 8 nucleotides at its 3' end are the same as the clamp attachment sequence. This allows the 40-nucleotide GC-clamp fragment to be attached to the end of the product of the first round of amplification. (*b*) The use of amplification to attach a chemi-clamp to one end of a test DNA fragment. One of the two primers used to amplify the test DNA fragment is synthesized with a psoralen derivative on its 5' end. After amplification, the sample is treated with UV light, which cross-links the psoralen-containing strand to its complementary strand, very near one end of the PCR-amplified fragment. This cross-link serves the same function as a GC-clamp fragment (i.e., it keeps one end of the fragments together during DGGE).

clamp because the DNA strands do not come apart during electrophoresis (Figure 2b) (Costes et al. 1993).

Use of Heteroduplexes

It was predicted and then demonstrated (Myers et al. 1988; Sheffield et al. 1989) that mismatched DNA fragments would have a greatly decreased mobility in denaturing gradient gels relative to the wild-type and mutant homoduplexes of the same DNA molecule. If equimolar amounts of two homoduplexes differing by a single base are melted and reannealed, four species of the DNA fragment (two homoduplexes and two heteroduplexes) are formed. These four forms of the fragment usually

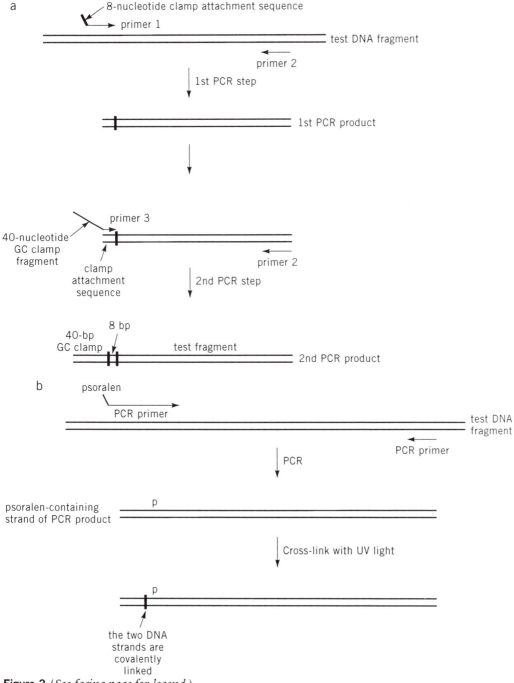

Figure 2 (*See facing page for legend.*)

separate from one another during DGGE, with the two heteroduplexes showing a retarded mobility relative to the two homoduplexes. This pattern of four fragments is easy to recognize, and the decreased mobility of heteroduplexes increases the resolution and the fraction of mutations that can be detected by DGGE.

Genomic DGGE

Although it is often desirable to use a screening technique such as DGGE at its maximum power, such that the highest fraction of base pairs in the target DNA fragments are screened for sequence variation, there are many occasions when a more rapid, but perhaps less thorough, approach is warranted. For example, it is often important to identify a polymorphic variation in a locus for which there is no genetic marker or for which the existing genetic markers are uninformative in the particular individuals being studied. For example, in the comparison of any two human chromosomes, identifying a single polymorphism may require screening up to several thousand base pairs of DNA at a locus. This amount of comparison is difficult with a DGGE approach, such as the PCR-based GC-clamping method described above, that examines single DNA fragments of a few hundred base pairs at a time. Indeed, any other method, including SSCP, RNase cleavage, CCM, and direct sequencing, that typically allows only a few hundred base pairs to be screened at a time must be performed multiple times to allow thousands of base pairs to be screened.

A variant of DGGE, known as gDGGE (Børresen et al. 1988; Burmeister et al. 1991), allows many DNA fragments to be examined in a single lane on the gel, such that many hundreds or several thousand base pairs can be screened for variation. Typi-cally, genomic DNA from any organism (including highly complex mammalian genomes) can be digested with one or more restriction enzymes that have 4-base recognition sites (e.g., *Alu*I, *Rsa*I, *Sau*3A, and *Hae*III), electrophoretically separated on two denaturing gradient gels with overlapping ranges of denaturant concentrations (e.g., 0–50% denaturants and 40–80% denaturants), and transferred onto a membrane by using an electroblotting apparatus. The DNA on the membrane can then be hybridized by using a labeled probe that recognizes a desired segment of DNA (usually 1–10 kb in length) and autoradiographed.

As with PCR-amplified or cloned samples detected by staining with ethidium bromide, variation in the genomic DNA samples is detected on such autoradiographs by mobilities of bands that vary from one sample to another. However, because complex genomic DNA is examined directly, there are no heteroduplexes when gDGGE is used, so the advantages of increased resolution and higher number of base pairs screened are lost. Nevertheless, because probes can be at least as long as 10 kb with this method, a large number of base pairs can be screened in a single hybridization. In addition, blots can be reprobed multiple times with probes that recognize different loci, which greatly increases efficiency when many loci must be examined.

In some cases, cloned or PCR-amplified DNA segments from two individuals or two different chromosome homologs from a single individual are available, and the principles of gDGGE can be applied to such samples to allow rapid screening for variation. However, unlike cases in which only genomic DNA is available, heteroduplexes can be formed between the two cloned samples, allowing for a larger fraction of the base pairs to be screened.

EQUIPMENT AND PROTOCOLS FOR DETECTING DNA VARIATION USING DGGE

Several preliminary considerations for using DGGE are discussed below, including a description of the equipment, a description of several ways of optimizing the gel conditions for particular target DNA fragments, and the methods for pouring and running the gels. These discussions are followed by protocols for performing DGGE with PCR-amplified or cloned DNA fragments (both with and without a GC clamp) and for performing gDGGE.

Equipment

The separation of DNA fragments by DGGE is sensitive to differences in temperature. It is therefore important that the temperature of the chemical denaturant gradient gel be maintained throughout electrophoresis. Because the T_m values of most naturally occurring DNAs are too high to achieve with chemical denaturants alone, DGGE is generally carried out at a fixed temperature of 60°C, which is a few degrees below the T_m of most DNA fragments. This fixed temperature is easily achieved by using a gel cassette and submerging the gel in an aquarium containing heated electrophoresis buffer (Figure 3). The cathode chamber in the gel cassette is separated from the heated buffer in the aquarium, which allows current to flow through the gel. Sources for the gel cassette, aquarium, heaters, and all required equipment include C.B.S. Scientific and Bio-Rad. This equipment has also been described in detail (Myers et al. 1987, 1988). Although the systems from different suppliers vary somewhat, the principles for preparing and running denaturing gradient gels with each are similar, and the protocols described here for parallel and perpendicular denaturing gradient gels can be adapted as suggested by the manufacturers.

Optimization of Gel Conditions for DGGE

Although a reasonable fraction of mutations are detected when DNA fragments are tested with two standard gel conditions in DGGE, the fraction can be increased by optimizing the conditions for each particular DNA fragment by using one of the following preliminary tests: the MELTMAP computer program, perpendicular DGGE, or empirical testing.

THE MELTMAP PROGRAM

The melting behavior of the target DNA fragment can be predicted by a computer program on the basis of the nucleotide sequence of the fragment as described in the protocol on pp. 315–317. A gel with the appropriate concentration range of denaturants can then be used to maximize separation of DNA fragments with such melting properties.

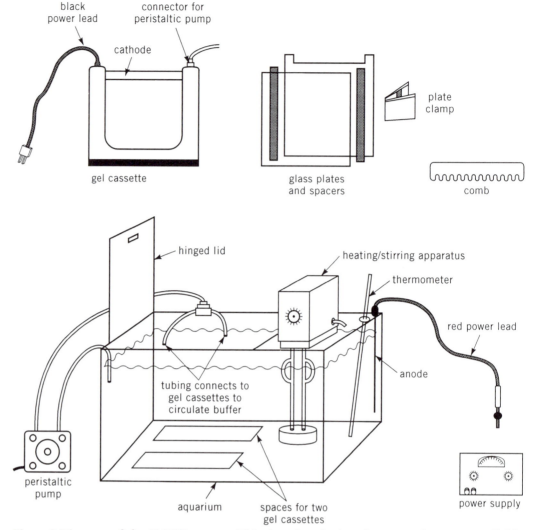

Figure 3 Diagram of the DGGE system. This drawing depicts the type of system available from C.B.S. Scientific, which is simpler to use and less expensive than the one described in Myers et al. (1987). Similar systems are available from Bio-Rad or can be built from drawings described in Myers et al. (1987). Both perpendicular and parallel denaturing gradient gels are cast in sealed and clamped gel plates. After polymerization of the gel, the sides of the plates are taped to prevent perpendicular electric fields from causing the samples in the outside lanes to migrate aberrantly. The taped gel plates are then clamped into the gel cassette and placed in an aquarium filled with electrophoresis buffer heated to 60°C. Electrodes are connected to the cassette and the aquarium. Tubing from a peristaltic pump is connected to the cassette. Electrophoresis is then performed at 150 V. For details of pouring gels, see text and Figure 5.

In cases where the DNA sequence of the test segment and the MELTMAP program are available, the most efficient way to use DGGE is to determine the melting behavior of a test DNA fragment before experimental manipulation and then use the gel conditions that are predicted to screen the maximum number of base pairs in this fragment on the basis of these melting calculations. Such preliminary work allows the investigator to determine the number, positions, and T_m values of the melting domains in the test DNA fragment and, in cases where a GC clamp is used, to determine at which of the two ends of the test fragment to place the GC clamp.

PERPENDICULAR DGGE

Electrophoresis of the target DNA sample can be performed on a gel containing a gradient of denaturants perpendicular to the direction of electrophoresis. This gradient ranges from a low concentration (usually 0% denaturants) on the left side of the gel to a high concentration (usually 80% denaturants) on the right. The gel is cast between two glass gel plates separated by spacers so that its top has a single large well into which the sample is loaded. The gradient solution is prepared by placing a different denaturant stock solution (one solution for the lowest and one for the highest desired denaturant concentration) in each of the two adjacent chambers of a gravity-fed gradient maker and then allowing the low-concentration denaturant solution to mix gradually with the high-concentration solution as it drips from the high-concentration chamber through a small gap in the spacers on the left side of the gel plates. After the gradient is poured and the gel polymerizes, the gel plates are clamped into a gradient gel cassette, and electrophoresis is performed with the gel submerged in an aquarium of electrophoresis buffer heated to a temperature near the T_m of most naturally occurring DNA fragments (generally 60°C). After electrophoresis, the gel is stained with ethidium bromide (or a blot is prepared and probed as described for gDGGE), and the melting profile of the target fragment is interpreted from the location of the DNA on the gel.

Perpendicular denaturing gradient gels can be used to determine empirically enough information about the melting properties of a target DNA fragment to predict optimal gel conditions of the fragment. Although perpendicular denaturing gradient gels do not indicate the location of melting domains in the target DNA fragment, they indicate the number of domains and the T_m of each domain. This information allows investigators to design the appropriate parallel denaturing gradient gel and to determine the optimum duration of electrophoresis for the fragment. A 0–80% perpendicular denaturing gradient gel is prepared, run, and interpreted in the protocols on pp. 318–326.

EMPIRICAL TESTING

If wild-type DNA and one or more known mutant DNA fragments are available for a target DNA fragment in cloned form, it is possible to estimate the best gel conditions by simple trial and error. Simply testing the ability of two gradient gels containing different denaturant concentration ranges to separate the wild-type and mutant forms of a target DNA fragment is often sufficient. Electrophoresis of the two samples should be performed for various lengths of time since no information is available about the number of melting domains.

The modifications required for loading and running parallel denaturing gradient gels to determine optimal gel conditions are provided in the protocol on p. 327.

Analysis of DNA Mutations by Parallel DGGE

The solutions and many of the technical details for preparing and running parallel denaturing gradient gels are the same as those for perpendicular denaturing gradient gels. The major difference in the parallel denaturing gradient gel is that the gradient of DNA denaturants (typically a mixture of urea and formamide) increases linearly from top to bottom and a comb is used to form multiple wells at the top of the gel as for most other types of gels. A typical gel has a 30% range of denaturant concentrations and 6.5% polyacrylamide. The basic protocol for preparing and running parallel denaturing gradient gels is provided on pp. 328–331.

A parallel denaturing gradient gel is used for almost all DGGE experiments, since it allows the examination of 20–30 samples on one gel. In general, appropriate melting conditions are first determined for the target DNA fragment by using the MELTMAP program, perpendicular DGGE, or empirical testing.

Although either homoduplexes or heteroduplexes can be analyzed for variations by parallel DGGE, this method is more efficacious when heteroduplexed DNA fragments containing mismatches at a single position are examined. Heteroduplexes can be prepared from cloned or PCR-amplified samples of the two variants of the target DNA fragment as described in the protocol on pp. 332–334.

Analysis by gDGGE is provided in the protocol on pp. 335–339. Various versions of protocols for performing gDGGE have been described previously (Børresen et al. 1988; Burmeister et al. 1991; Gray et al. 1991; Reindollar et al. 1992). In gDGGE, multiple DNA fragments are loaded in each lane. These fragments are typically restriction-enzyme-digested products from genomic DNA. (The principles of gDGGE can also be applied to digested products from a complex cloned sample [e.g., cosmid, BAC, bacteriophage P1] or from a mixture of different PCR-amplified samples.) After electrophoresis, the DNA is transferred onto a membrane by using an electroblotting apparatus and then hybridized by using a labeled, single-copy probe that recognizes up to 10 kb of the sample in the blot. After detection of signals, the blot can be stripped and rehybridized multiple times with additional probes.

Optimizing Gel Conditions Using Preliminary Melting Determinations from the MELTMAP Program

1. Enter the nucleotide sequence of the test DNA fragment in the computer. If a GC clamp is to be used, enter the test sequence in two separate files, one in which a 40-nucleotide GC-clamp sequence is attached to one end of the fragment and one in which the clamp is attached to the other end of the fragment.

   Notes: For experiments in which a GC clamp will be attached to the target fragment by PCR, the nucleotide sequence of a GC clamp that is known to be effective is: 5'CGCCC GCCGC GCCCC GCGCC CGCCC CGCCG CCCCC GCCCG3', where 18–25 nucleotides of unique DNA sequence corresponding to one end of the target DNA fragment continues at the 3'end of this clamp.

   Both melting theory and experiments indicate that most random GC-rich sequences will serve as effective GC clamps. The sequence above is recommended because it has been used successfully with a large number of different target DNA fragments. In the original publication describing PCR-based GC clamps (Sheffield et al. 1989), a GC clamp of almost identical sequence to that above was used, except that a T residue was placed at one position near the middle of the clamp. Because the melting calculations showed that this T residue decreased the T_m of the GC clamp enough to cause problems with GC-rich target segments, all subsequent experiments have been performed with the clamp consisting entirely G and C that is listed above.

   A description of the MELTMAP program and its use is published in Lerman and Silverstein (1987) and the program is available from L. Lerman (lerman@fang.mit.edu). Alternatively, Macintosh and PC versions of the program are available from A. Grow, L. Lerman, and R.M. Myers at myers@shgc.stanford.edu, and a Macintosh version is available commercially from Bio-Rad.

2. Run the MELTMAP program and generate an output file of the melting map.

   Note: An example showing the two melting maps for a typical target fragment with a GC clamp at its right or left end is shown in Figure 4.

3. Use the information from the melting profiles to determine the optimal placement of the GC clamp on the test DNA fragment.

   Note: In the example shown in Figure 4, MELTMAP indicates that the GC clamp attached to the right end of the test DNA fragment (upper panel) has the better of the two melting profiles, since there is a single melting domain for the test fragment with a T_m of 71°C and a single higher-temperature melting domain (>80°C) for the GC clamp. The test fragment with the GC clamp placed at its left end (lower panel) is predicted to have two melting domains in addition to the domain corresponding to the GC clamp; two different gel conditions would therefore be required to optimize separation of wild-type and mutant fragments (see below).

4. To design the appropriate parallel denaturing gradient gel(s) for the test DNA fragment, use the T_m values from the melting profiles and the follow-

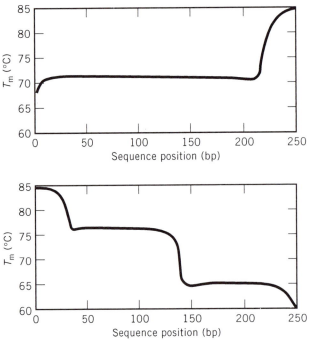

Figure 4 Melting maps. The output of the MELTMAP program showing the melting maps for a typical target DNA fragment with a GC clamp on its right end (*upper panel*) or left end (*lower panel*). At temperatures above the plotted line, the DNA at a sequence position is melted; at temperatures below the line, the DNA is double-helical. Flat segments on the plotted line represent melting domains. For example, the DNA fragment in the upper panel has one domain with a T_m of approximately 71°C from sequence positions 1–210 bp and a second domain (which corresponds to the GC clamp) with a T_m above 80°C.

ing relationship to determine the appropriate denaturant concentrations for the gel(s). When gels are run at 60°C,

denaturant concentration (in %) = 3.2 x (T_m [in °C] − 57)

For test DNA predicted to have one melting domain in addition to the domain corresponding to the GC clamp:
Prepare one parallel denaturing gradient gel with a denaturant concentration ranging from 15% above to 15% below the calculated percentage of denaturants. For example, the single domain in the test fragment in the upper panel of Figure 4 melts at a calculated denaturant concentration of 45%. Therefore, a gel with a denaturant concentration range of 30–60% is optimal.

For test DNA predicted to have two melting domains in addition to the domain corresponding to the GC clamp:
Use two different parallel denaturing gradient gels to maximize the ability to detect mutations in each. For each melting domain, prepare a gel with a denaturant concentration ranging from 15% above to 15% below the calculated percentage of denaturants. For example, the target DNA fragment

in the lower panel of Figure 4 has three melting domains. The one with the lowest T_m (66°C) melts at a calculated denaturant concentration of 29%; the next lowest (76°C) melts at a calculated denaturant concentration of 61%; the domain with the highest T_m corresponds to the GC clamp. For the first domain, a parallel denaturing gradient gel with 15–45% denaturants is optimal. For the second domain, a gel with 45–75% denaturants is optimal.

For test DNA predicted to have two melting domains in addition to the domain corresponding to the GC clamp and calculated denaturant concentrations within 10–15% of each other:
Use a single parallel denaturing gradient gel with denaturant concentrations ranging from approximately 10% above the higher calculated denaturant concentration to approximately 10% below the lower concentration.

Optimizing Gel Conditions Using Perpendicular DGGE

PREPARING PERPENDICULAR DENATURING GRADIENT GELS

1. Prepare the acrylamide:bisacrylamide and DNA denaturant stock solutions.

   Note: It is useful to understand a convention used in DGGE to describe the concentration range of denaturants. A solution of acrylamide containing 7 M urea and 40% formamide is referred to as a 100% denaturant stock solution, since these concentrations are near saturation in aqueous solution. In general, an 80% denaturant stock solution is the maximum required. For gels run at 60°C, only rare GC-rich target fragments require a denaturant concentration higher than approximately 80% to reach the T_m for the melting domain with the highest T_m.

Acrylamide:bisacrylamide (37.5:1) solution

Component and final concentration	Amount to add per 0.5 liter
40% **acrylamide** (high grade)	200 g
1.07% **bisacrylamide**	5.35 g
H_2O	to make 0.5 liter

Do not autoclave. Store at 4°C for several months.

0% Denaturant stock solution

Component and final concentration	Amount to add per 0.5 liter
6.5% acrylamide	81.25 ml of acrylamide: bisacrylamide (37.5:1) solution
1x TAE buffer	10 ml of 50x
H_2O	408.75 ml

Store at 4°C for up to 6 months.

80% Denaturant stock solution

Component and final concentration	Amount to add per 0.5 liter
6.5% acrylamide	81.25 ml of acrylamide: bisacrylamide (37.5:1) solution
32% deionized **formamide**	160 ml
5.6 M urea (ultrapure)	168 g
1x TAE buffer	10 ml of 50x
H_2O	to make 0.5 liter

Store at room temperature for up to 6 months.

acrylamide, bisacrylamide, formamide (see Appendix for Caution)

2. Fill the aquarium with 1x TAE buffer and turn on the heating/stirring apparatus to heat the buffer to 60°C.

   Notes: To minimize evaporation, keep the lid on the aquarium closed while heating. If no lid is available, use floating plastic balls or styrofoam chips to cover the surface of the buffer.

   For a discussion of the equipment used for DGGE, see p. 311; see also Figures 3 and 5.

3. Clean the glass gel plates. Place the spacers at each edge of the notched plate, but leave a 2–3-cm gap at the bottom of the left side of the plate. Also place a single large spacer at the top of the glass plate to form a large loading well (and to keep the gel solution within the plates). Carefully place the unnotched glass plate on top of the spacers. Tape all edges (except the 2–3-cm gap) of the plates together carefully with waterproof tape; be particularly careful to seal the bottom of the gel and to avoid creating bubbles in the tape.

   Notes: A bottom spacer is also available for some systems. The use of this spacer helps minimize leaking, but tape should also still be used in addition to the spacer.

   For the system available from C.B.S. Scientific, sealing gaskets called Gel-Wrap allow gels to be poured without taping the plates.

4. Place clamps as close together as possible around the taped edges. Stand the taped and clamped plates on their right edge so that gel solution can be introduced through the gap on the left side.

5. Determine the precise volume of liquid required to fill the space between the gel plates. Transfer 0% denaturant stock solution into a test tube to provide half of the required volume, and transfer 80% denaturant stock solution into a separate test tube to provide the other half of the volume. Place the tubes on ice.

   Note: The precise volume of liquid required to fill the space between the gel plates should be determined empirically by preparing a set of plates as described above and measuring the volume of liquid required to fill them. Most spacer/gel plate combinations require approximately 22 ml. Therefore, approximately 11 ml of each of the two denaturant stock solutions is required to pour the gradient.

6. Add 10 μl of TEMED and 100 μl of 10% ammonium persulfate to each of the two tubes. Mix thoroughly by inverting each tube several times and place on ice.

   Note: These volumes of TEMED and ammonium persulfate are appropriate for approximately 10–12 ml of denaturant solution.

7. With the stopcocks closed, pour the 80% denaturant stock solution into the chamber of the gradient maker that contains the exit port (for details, see Figure 5). Briefly open the stopcock between the two chambers to allow liquid to flow into the small channel connecting the chambers. Use a pasteur pipette to transfer the solution that flows into the empty chamber back into the chamber containing the 80% denaturant, but be careful to leave liquid (no air bubbles) in the small channel.

8. Pour the 0% denaturant stock solution into the other chamber of the gradient maker.

Note: Visually confirm that the levels of the solutions in the two chambers are approximately the same and that there are no air bubbles in the channel between the chambers.

9. Open the stopcock between the chambers of the gradient maker so that the solutions in the two chambers are connected. Use a vibrating mixer or a magnetic stirrer to begin mixing the solution containing the higher concentration of denaturants.

10. Allow the mixed solution to drip from the exit port of the gradient maker through thin-walled tubing and a 20-gauge needle into the taped glass plates through the gap on the left side.

   Note: As the 80% denaturant stock solution leaves its chamber, an equal volume of the 0% solution enters that chamber. Make sure that the levels of both denaturant solutions diminish at the same rate. Mixing of the two solutions should continue throughout the time the gel is being poured. The flow rate should be such that approximately 5 minutes is required to fill the space between the plates.

Figure 5 Preparing perpendicular and parallel denaturing gradient gels. The total volume of acrylamide/denaturant solution required to just fill the gel plates assembled with a particular set of spacers is determined empirically. (*a*) For pouring a perpendicular denaturing gradient gel, the spacer on the left side of the gel should be placed approximately 2–3 cm from the bottom of the plates to allow the solution to enter the gel, and the plates should be stood on their right edge. Two solutions, each equal to half of the required total volume, are prepared with denaturant concentrations representing the extremes of the range desired. After polymerization catalysts are added to both solutions, the solution with the higher concentration of denaturants is poured into the right side of the gradient maker, which contains the exit port. The stopcocks that control the exit port and that connect the left and right chambers of the gradient maker are kept in their closed positions (perpendicular to the direction of liquid flow) at this time. The stopcock connecting the chambers is opened briefly to allow liquid to flow into the small channel connecting the two chambers, and the solution that flows into the left chamber is transferred back into the right chamber with a pasteur pipette. The purpose of this step is to avoid the presence of an air bubble in the small channel, which would prevent the solution on the left from mixing with the solution on the right during the gel pouring process. The solution with the lower concentration of denaturants is then poured into the left chamber, and a small stirring bar or mixer is placed in the right chamber. The gradient maker is then raised above the gel plates on a ring stand, and, while a vibrating mixer or a magnetic stirrer mixes the solution on the right side, the two stopcocks are carefully opened. Liquid is allowed to leave the exit port through thin-walled tubing and enter into the gap in the gel plates through a 20-gauge needle. The liquid should drip into the plates in a thin, continuous stream. The levels of the two solutions in the right and left chambers should gradually decrease at the same rate, which is the result of appropriate mixing. (*b*) The same procedure is used to pour a parallel denaturing gradient gel, except that the two side spacers are placed at the bottom edge (i.e., there should be no gap since the gel liquid is poured into the top) of the gel plates, a comb is used instead of the top spacer, and the plates are placed in their normal running orientation during the pouring of the gel. For both types of gels, the plates should be placed almost flat on a bench top to allow polymerization to occur after the gel is poured. After polymerization, the sides of the gel plates are taped if they were not already taped, and the Gel-Wrap or tape is removed from the bottom of the gel to allow electrophoresis to occur.

11. Allow the gel to polymerize at room temperature in a horizontal position, with the left side of the plates slightly propped up to prevent the liquid from dripping out.

   Note: Because relatively low concentrations of TEMED and ammonium persulfate are used to allow extra time for pouring a gradient gel, polymerization may require longer than it does for other types of gels. Typically, approximately 30 minutes is sufficient.

12. Remove the tape (and spacer, if used) from the bottom of the polymerized gel. Clamp the gel/plates into the gel cassette and place in the aquarium filled with 1x TAE buffer heated to 60°C. Adjust the level of buffer so that the entire gel is submerged but the upper electrophoresis chamber is above the buffer level.

   Notes: The upper electrophoresis chamber is created when the gel/plates are clamped into the gel cassette.

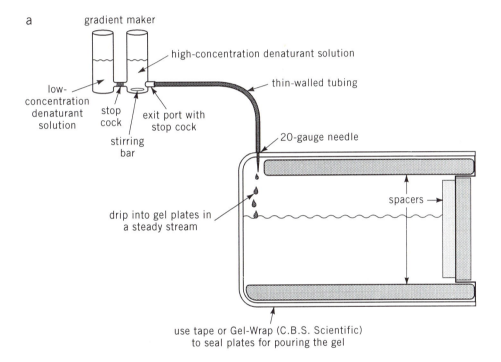

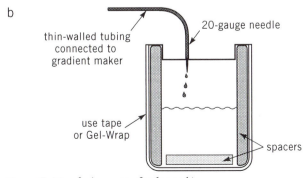

Figure 5 (*See facing page for legend.*)

If the system available from C.B.S. Scientific is being used, remove the Gel-Wrap and tape the sides of the plates with waterproof tape to prevent electric fields from forming perpendicular to the gel.

13. Attach the tubes from the peristaltic pump to the gel cassette, and begin circulating buffer from the aquarium into the upper electrophoresis chamber.

14. Remove the large spacer "comb" from the top of the gel plates. Use a pasteur pipette to rinse the single large well with electrophoresis buffer. The gel is now ready for loading the sample.

RUNNING A PERPENDICULAR DENATURING GRADIENT GEL TO DETERMINE MELTING PROPERTIES OF A TEST DNA FRAGMENT

1. Generate the target DNA sample for the wild-type fragment (~100–600 bp) by PCR or by digesting a cloned DNA fragment with restriction enzymes. If a GC clamp is to be attached to the target fragment by PCR, use one primer with a 5′ GC-rich segment (for a description the segment, see p. 315).

Notes: If cloned DNA is used, the target fragment should be excised from the vector with the appropriate restriction enzymes. However, there is no need to purify the insert from the vector.

The minimum amount of target DNA fragment required for detection by staining with ethidium bromide is approximately 4 µg. The reason such a large amount is needed is that the test DNA fragment will be spread across 10–20 cm of the gel. Much smaller amounts of the test fragment (as little as a few nanograms or up to a few micrograms of cloned or PCR-amplified DNA) can be loaded if the gel is to be blotted and probed by hybridization as described in the gDGGE protocol on pp. 335–339.

2. If desired, analyze an aliquot (50–200 ng of target DNA) of the sample and appropriate molecular-weight markers (100–1000 bp) on a 1% agarose minigel in 1x TAE buffer. Estimate the size and concentration of the DNA fragment. Also confirm that amplification or digestion was successful and then proceed with step 3. Keep the remainder of the DNA on ice while the gel is running.

Note: If digestion was incomplete, add additional enzyme and continue incubating until digestion is complete. If amplification was unsuccessful, prepare PCR products again.

3. Precipitate the target DNA with 0.1 volume of 3 M sodium acetate (pH 5.2) and 2–2.5 volumes of absolute ethanol and dry the pellet under vacuum (see Appendix).

4. Dissolve the target DNA in approximately 100 µl of nondenaturing gel-loading buffer. Carefully load the sample across the well of the gel prepared in the protocol above. Make sure that the sample is distributed evenly.

Note: It is possible to examine multiple target DNA fragments in a single perpendicular denaturing gradient gel if their lengths are known. Although the shapes of melting profiles are not predicted on the basis of fragment length, each fragment in a mixture can be identified by its location on the left (i.e., nondenaturant) side of the gel, where mobility is dependent only on molecular weight.

Nondenaturing gel-loading buffer

Component and final concentration	Amount to add per 10 ml
20% Ficoll (Type 400) or sucrose	2 g
10 mM Tris-Cl	100 μl of 1 M (pH 8.0 at room temperature)
1 mM EDTA	20 μl of 0.5 M (pH 8.0)
~0.1% dye (orange G, xylene cyanole FF, or bromophenol blue)	~10 mg
H$_2$O	to make 10 ml

Store at −20°C for up to 1 year.

5. Connect the electrodes and perform electrophoresis at 150 V (~35–40 mA, constant voltage) for 5–10 hours.

   Notes: Because the DNA sample is loaded across the entire width of a perpendicular denaturing gradient gel and molecules enter all concentrations of denaturant at the beginning of electrophoresis, the duration of electrophoresis is less critical than for parallel denaturing gradient gels. If the gel is run for too short a time for a particular DNA fragment, there may be a less dramatic mobility transition at the point in the gel where a domain melts. This is usually not a problem because it is still possible to determine the positions of the transitions and therefore the number and T_m values of the melting domains. If the gel is run too long for a particular fragment, the DNA may run off the gel at the left side of the gradient where the concentration of denaturants is low. However, since it is likely that DNA will still be present at the right (i.e., high denaturant concentration) side of the gel and at the points in the gradient where the melting transitions for each domain are seen, the T_m values for those domains can still be assessed.

   Because the mobility of a DNA fragment has some dependence on length in DGGE, it is a good idea to run gels longer for large DNA fragments than for small DNA fragments on perpendicular denaturing gradient gels. Table 2 provides a useful starting point for fragments based on size, but the duration of electrophoresis should be adjusted on the basis of results for each test fragment.

6. Turn off the power supply, the heating/stirring apparatus, and the peristaltic pump. Remove the gel plates from the gel cassette. Remove the gel from the plates, and stain the DNA with **ethidium bromide** solution (0.1–0.5 μg/ml) for 15–30 minutes. Examine and photograph the stained gel (with a ruler placed across the top or bottom of the gel) under **UV** transillumination.

Table 2 Approximate Duration of Electrophoresis for DNA Fragments as a Function of Length for Perpendicular DGGE

DNA fragment length (bp)	Duration of electrophoresis (hr)
50–150	4
150–300	6
300–500	8
500–1000	11

These values are estimated for gels containing 6.5% acrylamide that are run at 60°C at 150 V.

Notes: These thin polyacrylamide gels are somewhat flimsy and require some care in handling to avoid tearing. To aid in staining and viewing by transillumination, move the gel from the staining vessel to the UV light box on a glass plate, and use a stream of water from a squirt bottle to help make the gel slide onto the viewing surface.

If the staining with ethidium bromide appears too faint or too heavy, alter the staining conditions to a higher or lower concentration of ethidium bromide, respectively.

ethidium bromide, UV radiation (see Appendix for Caution)

7. Determine the number and T_m of the melting domains in the test DNA fragment on the basis of the pattern on the gel as described in the protocol below.

CALCULATING THE T_m OF DNA FRAGMENTS FROM PERPENDICULAR DENATURING GRADIENT GELS

1. To determine the T_m of the first melting domain in a test DNA fragment with two melting domains, use a ruler to draw a vertical line from the midpoint of the melting transition to the bottom of the gel.

   Note: Figure 6 shows a melting map of a DNA fragment with two melting domains and indicates how to estimate the denaturant concentration corresponding to the T_m for the domain with the lower T_m.

2. Determine the distance from the midpoint to the left side of the gel (where the denaturant concentration was 0%), and divide it by the entire width of the gel. (The distance from left and right corresponds to the distance between the positions containing denaturant concentrations of 0% and 80%, respectively.) Multiply this fraction by 80% (i.e., the range of denaturant concentrations on the gel) to obtain the denaturant concentration at the position of the melting transition, which corresponds to the T_m.

   Notes: For the example in Figure 6, the denaturant concentration at the position of the melting transition is:

   6 cm/16 cm = 0.375
   0.375 x 80% = 30% denaturants

   Similar to the results obtained with MELTMAP, these results indicate that a parallel denaturing gradient gel with a denaturant concentration range of 15–45% should be used for this test DNA fragment.

3. For gels run at 60°C, calculate the T_m for the melting domain with the lower T_m for a test DNA fragment that melts at a denaturant concentration of 30% as follows:

   T_m (in °C) = (denaturant concentration [in %] / 3.2) + 57
   T_m = (30%/3.2) + 57 = 66°C

   Notes: In many cases, the T_m does not need to be calculated; simply use the experimentally determined denaturant concentration to design the subsequent parallel denaturing gradient gels for the test DNA fragment. However, it is often useful to calculate the T_m if the results of the perpendicular denaturing gradient gel are to be compared to those obtained with the MELTMAP program.

Figure 7 shows the determination of the denaturant concentrations corresponding to the T_m values for a DNA fragment that melts in three domains. As in the approach outlined above, the denaturant concentrations can be determined by measuring the positions of the midpoints of each transition. In this example, the first domain melts at 27% denaturants and the second domain melts at 60% denaturants, so parallel denaturing gradient gels with denaturant concentration ranges of 10–40% and 45–75% denaturants should be used.

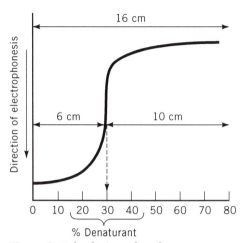

Figure 6 Calculating the denaturant concentration corresponding to the T_m of a DNA fragment with two melting domains in a perpendicular denaturing gradient gel. Shown here is a drawing of an ethidium-bromide-stained perpendicular denaturing gradient gel containing a test DNA fragment with two melting domains. The steep decrease in mobility in the middle of the gel is due to the cooperative melting of the domain with the lower T_m. A ruler is placed across the top or bottom of the gel when it is photographed; this allows the relative position of the mobility transition midpoint (marked with the dotted vertical line) to be determined. The fraction of the left-to-right distance on the gel at which this transition occurs corresponds to the percentage denaturants at which the domain melts. In this example, the midpoint occurs 6 cm from the left (i.e., 0% denaturants) side of the 16-cm-wide gel (not shown to scale). Thus, 6/16, or 0.375, is the fraction of the total distance at which the mobility transition occurs. Because the range of denaturant concentrations in this gel is 80%, the transition occurs at 0.375 x 80%, or 30% denaturants. On the basis of this information, a parallel denaturing gradient gel with a denaturant concentration range of approximately 15–45% (with 30% denaturants approximately halfway down the gel) would be optimal for analyzing this fragment (see text for details).

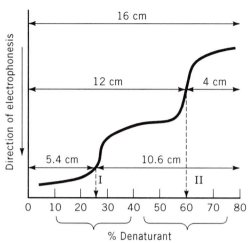

Figure 7 Calculating the denaturant concentrations corresponding to the T_m values for a DNA fragment with three melting domains in a perpendicular denaturing gradient gel. Shown here is a drawing of an ethidium-bromide-stained perpendicular denaturing gradient gel containing a test DNA fragment with three melting domains (not shown to scale). The two steep transitions correspond to the two domains with lower T_m values. As in Figure 6, the denaturant concentrations corresponding to the two mobility transitions of these two domains are calculated as 27% for the first domain (labeled I) and as 60% for the second domain (labeled II). Although it would be possible to detect DNA variants in both domains in a single parallel denaturing gradient gel with a denaturant concentration range of approximately 15–75%, better resolution would be obtained with two different parallel denaturing gradient gels, one with 10–40% denaturants for domain I variants and one with 45–75% denaturants for domain II variants.

PROTOCOL

Optimizing Gel Conditions for DGGE by Empirical Testing

1. Prepare two parallel denaturing gradient gels, one with a denaturant concentration range of 10–50% and the other with a range of 40–80%, as described on pp. 328–330, steps 1–14.

   Note: If wild-type DNA and one or more known mutant DNA fragments are available for a target DNA fragment in cloned form, this protocol can be used to estimate the best gel conditions by simple trial and error. Two gradient gels containing different denaturant concentration ranges are used to separate the wild-type and mutant forms of the target DNA fragment. For each gel, electrophoresis of the two samples is performed for various lengths of time since no information is available about the number of melting domains.

2. Follow the protocol on pp. 330–331, steps 1–7, but use the following modifications to load and run each of the parallel denaturing gradient gels:

   a. Load an aliquot of the wild-type form and an aliquot of a known mutant form of the target DNA fragment in adjacent lanes of each gel. Mark the glass plates with a felt-tipped pen to indicate the positions of the wells that were loaded. Perform electrophoresis for 2 hours.

   b. Turn off the power supply. Clean out the two empty wells adjacent to the previously loaded ones by squirting electrophoresis buffer into them with a pasteur pipette.

   c. Repeat step a, loading the aliquots in the clean wells.

   d. Repeat steps b and c until five to seven more pairs of samples have been loaded and run on each gel.

   Note: At the end of the experiment, the pairs of samples should have been run for 2, 4, 6, 8 hours and so on.

3. Determine which range of denaturant concentrations and which period of electrophoresis, if any, resulted in the maximum degree of separation between the wild-type and mutant forms of the target DNA fragment. Use these optimal conditions for testing additional unknown variants of the target DNA fragment.

PROTOCOL

Detecting DNA Variation Using Parallel DGGE

PREPARING PARALLEL DENATURING GRADIENT GELS

1. Prepare the acrylamide:bisacrylamide and DNA denaturant stock solutions as described on p. 318, step 1.

2. Fill the aquarium with 1x TAE buffer and turn on the heating/stirring apparatus to heat the buffer to 60°C.

   Notes: To minimize evaporation, keep the lid on the aquarium closed while heating. If no lid is available, use floating plastic balls or styrofoam chips to cover the surface of the buffer.
   For a discussion of the equipment used for DGGE, see p. 311; see also Figures 3 and 5.

3. Clean the glass gel plates. Place the spacers at each edge of the notched plate. Carefully place the unnotched glass plate on top of the spacers. Tape the edges (except for the top) of the plates together carefully with waterproof tape; be particularly careful to seal the bottom of the gel and to avoid creating bubbles in the tape.

   Notes: A bottom spacer is also available for some systems. The use of this spacer helps minimize leaking, but tape should also still be used in addition to the spacer.
   For the system available from C.B.S. Scientific, sealing gaskets called Gel-Wrap allow gels to be poured without taping the plates.

4. Place clamps as close together as possible around the taped edges. Stand the taped and clamped gel plates upright (on their bottom edge in their normal running orientation) adjacent to the gradient maker.

   Note: The clamps should be placed on the two sides such that they apply pressure directly over the spacers holding the two plates apart.

5. Determine the precise volume of liquid required to fill the space between the gel plates. Mix the appropriate volumes of 0% and 80% denaturant stock solutions to prepare two solutions containing the highest and lowest denaturant concentrations desired for the gradient (as determined from the T_m). (The volume of each solution should be half the volume required to fill the gel plates.) Place the solutions on ice, and label them L (low) and H (high).

   Notes: The precise volume of liquid required to fill the space between the gel plates should be determined empirically by preparing a set of plates as described above and measuring the volume of liquid required to fill them. Most spacer/gel plate combinations require approximately 22 ml. Therefore, approximately 11 ml of each of the two denaturant stock solutions is required to pour the gradient.
   A typical denaturant concentration range for parallel denaturing gradient gels is approximately 30% denaturants. (Optimal concentration ranges can be determined empirically as

Table 3 Proportions of 0% and 80% Denaturant Stock Solutions Needed to Prepare 11 ml of a Desired Denaturant Solution

Desired denaturant solution (%)	Volume of stock solution required (ml)	
	0%	80%
0	11	0
10	9.6	1.4
20	8.2	2.8
30	6.9	4.1
40	5.5	5.5
50	4.1	6.9
60	2.8	8.2
70	1.4	9.6
80	0	11

These values are estimated for gels containing 6.5% acrylamide that are run at 60°C at 150 V.

described on p. 327.) Thus, for a DNA fragment with a melting domain of 35%, prepare a parallel denaturing gradient gel with 20–50% denaturants.

Table 3 shows the proportions of 0% and 80% denaturant stock solutions needed to prepare 11 ml of particular denaturant solutions.

6. Add 10 μl of TEMED and 100 μl of 10% ammonium persulfate to each of the two solutions (L and H). Mix thoroughly by inverting and place on ice.

   Note: These volumes of TEMED and ammonium persulfate are appropriate for approximately 10–12 ml of denaturant solution.

7. With the stopcocks closed, pour the solution containing the higher concentration of denaturants into the chamber of the gradient maker that contains the exit port (for details, see Figure 5). Briefly open the stopcock between the two chambers to allow liquid to flow into the small channel connecting the chambers. Use a pasteur pipette to transfer the solution that flows into the empty chamber back into the chamber containing the denaturant solution, but be careful to leave liquid (no air bubbles) in the small channel.

8. Pour the solution containing the lower concentration of denaturants into the other chamber of the gradient maker.

   Note: Visually confirm that the levels of the solutions in the two chambers are approximately the same and that there are no air bubbles in the channel between the chambers.

9. Open the stopcock between the chambers of the gradient maker so that the solutions in the two chambers are connected. Use a vibrating mixer or a magnetic stirrer to begin mixing the solution containing the higher concentration of denaturants.

10. Allow the mixed solution to drip from the exit port of the gradient maker through thin-walled tubing and a 20-gauge needle into the top left-hand side of the taped glass plates.

Note: As the higher-concentration denaturant solution leaves its chamber, an equal volume of the lower-concentration solution enters that chamber. Make sure that the levels of both denaturant solutions diminish at the same rate. Mixing of the two solutions should continue throughout the time the gel is being poured. The flow rate should be such that approximately 5 minutes is required to fill the space between the plates.

11. Insert a gel comb into the top of the glass plates. Allow the gel to polymerize at room temperature in a horizontal position, with the comb end of the plates slightly propped up to prevent the liquid from dripping out.

Note: Because relatively low concentrations of TEMED and ammonium persulfate are used to allow extra time for pouring a gradient gel, polymerization may require longer than it does for other types of gels. Typically, approximately 30 minutes is sufficient.

12. Remove the tape (and spacer, if used) from the bottom of the polymerized gel. Clamp the gel/plates into the gel cassette and place in the aquarium filled with 1x TAE buffer heated to 60°C. Adjust the level of buffer so that the entire gel is submerged but the upper electrophoresis chamber is above the buffer level.

Notes: The upper electrophoresis chamber is created when the gel/plates are clamped into the gel cassette.

If the system available from C.B.S. Scientific is being used, remove the Gel-Wrap and tape the sides of the plates with waterproof tape to prevent electric fields from forming perpendicular to the gel.

13. Attach the tubes from the peristaltic pump to the gel cassette, and begin circulating buffer from the aquarium into the upper electrophoresis chamber.

14. Carefully remove the comb from the top of the gel plates. Use a pasteur pipette to rinse the wells with electrophoresis buffer. The gel is now ready for loading the samples.

RUNNING PARALLEL DENATURING GRADIENT GELS

1. Generate the target DNA samples for the wild-type and mutant test fragments (~100–600 bp) by PCR or by digesting cloned DNA fragments with restriction enzymes. If a GC clamp is to be attached to the target fragment by PCR, use one primer with a 5′ GC-rich segment (for a description the segment, see p. 315).

Note: If cloned DNA is used, the target fragments should be excised from the vector with the appropriate restriction enzymes. However, there is no need to purify the insert from the vector.

2. Analyze an aliquot (50–200 ng of target DNA) of each sample and appropriate molecular-weight markers (100–1000 bp) on a 1% agarose mini-gel in 1x TAE buffer. Estimate the size and concentration of the DNA fragments. Also confirm that amplification or digestion was successful and then proceed with step 3. Keep the remainder of the DNA on ice while the gel is running.

Note: If digestion was incomplete, add additional enzyme and continue incubating until digestion is complete. If amplification was unsuccessful, prepare PCR products again.

3. If fragments of the expected size are seen, precipitate the DNA in each sample with 0.1 volume of 3 M sodium acetate (pH 5.2) and 2–2.5 volumes of absolute ethanol and dry the pellets under vacuum (see Appendix).

4. Dissolve each DNA pellet in nondenaturing gel-loading buffer (for preparation, see p. 322) such that approximately 3 μl contains a minimum of 100 ng of target DNA fragment.

   Note: If the gel is to be blotted and the target DNA fragment detected by hybridization, much smaller amounts (as little as 0.1 ng of each fragment) of DNA can be loaded on the gel.

5. Load each DNA sample (~3 μl) in a separate lane of the parallel denaturing gradient gel prepared above.

6. Connect the electrodes and perform electrophoresis at 150 V (~35–40 mA, constant voltage) for 5–12 hours, depending on the sizes and melting properties of the test DNA fragments.

   Note: If an empirical test has been performed to determine optimal electrophoresis conditions (see p. 327), use the optimized period of electrophoresis to run the samples. In cases where such a test has not been done, loading the target fragment at multiple times (as described in the empirical test) is recommended in initial experiments. Although wide variability in appropriate periods of electrophoresis is seen for different target fragments of the same length (because of different melting behaviors), there is some effect of length on final position in the gel. Thus, DNA fragments in the size range of 100–300 bp usually require shorter periods of electrophoresis (5–8 hours) than do fragments of 300–600 bp, which typically require 8–12 hours of electrophoresis.

7. Turn off the power supply, the heating/stirring apparatus, and the peristaltic pump. Remove the gel plates from the cassette. If the test DNA samples are present in sufficient amounts to detect without probing, remove the gel from the plates, stain the DNA with **ethidium bromide** solution (0.1 μg/ml) for 15–30 minutes, and then examine and photograph the stained gel under **UV** transillumination. If the test DNA samples are not present in sufficient amounts, blot the gel and probe by hybridization as described in the gDGGE protocol on pp. 335–339.

   Notes: These thin polyacrylamide gels are somewhat flimsy and require some care in handling to avoid tearing. To aid in staining and viewing by transillumination, move the gel from the staining vessel to the UV light box on a glass plate, and use a stream of water from a squirt bottle to help make the gel slide onto the viewing surface.
     If the staining with ethidium bromide appears too faint or too heavy, alter the staining conditions to a higher or lower concentration of ethidium bromide, respectively.

   ethidium bromide, UV radiation (see Appendix for Caution)

8. Interpret the DNA patterns. A shift in mobility of a test DNA fragment relative to the wild-type fragment indicates that the test fragment contains a variation in its DNA sequence. Variation that leads to a higher G + C content generally results in increased mobility in gels, whereas variation that increases A + T content leads to decreased mobility.

PROTOCOL
___

Formation of Heteroduplexes and Analysis by Parallel DGGE

1. Prepare two parallel denaturing gradient gels, one with a denaturant concentration range of 10–50% and the other with a range of 40–80%, as described on pp. 328–330.

2. Prepare the target DNA samples for the wild-type and mutant test fragments as described on p. 330, steps 1–3.

3. Dissolve each precipitated sample in TE (pH 7.8). Place equal amounts (1–2 µg) of each of the two DNA samples (wild-type and mutant) in a single 0.5-ml microcentrifuge tube (tube A) and adjust the volume to 20 µl with TE (pH 7.8). As controls, prepare similar samples containing wild-type DNA alone (tube B) and mutant DNA alone (tube C).

   Notes: The reason that such a large amount of DNA is used in this test is that the samples will be used for several different lanes on two different gels so that optimal separation conditions can be determined. Once this preliminary test is done, smaller amounts (100–200 ng) of each sample can be used for this type of analysis.

   Four DNA species, each at approximately the same concentration, will be formed when this experiment is performed (Figure 8). The two starting homoduplexes will be regenerated and two different heteroduplexes will be formed, one heteroduplex with the upper strand of the wild-type homoduplex plus the lower strand of the mutant homoduplex and one with the lower strand of the wild-type homoduplex plus the upper strand of the mutant homoduplex.

4. Place the samples in a heating block filled with H_2O or in a boiling-water bath and heat at 95–98°C for 10 minutes to separate the strands. Centrifuge briefly to collect the liquid at the bottom of the tubes.

5. Add 2 µl of 5 M NaCl to each tube.

   Note: After the fragments are mixed and heated in a solution of low ionic strength, the strands are allowed to reanneal in the presence of a salt concentration greater than 300 mM.

6. Place the samples in a heating block filled with H_2O or in a water bath and heat at 55°C for 30 minutes to reanneal the strands. Centrifuge briefly to collect the liquid at the bottom of the tubes.

7. Precipitate the DNA in each tube with 60 µl of absolute ethanol and dry the pellets under vacuum (see Appendix).

8. Dissolve each DNA pellet in 5–10 µl of nondenaturing gel-loading buffer (for preparation, see p. 322).

9. Load and run each of the parallel denaturing gradient gels as follows:

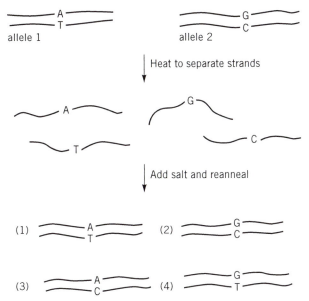

Figure 8 Formation of a mixture of homoduplexes and heteroduplexes from two alleles of a DNA fragment. Two different homoduplexes comprising two alleles of a DNA fragment are mixed, heated in a solution of low ionic strength, and then allowed to reanneal in the presence of a salt concentration greater than 300 mM. If the two original homoduplexes are at the same concentration (which is the case either when diploid, heterozygous genomic DNA or PCR products made from such DNA are adjusted to equal concentrations or when cloned or PCR-amplified homoduplexes are adjusted to equal concentrations), this denaturation and renaturation experiment will yield four species of the DNA fragment. These species are the two homoduplexes (species 1 and 2 in the figure) and two different heteroduplexes, one composed of the upper strand of allele 1 annealed to the lower strand of allele 2 (species 3 in the figure) and the other composed of the upper strand of allele 2 annealed to the lower strand of allele 1 (species 4 in the figure). All four species of most DNA fragments can be separated by DGGE, the two heteroduplexes showing a retarded mobility relative to the two homoduplexes (see Figure 9).

a. Load approximately 10% of each sample (A, B, and C) in adjacent lanes of each gel. Mark the glass plates with a felt-tipped pen to indicate the positions of the wells that were loaded.

b. Connect the electrodes and perform electrophoresis at 150 V (~35–40 mA, constant voltage) for 2 hours.

c. Turn off the power supply. Clean out the three empty wells adjacent to the previously loaded ones by squirting electrophoresis buffer into them with a pasteur pipette.

d. Repeat steps a and b, loading the aliquots in the clean wells.

e. Repeat steps c and d until five to seven sets of samples have been loaded and run on each gel.

Note: At the end of the experiment, the sets of samples should have been run for 2, 4, 6, 8 hours and so on.

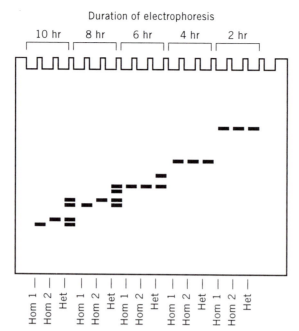

Figure 9 Effect of the duration of electrophoresis on the mobility of homoduplexes and heteroduplexes on a parallel denaturing gradient gel. Shown here is a diagram of typical gel mobility patterns for homoduplexes and heteroduplexes as a function of the duration of electrophoresis. In this example, the two homoduplexes for different alleles shown in the first two lanes of each set of three do not separate under these gel conditions for the first 6 hours of electrophoresis. However, the two heteroduplexes begin to separate from the homoduplexes after 6 hours of electrophoresis, and both the homoduplexes and the heteroduplexes separate from each other after 8 hours of electrophoresis. In other cases, the two heteroduplexes may separate from the homoduplexes but not from each other, or the homoduplexes may not separate from each other but the heteroduplexes do.

10. Turn off the power supply, the heating/stirring apparatus, and the peristaltic pump. Remove the gel plates from the cassettes. Remove the gels from the plates, stain the gels with **ethidium bromide** solution (0.1 μg/ml) for 15–30 minutes, and then photograph the stained gels under **UV** transillumination.

   ethidium bromide, UV radiation (see Appendix for Caution)

11. Determine which range of denaturant concentrations and which period of electrophoresis resulted in the maximum degree of separation of the different alleles (an example is shown in Figure 9). Use these optimal conditions for testing additional unknown variants of the target DNA fragment (see step 3, Note).

PROTOCOL

Detecting DNA Variation Using gDGGE

1. Prepare target DNA samples by one or more of the methods below.

   Note: Although gDGGE is performed with genomic DNA, the principles of gDGGE can be applied to a complex cloned sample or to a mixture of different PCR-amplified samples as described here.

   For analysis of genomic DNA:

   a. In separate reaction mixtures, digest genomic DNA from two or more individuals with two or three restriction enzymes that have 4-base recognition sites (e.g., *Hae*III, *Sau*3A, *Dde*I, and *Rsa*I).

   Notes: Genomic DNA can be prepared and digested as described in Wolff and Gemmill (1997).

   The amount of digested genomic DNA required for detection by radiolabeled DNA probes varies with the complexity of the genomic DNA source. For human and other similarly complex genomes, approximately 10 µg per lane is recommended. For less complex genomes, 1–10 µg can be used.

   Because staining with ethidium bromide is used in step b to analyze 2–5 µg of the digested genomic DNA to ensure that the digestion was complete, it is recommended that at least this much extra genomic DNA be digested for each sample.

   b. Analyze an aliquot (2–5 µg) of each DNA sample and appropriate molecular-weight markers (100–2000 bp) by electrophoresis on a 1% agarose gel and staining of the gel with **ethidium bromide**. Confirm that digestion is complete and then proceed with step c. Keep the remainder of the DNA on ice while the gel is running.

   Notes: Incomplete digestion of genomic DNA generally appears as a diffuse smear near the wells. When digestion is complete, the smear appears broadly throughout the lane, with very little stained material near the well.

   If digestion is incomplete, add additional enzyme and continue incubating until digestion is complete.

   ethidium bromide (see Appendix for Caution)

   c. Precipitate the DNA in each sample with 0.1 volume of 3 M sodium acetate (pH 5.2) and 2–2.5 volumes of absolute ethanol and dry the pellets under vacuum (see Appendix).

   d. Dissolve each DNA pellet in 5 µl of nondenaturing gel-loading buffer (for preparation, see p. 322).

   e. Proceed with step 2.

   For analysis of homoduplexes from cloned DNA:

   a. Follow steps a–d for preparing genomic DNA samples, but start with 1–2 µg of cloned DNA in each digestion and analyze only 100–200 ng of the DNA by staining with ethidium bromide to ensure that the digestion was complete. Prepare a wild-type cloned DNA as a control and cloned DNA test samples.

b. Proceed with step 2.

For analysis of heteroduplexes from cloned DNA:

a. Follow steps a–c for preparing genomic DNA samples, but start with 1–2 μg of cloned DNA in each digestion and analyze only 100–200 ng of the DNA by staining with ethidium bromide to ensure that the digestion was complete. Prepare a wild-type cloned DNA and cloned DNA test samples.

b. Dissolve each precipitated sample in TE (pH 7.8). To prepare heteroduplexes from each test sample, place equal amounts (~1–2 μg) of each of the two digested DNA samples (wild-type and test DNA digested with the same enzyme) in a single 0.5-ml microcentrifuge tube, adjust the volume to 20 μl with TE (pH 7.8), and then follow steps 4–8 on p. 332. As controls, prepare similar samples containing wild-type DNA alone and test DNA alone.

c. Proceed with step 2.

For analysis of multiple PCR samples in a single lane:

a. For multiple individuals, amplify as many as 10–30 target loci from each in separate PCR assays. If a GC clamp is to be attached to the target fragment by PCR, use one primer with a 5' GC-rich segment (for a description the segment, see p. 315).

b. Analyze an aliquot (~50–200 ng) of each amplified product and appropriate molecular-weight markers (100–1000 bp) on a 1% agarose gel. Estimate the size and concentration of the DNA fragments. Also confirm that amplification was successful and then proceed with step c. Keep the remainder of the DNA on ice while the gel is running.

Notes: If amplification was unsuccessful, prepare PCR products again.

These samples will be detected by hybridization on a blot. Additional, spurious bands therefore do not usually interfere with detection of specific target fragments.

c. Using PCR products from a single individual, mix samples from as many as 30 different loci in microcentrifuge tubes. Precipitate the DNA in each tube with 0.1 volume of 3 M sodium acetate (pH 5.2) and 2–2.5 volumes of absolute ethanol and dry the pellets under vacuum (see Appendix).

Note: If PCR has been performed with diploid DNA from an individual heterozygote at one or more positions at a locus, a mixture of heteroduplexes and homoduplexes will be present in the mixture.

d. Dissolve each pellet in 5 μl of nondenaturing gel-loading buffer (for preparation, see p. 322).

e. Proceed with step 2.

2. Prepare two parallel denaturing gradient gels, one with a denaturant concentration range of 10–50% and the other with a range of 40–80%, as described on pp. 328–330, steps 1–14.

3. Heat the samples (from one of the preparations in step 1) at 65°C for 5 minutes and then load half of each sample in separate lanes on each gel.

4. Connect the electrodes and perform electrophoresis at 150 V (~35–40 mA, constant voltage) for 9 hours (10–50% gel) or 11 hours (40–80% gel).

5. Turn off the power supply, heating/stirring apparatus, and the peristaltic pump. Remove the gel cassette from the aquarium, and remove the gel plates from the cassette. Mark the orientation of the gel by cutting off one corner, and then place the gel on a piece of used, developed X-ray film to support it.

   Note: These thin polyacrylamide gels are somewhat flimsy and require some care in handling to avoid tearing.

6. Place the gel and its support in 200 ml of 0.5 N **NaOH** at room temperature for 5 minutes.

   NaOH (see Appendix for Caution)

7. Transfer the gel and support into 200 ml of 0.5 M Tris-Cl (pH 8.0 at room temperature) at room temperature for 5 minutes.

8. Transfer the gel and support into 200 ml of transfer buffer at room temperature for 10 minutes.

   Transfer buffer

Component and final concentration	Amount to add per 2 liters
20 mM Tris-Cl	40 ml of 1 M (pH 8.0 at room temperature)
1 mM EDTA	4 ml of 0.5 M (pH 8.0)
H$_2$O	1.956 liters

   Store at room temperature indefinitely.

9. Transfer the gel onto a sheet of Whatman 3MM paper wetted with transfer buffer by laying the wetted paper on top of the gel and then carefully turning the stack over. Remove the used X-ray film and place the paper on the bench top, gel side up.

10. Cut a sheet of Hybond N$^+$ nylon membrane (Amersham) to the same dimensions as the gel and wet it with transfer buffer. Carefully place the membrane on top of the gel, avoiding the formation of air bubbles between the gel and the membrane.

11. Sandwich the gel between four additional sheets of Whatman 3MM paper, and then clamp it between two sheets of the sponge mesh provided with a commercial electroblotting apparatus (e.g., Bio-Rad).

    Note: The gel sandwich should now be one sheet of sponge mesh, the two new sheets of Whatman 3MM paper, the original sheet of 3MM paper, the gel, the membrane, the other two new sheets of 3MM paper, and then the second sheet of sponge mesh.

12. Fill the electroblotting apparatus with transfer buffer at room temperature, and place the gel sandwich in the apparatus. Be sure to orient the gel so that the nylon membrane lies between the gel and the positive electrode of the apparatus.

13. Set the electroblotting apparatus at 30 V (which should result in a current of approximately 600 mA) for 3–5 hours.

14. Turn off the power supply and remove the gel sandwich from the apparatus.

15. Remove the nylon membrane from the gel sandwich and place the membrane on top of a piece of Whatman 3MM paper soaked in 0.4 N NaOH. Allow to stand for 20 minutes.

    Note: This step denatures the DNA bound to the membrane.

16. Place the membrane in 100 ml of 2x SSC at room temperature for 2 minutes.

17. Place the membrane in a sealable bag containing 10 ml of hybridization buffer and prehybridize at 65°C for 2 hours.

    Hybridization buffer for gDGGE

Component and final concentration	Amount to add per 0.5 liter
0.5 M sodium phosphate	250 ml of 1 M buffer (pH 7.2; see Appendix)
2 mM EDTA	2 ml of 0.5 M (pH 8.0)
7% **SDS**	175 ml of 20%
H_2O	73 ml

    Store at room temperature indefinitely.

    SDS (see Appendix for Caution)

18. Add **radiolabeled probe** to the bag (~10^6 dpm/ml for PCR or cloned samples and ~10^8 dpm/ml for complex genomic samples). Hybridize cloned or PCR-amplified target DNA at 65°C for 2–4 hours. Hybridize genomic target DNA for 12–18 hours.

    Notes: In gDGGE, the segment of DNA recognized by the labeled probe is usually 1–10 kb in length.

    Probes should be labeled to a specific activity of 10^7 dpm/µg for PCR-amplified or cloned targets and to 10^9 dpm/µg for complex genomic DNA targets. If the probe has or is thought to have repetitive sequences in it, competitive preannealing of the probe with unlabeled repeat-containing DNA is necessary. For protocols for probe preparation and preannealing, see Wolff and Gemmill (1997).

radioactive substances (see Appendix for Caution)

19. Wash the blot in several changes of washing buffer at 65°C for a total of 30–45 minutes.

Washing buffer for gDGGE

Component and final concentration	Amount to add per 0.5 liter
40 mM sodium phosphate	20 ml of 1 M buffer (pH 7.2)
1% SDS	25 ml of 20%
H$_2$O	455 ml

Store at at room temperature indefinitely.

20. Wrap the moist blot in plastic wrap and place on X-ray film. Expose at room temperature without an intensifying screen or at –80°C with one and develop the film.

Notes: Typically, exposure will need to be for 12–24 hours with an intensifying screen when complex genomic DNA samples (e.g., those from mammalian cells) are used. Exposure can be much shorter and an intensifying screen is not necessary when the test DNA is derived from cloned or PCR-amplified fragments or from simpler genomes.

Blots can be stripped by incubating in 0.1x SSC at 65°C for 30 minutes and reprobed several times if they are not permitted to dry out completely.

21. Interpret the bands on the autoradiograph as described for ethidium-bromide-stained gels on p. 331, step 8.

RNase Cleavage at Mismatches

THEORETICAL BASIS OF THE RNase CLEAVAGE METHOD

The RNase cleavage method is one of several approaches in which single-base mutations and polymorphisms are detected by forming heteroduplexes between wild-type and mutant DNA or RNA strands and then treating the heteroduplexes with reagents that selectively cleave the nucleic acid backbone at mismatches. Investigators found that RNase A efficiently recognizes and cleaves at many mismatched positions in RNA:DNA (Myers et al. 1985a) and RNA:RNA heteroduplexes (Winter et al. 1985), and this observation led to the development of the following experimental paradigm. First, a uniformly labeled RNA transcript is prepared by linearizing a plasmid that contains the test fragment clone adjacent to a bacteriophage SP6, T7, or T3 RNA polymerase promoter and then transcribing with the appropriate RNA polymerase to the end of the restriction fragment (this is called a "run-off" transcript). This RNA probe is annealed to the test nucleic acid (which can be genomic DNA, PCR-amplified DNA, cloned DNA, or RNA) to form hybrid double-stranded molecules. If a single-base change is present in the test fragment, the hybrids will contain a single-base mismatch. The hybrids are then treated with RNase A, which cleaves the RNA strand at the position of a mismatch in RNA:DNA and RNA:RNA hybrids. After treatment with the enzyme, the double-stranded hybrids are denatured and analyzed on a denaturing polyacrylamide gel, which is then autoradiographed. By analyzing the sizes of the labeled probe fragments, it is possible to determine whether or not a mismatch was present and the approximate position of the mismatch relative to one of the ends of the test fragment.

RNase A cleaves at the 3′end of pyrimidine residues in RNA, which might lead to the expectation that approximately 50% of all single-base changes in an RNA:DNA duplex would be detected with the RNase cleavage method if all possible mismatches containing a pyrimidine in the RNA strand were cleaved. However, analysis of a large number of single-base mismatches in many different sequence contexts indicates that only about 30–40% of the mismatches containing a pyrimidine in the RNA strand are cleaved efficiently (efficiently is defined as the cleavage of at least 50% of the molecules in a sample with a particular mis-

match containing a pyrimidine in the RNA strand). For some reason, perhaps because the sequence context of each mismatch determines the accessibility of the enzyme, some pyrimidine mismatches are not cleaved well. Surprisingly, some base changes caused by mismatches with a purine in the RNA strand are detected with this procedure because the RNase A cleaves at pyrimidines that lie adjacent to the mismatched purine in the RNA strand. Why some mismatches at purines in the RNA strand are detected and others are not is again presumably due to different local structures of the mismatches that affect accessibility of the enzyme.

The fraction of all possible single-base changes that can be detected by this procedure can be doubled to 60–80% by using, in separate assays, both strands of the test fragment as the RNA probe. Those mismatches not detected when one strand is used as the RNA probe are almost always recognized when the opposite strand is used as the probe. For example, a single-base change causing an A:C mismatch (with A in the RNA probe strand and C in the DNA strand) may not be cleaved by RNase A, whereas the U:G mismatch formed with the opposite strand RNA probe (with U in the RNA probe strand and G in the DNA strand) will be cleaved efficiently.

EXPERIMENTAL CONSIDERATIONS FOR USING THE RNase CLEAVAGE METHOD

Radiolabeled RNA probes for the RNase cleavage method are prepared by run-off transcription of cloned template DNA. The template containing the bacteriophage promoter and target sequence is truncated by cleavage with restriction enzymes so that a transcript synthesized in vitro by the RNA polymerase begins at the promoter, extends through the target sequence, and ends at the end of the target sequence because the polymerase cannot extend past the end of the template. RNA probes are prepared with either low specific radioactivity or high specific radioactivity, depending on the nature of the test samples. Low-specific-radioactivity probes generated from cloned wild-type DNA are used for testing cloned or PCR-amplified samples, whereas high-specific-radioactivity probes are necessary when the test sample is more complex (e.g., genomic DNA).

The advantages of testing PCR-amplified or cloned samples instead of genomic DNA with this

Figure 10 Diagram of an autoradiograph showing typical results of an RNase cleavage experiment. Lane 1 contains uncleaved RNA probe with no test DNA; lane 2 is RNase-treated RNA probe with no test DNA; lanes 3–6 show the results obtained when labeled RNA probe is annealed to wild-type and mutant test DNA samples and then treated with RNase; lane 7 contains markers. Lane 3 contains the fragment produced when the wild-type probe is annealed to wild-type test DNA. Note that the fragment is slightly smaller than the intact probe because the nonhybridizing tail of the probe (see text for description) is trimmed. Lanes 4 and 5 contain the two fragments produced by mutant test DNAs annealed to the wild-type probe. In these cases, the cleavage was complete, and the sum of the lengths of the two fragments in each lane approximately equals the length of the wild-type full-length fragment. The sizes of the two fragments can be used to deduce the distance the mutation is located from one or the other end of the test fragment. Lane 6 shows the results of a mutant test DNA fragment annealed to the wild-type probe where the cleavage by RNase was not complete and some full-length fragment remains. As described in the text, it

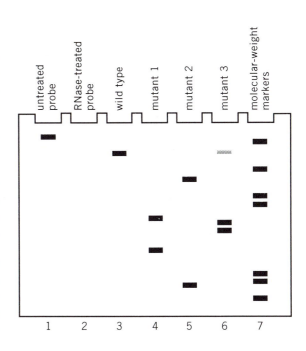

is often possible to drive such partial cleavage reactions to completion by treating with RNase for longer periods of time. When diploid test DNA samples are used in the RNase cleavage method, the cases where incomplete cleavage with RNase occurs can lead to confusing results, particularly when the samples are from compound heterozygotes (i.e., samples containing either one wild-type allele and one mutant allele or two different mutant alleles). Because most sequence variants show complete cleavage, most samples that have 50% of the signal at the position of the wild-type band and 50% of the signal in two smaller bands are samples with one wild-type and one mutant allele. In cases where the wild-type band is signficantly more or less than 50% of the signal, the samples could be heterozygous or the results could be due to incomplete cleavage. If such results are obtained, treatment with RNase for a longer period of time is recommended.

method are not only a large increase in sensitivity (signal-to-noise ratio) and the much smaller amounts of DNA needed when PCR is used but also the fact that uniformly labeled low-specific-radioactivity RNA probes are easier and safer to handle than high-specific-radioactivity probes. Synthesis of low-specific-radioactivity RNA probes requires handling lower amounts of radioactive nucleotides (10 μCi of [^{32}P]NTP instead of the 80–100 μCi required for high-specific-activity probes). In addition, low-specific-radioactivity RNA probes do not degrade as rapidly and can therefore be used a week or two after they are synthesized instead of only a day or two.

Although RNA probes several thousand nucleotides in length can be synthesized by in vitro run-off transcription, it is best in practice to limit the length of RNA probes used in the RNase cleavage method to approximately 500–600 nucleotides.

The reason for this limitation has to do with the signal-to-noise ratio; longer RNA:DNA (or RNA:RNA) hybrids generally produce more nonspecific cleavage by RNase at paired bases in the duplex, which results in radioactive background in the autoradiographs and obfuscation of signals.

The reason the RNase cleavage method works is that the enzyme cuts at mismatched bases at a much higher rate than at paired bases. However, there is some cleavage at perfectly paired bases, presumably because of the temporary formation of unpaired regions in the duplexes during the RNase reaction. The longer the duplex, the more frequently such single-stranded regions will occur in the molecule and the more likely RNase will cleave at nonmismatched positions. This problem is minimized to some extent by performing the cleavage reactions at 25ºC in 300 mM salt (NaCl plus LiCl in the protocol here). Lower temperature and higher salt concentration minimize cleavage at non-mismatched positions even more significantly, but only at the expense of cleavage at mismatches by the enzyme. Thus, cleavage at 25°C in 300 mM salt is a compromise that results in the maximum amount of the desired cleavage and the minimum amount of undesired cleavage within the size constraints of a few hundred base pairs.

To perform the RNase cleavage method, a double-stranded DNA fragment corresponding to the segment to be tested should be cloned into the multiple cloning site of a plasmid vector that contains one or more of the bacteriophage promoters for which specific RNA polymerases are available. Examples are the pSP72 series (Krieg and Melton 1987), the pGEM series (Promega), and Bluescript and Bluescribe (Stratagene). Since most of these vectors contain two different bacteriophage RNA polymerase promoters, one on each side of the multiple cloning site, a single cloning step can provide a plasmid that can be used to produce two different RNA probes. The plasmid can be linearized by digestion at either end of the multiple cloning site; this generates two templates from which RNA transcripts of each strand of the test fragment can be made by using the two polymerases in separate reactions.

The multiple cloning sites in these vectors provide an additional feature that is useful in RNase cleavage—a 5' and/or 3' tail on the RNA probe that does not hybridize to the test DNA or RNA segment. The activity of the RNase can be monitored during the procedure by showing that the tail is completely removed from a probe annealed to a complementary control template (see Figure 10).

PROTOCOLS FOR DETECTING DNA VARIATION USING RNase CLEAVAGE

Probe Preparation

In most cases, the RNase cleavage method uses low-specific-radioactivity RNA probes (10 Ci/mmole in one of the nucleotides). Preparation of such a probe is provided in the protocol on pp. 344–346 (Myers et al. 1985a). Alternative protocols provided with commercially available kits (e.g., Stratagene, Promega, or Epicentre Technologies) are also suitable for generating RNA probes. Preparation and subsequent usage of the probe require the observance of several precautions to avoid breakdown of the RNA probe by contaminating RNase. These precautions include using DEPC-treated H_2O for preparing solutions (except those containing Tris), wearing gloves, and using sterile plasticware instead of glasswear (for further discussion, see pp. 67–68).

Cleavage at Mismatches

Although the RNase cleavage method works best with PCR-amplified or cloned target DNA fragments (see above), it can also be used with genomic DNA or mRNA. In such cases, a molar excess of labeled RNA probe over target DNA must be used to obtain reasonable signals on autoradiography. The excess is necessary for kinetic reasons (so that the annealing reaction can occur in an overnight reaction) and for thermodynamic reasons (to ensure that enough labeled hybrids are formed to yield a signal). However, in most applications of the method that use PCR-amplified or cloned target DNA, complete hybridization of the probe can be obtained if the probe is used in smaller amounts than the target.

It is helpful to include a wild-type and, if available, a mutant control target DNA fragment in the RNase cleavage method when unknown samples are being tested. The wild-type control can be the same plasmid DNA that was used to generate the labeled RNA probe; a similar plasmid containing a single-base change is a useful control for cleavage at a mismatch. A protocol for the annealing and cleavage reactions for the RNase cleavage method is provided on pp. 347–350.

Analysis of RNase Cleavage Products

To determine whether or not a single-base change is present in the test DNA sample, the radiolabeled RNase cleavage products are separated from their complementary DNA strands and loaded on a denaturing polyacrylamide sequencing gel to separate the RNA fragments by size. Standard sequencing gels and apparatuses are appropriate for this purpose. To examine RNA fragments in the size range of 500–600 nucleotides, a 6% polyacrylamide gel should be used, whereas a 10% or 12% gel works best when the RNA fragments are shorter than 500 nucleotides. A protocol for analyzing RNase cleavage products on a sequencing gel is provided on p. 351.

PROTOCOL

Preparing Uniformly Labeled RNA Probes

Be sure to observe the special considerations for working with RNA throughout this procedure (see p. 343).

1. Digest 5–10 µg of the plasmid containing the test sequence as specified by the manufacturer of the restriction enzyme. Use a restriction enzyme that cleaves near one end of the test sequence. If both strands of the test sequence are to be used as probes, perform two different digestions with restriction enzymes appropriate for leaving a tail of at least 25 nucleotides at one or both ends of the segment to be tested, and process the products separately throughout the rest of this procedure.

   Notes: The double-stranded DNA fragment corresponding to the segment to be tested should be cloned into the multiple cloning site of a plasmid vector that contains one or more of the bacteriophage promoters for which specific RNA polymerases are available (e.g., the pSP72 series [Krieg and Melton 1987], the pGEM series [Promega], or Bluescript and Bluescribe [Stratagene]).

   A single in vitro run-off transcription reaction requires approximately 1 µg of template. It is useful to prepare enough restriction-enzyme-digested template for several probe-labeling reactions. Digested template can be stored at –20°C for several years for later use.

2. Analyze an aliquot (100–200 ng) of the DNA template and appropriate molecular-weight markers (1–14 kb) on a 1% agarose gel. Confirm that digestion is complete and then proceed with step 3. Keep the remainder of the DNA template on ice while the gel is running.

   Notes: If digestion is complete, the DNA template should be seen as a single band that migrates at a size equal to that of the vector plus that of the insert.

   If digestion is incomplete, add additional enzyme and continue incubating until digestion is complete.

3. Extract the digested DNA template with 1 volume of **phenol** and then with 1 volume of phenol:**chloroform** (1:1) (see Appendix). Use phenol saturated with **DEPC**-treated H_2O.

   phenol, chloroform, DEPC (see Appendix for Caution)

4. Precipitate the DNA template in the aqueous phase with 0.1 volume of 3 M sodium acetate (pH 5.2) and 2–2.5 volumes of absolute ethanol and dry the pellet under vacuum (see Appendix).

5. Dissolve the DNA template in TE (pH 7.8) at a final concentration of 1 mg/ml.

6. Combine the following (in the order listed) in a 1.5-ml microcentrifuge tube on ice:

H$_2$O	9 μl
digested template DNA (1 mg/ml)	1 μl
2 mM unlabeled GTP	1 μl
CTP/ATP/UTP mixture (each at 10 mM)	1 μl
[α-^{32}P]GTP (10 mCi/ml, 400 Ci/mmole)	1 μl
1 M DTT	2 μl
BSA (2 mg/ml)	1 μl
placental RNase inhibitor	2–5 units
10x transcription buffer	2 μl

Notes: Nucleotide solutions are relatively unstable and should be stored at –80°C for up to 1 year. Working stock solutions can be stored at –20°C for up to 3–6 months, but frequent thawing should be avoided.

   If high-specific-activity probes (200–400 Ci/mmole in one of the nucleotides) are needed for analysis of genomic DNA, use 80–100 μCi of [^{32}P]NTP in the reaction mixture.

10x Transcription buffer

Component and final concentration	Amount to add per 1 ml
400 mM Tris-Cl	400 μl of 1 M (pH 7.5 at room temperature)
60 mM MgCl$_2$	60 μl of 1 M MgCl$_2$
20 mM spermidine	20 μl of 1 M
H$_2$O	0.52 ml

Store at –20°C or at –80°C for up to 1 year.

radioactive substances (see Appendix for Caution)

7. Add 2–5 units of the appropriate RNA polymerase (bacteriophage SP6, T7, or T3).

   Notes: The polymerase's promoter should be opposite the site of cleavage.
      RNA polymerase from any manufacturer can be used.

8. Incubate at 40°C for 1 hour.

9. Add 1 μl of RNase-free DNase I (1 mg/ml) and incubate at 37°C for 30 minutes to destroy the DNA template.

   Notes: The DNase solution must be highly purified and RNase-free so that the RNA probe remains intact. Several commercial sources provide DNase I of sufficient quality to use in synthesizing RNA probes (e.g., Worthington Biochemical Corp. or Promega).
      DNase I has a relatively short half-life at –20°C in aqueous solutions (~6 months) and loses activity with frequent thawings.

10. Add 80 μl of 2 M ammonium acetate and 10 μg of carrier tRNA (yeast or *E. coli* tRNA; e.g., Sigma, Promega, or Boehringer Mannheim).

11. Extract the sample with 1 volume of phenol:chloroform (1:1). Use phenol saturated with DEPC-treated H$_2$O.

12. Precipitate the RNA probe in the aqueous phase with 2.5 volumes of absolute ethanol and dry the pellet under vacuum.

13. Dissolve the pellet in 100 μl of 2 M ammonium acetate and precipitate with ethanol again.

14. Dissolve the RNA probe in 200 μl of TE (pH 7.8) containing 0.1% **SDS**. Store at –20°C until needed for the annealing and cleavage reactions on pp. 347–350.

Note: Low-specific-activity probes (10 Ci/mmole in one of the nucleotides) are stable for up to 1–2 weeks. High-specific-activity probes can be used for only a day or two.

SDS (see Appendix for Caution)

PROTOCOL

Annealing and Cleavage Reactions

Be sure to observe the special considerations for working with RNA throughout this procedure (see p. 343).

1. Combine the components below in a 1.5-ml microcentrifuge tube and mix by vortexing. Prepare this 32-μl hybridization mixture for each control and each sample to be tested. Also prepare two tubes containing all of the components except the test DNA. Process each mixture separately throughout the rest of this procedure (except as noted in step 5).

hybridization buffer	30 μl
radiolabeled RNA probe (from pp. 344–346)	1 μl
test DNA (1–100 ng of PCR-amplified or cloned DNA)	1 μl

   Notes: Target DNA samples for the wild-type and mutant test fragments should be generated by PCR or by digesting cloned DNA fragments with restriction enzymes. If cloned DNA is used, the target fragment should be excised from the vector with the appropriate restriction enzymes. However, there is no need to purify the insert from the vector.

   If complex genomic DNA is being used as the target, use approximately 10 μg if the DNA is mammalian or from similarly complex source; use 1–10 μg if the DNA is from yeast, *Drosophila*, or similarly complex source. Also use high-specific-activity probes (200–400 Ci/mmole in one of the nucleotides). Genomic DNA can be prepared as described in Wolff and Gemmill (1997).

   It is helpful to include a wild-type and, if available, a mutant control target DNA fragment. The wild-type control can be the same plasmid DNA that was used to generate the labeled RNA probe; a similar plasmid containing a single-base change is a useful control for cleavage at a mismatch.

Hybridization buffer for RNase cleavage

Component and final concentration	Amount to add per 10 ml
80% deionized **formamide**	8 ml
50 mM PIPES	1 ml of 0.5 M PIPES buffer (pH 6.4)
0.5 M NaCl	1 ml of 5 M
1 mM EDTA	20 μl of 0.5 M (pH 8.0)

Do not autoclave. Store at –80°C for up to 6 months.

0.5 M PIPES buffer (pH 6.4)
Dissolve 173 g of PIPES (disodium salt) in approximately 0.9 liter of H_2O. Ajust the pH to 6.4 with 1 N **HCl** and then adjust the final volume to 1 liter with H_2O. Store at room temperature for 6 months (discard if precipitation is seen).

radioactive substances, formamide, concentrated acids (see Appendix for Caution)

2. Heat the hybridization mixture at 95–100°C for 10 minutes to denature the DNA strands.

3. Centrifuge briefly to collect the liquid at the bottom of the tube.

4. Incubate at 55°C for 30 minutes to allow the labeled RNA probe to anneal to the denatured DNA.

   Note: If genomic DNA is being used as the target DNA, the annealing reaction should be allowed to proceed for 12–16 hours.

5. To one of the tubes that has no test DNA and to each control and test sample, add 350 μl of RNase reaction buffer/RNase A/carrier tRNA mixture. Do not add RNase to the other tube that has no test DNA. Incubate all mixtures at 25°C for 1 hour.

   Notes: A 40 μg/ml concentration of RNase A is generally appropriate for RNase cleavage reactions. However, some batches of the enzyme have been found to be more active than others, such that the use of 40 μg/ml causes damage to the RNA:DNA hybrids. In these cases, lowering the enzyme concentration five- to tenfold often corrects the problem. If a new batch of RNase gives poor yields or smeared bands in the cleavage reactions, it may be possible to solve the problem by titrating control reactions of mutant and wild-type test DNA samples with lower concentrations of the enzyme.

   It is often possible to drive partial cleavage reactions to completion by treating with RNase for longer periods of time (for further discussion and an example of the results seen in cases of partial digestion, see Figure 10).

RNase reaction buffer/RNase A/carrier tRNA mixture

Component and final concentration	Amount to add per 10 reactions
~1x RNase reaction buffer	3.5 ml
40 μg/ml RNase A	70 μl of 2 mg/ml
20 μg/ml tRNA	15 μl of 5 mg/ml

Prepare enough RNase reaction buffer/RNase/tRNA mixture to perform the appropriate number of reactions. Place on ice until needed.

RNase reaction buffer

Component and final concentration	Amount to add per ~100 ml
20 mM Tris-Cl	2 ml of 1 M (pH 7.5 at room temperature)
200 mM NaCl	4 ml of 5 M
100 mM LiCl	10 ml of 1 M
1 mM EDTA	200 μl of 0.5 M (pH 8.0)
H$_2$O	84 ml

Sterilize by autoclaving. Divide into aliquots and store at room temperature for indefinitely.

RNase A solution (2 mg/ml)
Purchase purified RNase A in crystallized form. One well-characterized source is Sigma (R 5125). To avoid contamination of the workspace with the enzyme, carefully open the bottle in a chemical fume hood and add H_2O to make an enzyme concentration of 2 mg/ml. Use a disposable pipette to transfer the solution into a screwcap plastic tube. Cap the tube tightly and place it in a boiling-water bath for 10--15 minutes to destroy DNases. Allow to cool to room temperature, divide into aliquots in screwcap tubes, and store at −20°C or at −80°C for at least 1 year.

Carrier tRNA solution (5 mg/ml)
Carrier tRNA (yeast or *E. coli*) is available from several sources (e.g., Sigma, Promega, or Boehringer Mannheim). Prepare a 5 mg/ml solution in H_2O. Store at −20°C indefinitely.

6. Stop the RNase reaction by adding 25 µl of 10% **SDS** (ultrapure) and 5 µl of proteinase K solution (20 mg/ml) and incubating at 37°C for 30 minutes.

    SDS (see Appendix for Caution)

7. Extract with 1 volume of **phenol** (see Appendix). (Use phenol saturated with TE [pH 8.0].) After the phases have been separated, carefully remove only 300 µl of the aqueous (upper) phase and transfer it into new microcentrifuge tube.

    Note: Carefully removing only part of the aqueous phase prevents transfer of any residual RNase from the interface.

    phenol (see Appendix for Caution)

8. Extract the 300-µl aqueous phase with 1 volume of phenol:**chloroform** (1:1). Use phenol saturated with TE (pH 8.0).

    chloroform (see Appendix for Caution)

9. Precipitate the nucleic acids in the aqueous phase with 0.1 volume of 3 M sodium acetate (pH 5.2) and 2–2.5 volumes of absolute ethanol and dry the pellet under vacuum (see Appendix).

10. Add 10–20 µl of 1x formamide/TBE gel-loading buffer to the pellet, and dissolve by pipetting up and down with a micropipette. Store at −20°C until needed for the sequencing gel on p. 351.

    Note: Except in cases where a high-specific-radioactivity probe is used (e.g., when non-amplified genomic DNA is used as the test DNA), the samples can be stored for a few days before analysis is necessary. However, since some breakdown of the radioactivity will occur, it is advisable to load the samples on the gel soon after the reactions are completed.

1x Formamide/TBE gel-loading buffer

Component and final concentration	Amount to add per 10 ml
80% deionized **formamide**	8 ml
1x TBE buffer	1 ml of 10x
0.01% xylene cyanole FF	100 µl of 1%
0.01% bromophenol blue	100 µl of 1%
H_2O	0.8 ml

Store at –20°C or at –80°C for up to 6 months.

formamide (see Appendix for Caution)

PROTOCOL

Analysis of RNase Cleavage Products on Sequencing Gels

Be sure to observe the special considerations for working with RNA throughout this procedure (see p. 343).

1. Prepare a sequencing gel (see, e.g., Sambrook et al. 1989) and run at 15–20 W for 30–60 minutes before loading the samples.

   Notes: To examine RNA fragments in the size range of 500–600 nucleotides, a 6% denaturing polyacrylamide gel should be used, whereas a 10% or 12% gel works best when the RNA fragments are shorter than 500 nucleotides.

   Most standard electrophoresis combs for sequencing apparatuses will work well. In general, clear bands can be seen with the samples loaded below if the lanes are 4–6 mm wide.

   When the gel is ready for loading samples, it should feel uniformly warm to the touch.

2. Heat the pellets dissolved in formamide/TBE gel-loading buffer (from pp. 347–350) at 100°C for 5 minutes to separate the RNA from the DNA strands. Centrifuge briefly to collect the liquid at the bottom of the tube.

3. Turn off the power supply to the gel, disconnect the electrodes, and load approximately 5 μl of each sample and appropriate molecular-weight markers (e.g., 50–1000 bp) in separate lanes of the gel.

4. Reconnect the electrodes and perform electrophoresis at 15–20 W for the appropriate length of time.

   Note: In most cases, the duration of electrophoresis should be adjusted so that the bromophenol blue dye just runs off the gel.

5. When electrophoresis is completed, transfer the gel onto a sheet of filter paper, cover with plastic wrap, and dry in a gel dryer.

6. Place the dried gel on X-ray film. Expose and develop the film.

   Note: Appropriate exposure varies from experiment to experiment. Under the conditions described here, reasonable RNA signals can generally be seen after exposing the film at room temperature for a few hours without an intensifying screen.

7. Analyze the sizes of the bands on the autoradiograph as described in Figure 10. Determine whether or not a mismatch was present and the approximate position of the mismatch relative to one of the ends of the test fragment.

The CCM Method

The CCM method (Cotton et al. 1988; Saleeba and Cotton 1993) detects mutations and polymorphisms by the same principle as other cleavage methods: Single-base mismatches in a heteroduplex nucleic acid are preferentially cleaved under conditions that leave the perfectly paired positions in the duplex intact, and the resulting cleaved products are observed by denaturing gel electrophoresis. In this case, however, the mismatches are not recognized by an enzyme but by hydroxylamine and osmium tetroxide, chemicals that damage C residues and T residues, respectively, in DNA and have a strong preference for single-stranded DNA. After reaction of the chemicals with mismatched bases, the DNA backbone is then cleaved at positions containing damaged bases, and the products are analyzed by electrophoresis. Because these two chemicals recognize mixmatched C residues and T residues efficiently, almost all single-base mutations can be detected with the method if both DNA strands are used as labeled probes (either separately or together). The reason that this is the case is the complementarity of double-stranded DNA; for example, a G on one strand, which does not react with either chemical, will be represented by a C on the other strand, which is recognized by one of the chemicals.

Despite the similarities, the CCM method is different in several ways from the RNase cleavage method. First, by using each of the two strands of the test sequence as a probe in separate reactions, it is possible to detect a higher percentage of all possible single-base changes by CCM (Saleeba and Cotton 1993). Second, although problems with the signal-to-noise ratio can occur for the same reasons in the two methods, it appears that longer probes can be used effectively with CCM, perhaps because the activity of the chemicals provides greater discrimination between mismatched and perfectly paired positions than there is with RNase A. Third, the CCM method can use DNA probes that are end-labeled instead of uniformly labeled, which is often more convenient because the probe segments do not have to be specially cloned into vectors containing bacteriophage promoters and because end-labeled probes last longer. Despite these positive features of the CCM method, the method has the disadvantage of requiring the handling of dangerous chemicals (hydroxylamine, osmium tetroxide, piperidine, and pyridine). Proper precautions (see below) are very important during the use of this method.

PROTOCOLS FOR DETECTING DNA VARIATION USING CCM

The CCM method on pp. 354–360 (Cotton et al. 1988; Saleeba and Cotton 1993) can be divided into five steps:

- preparation of labeled probe
- formation of heteroduplexes
- modification of mismatched bases with hydroxylamine or osmium tetroxide
- cleavage at modified bases
- analysis of cleaved products by denaturing gel electrophoresis

Either single-stranded or double-stranded wild-type DNA fragments, labeled at one or both ends with ^{32}P, can be used as probes in CCM. These fragments are generally cloned DNA segments 100–1500 bp in length (500–1000-bp segments are ideal) that are derived from a wild-type reference allele. The simple method on pp. 354–355 is for labeling both 5' ends of a double-stranded DNA fragment generated by PCR.

Throughout the procedures in this section, observe proper precautions for handling the dangerous chemicals required. Work in a well-functioning chemical fume hood. Carefully separate, store, and dispose of chemical wastes. Make sure all personnel have appropriate training.

PROTOCOL

Preparation of End-labeled Probe

1. Amplify the probe fragment from a single cloned wild-type reference allele by PCR according to standard procedures.

   Notes: Since the primers will contain 5′-hydroxyl groups, the standard polynucleotide kinase reaction conditions below can be used. The length of the probe is limited by the resolution of the denatured DNA fragments in the polyacrylamide gel (generally up to approximately 1.5 kb). As with the RNase cleavage method, smaller probes typically give cleaner results; 500–1000-bp probes are ideal.

   For DNA fragments generated by cleavage with restriction enzymes, 5′-phosphate groups must be removed with phosphatase or the kinase exchange reaction must be used (Sambrook et al. 1989, pp. 10.64–10.65 or 10.66–10.67).

2. Remove the DNA polymerase used during amplification as follows:

   a. Extract the sample with 1 volume of **phenol:chloroform** (1:1) (see Appendix). Use phenol saturated with TE (pH 8.0).

      phenol, chloroform (see Appendix for Caution)

   b. Precipitate the DNA in the aqueous phase with 0.1 volume of 3 M sodium acetate (pH 5.2) and 2–2.5 volumes of absolute ethanol and dry the pellet under vacuum (see Appendix).

3. Dissolve the DNA in TE (pH 8.0) at a concentration of 200 μg/ml.

4. Prepare a 20-μl reaction mixture by combining the components below (in the order listed):

DNA (200 ng/μl)	1 μl
10x bacteriophage T4 polynucleotide kinase buffer (pH 7.6; for preparation, see p. 297)	2 μl
H$_2$O	14 μl
[γ-^{32}P]ATP (10 μCi/μl, 4000–7000 Ci/mmole)	2 μl
bacteriophage T4 polynucleotide kinase (5 units/μl)	1 μl

   radioactive substances (see Appendix for Caution)

5. Incubate at 37°C for 1 hour.

6. Precipitate the DNA with 0.2 volume of 10 M ammonium acetate (pH 7.0–7.4) and 2–2.5 volumes of absolute ethanol (see Appendix).

   Note: There is no need to dry the pellet at this point.

7. Dissolve the pellet in 100 μl of H$_2$O. Precipitate the DNA again as described in step 6 and dry the pellet under vacuum.

Note: Two precipitations are performed with ammonium acetate as the salt. This removes most of the unincorporated [γ-^{32}P]ATP.

8. Dissolve the probe pellet in TE (pH 7.8) at a concentration of 2 ng/μl. Store the probe at –80°C until needed for forming heteroduplexes as described on p. 356.

   Note: Probes are stable for up to 2 weeks.

Formation of Heteroduplexes for Analysis by the CCM Method

1. For each test DNA and wild-type control, combine the components below in a 1.5-ml microcentrifuge tube and mix.

test DNA (or control)	250 ng
end-labeled probe (4 ng/2 μl; from pp. 354–355)	2 μl
TE (pH 7.8) containing 100 mM NaCl	to a final volume of 25 μl

   Notes: Target DNA samples for the wild-type and mutant test fragments should be generated by PCR or by digesting cloned DNA fragments with restriction enzymes. If cloned DNA is used, the target fragment should be excised from the vector with the appropriate restriction enzymes. However, there is no need to purify the insert from the vector.

   The wild-type control tube for the cleavage reactions should be prepared so that probe anneals to the DNA from which it is derived (instead of test DNA).

   There is a large molar excess of test DNA over probe in this reaction. This drives all of the probe to form duplexes with the test DNA.

   Enough heteroduplexes for six reactions must be prepared for each DNA sample so that multiple reaction times can be used for treating each heteroduplex with hydroxylamine and osmium tetroxide in the protocol on pp. 357–358.

   Each of the two strands of the probe can be used in separate reactions to detect a higher percentage of all possible single-base changes (Saleeba and Cotton 1993).

   radioactive substances (see Appendix for Caution)

2. Place the tubes in a boiling-water bath or heating block and heat at 100°C for 10 minutes. Centrifuge briefly to collect the liquid at the bottom of the tubes.

3. Place the tubes in a water bath or incubator and heat at 55°C for 30 minutes. Centrifuge briefly to collect the liquid at the bottom of the tubes.

4. Precipitate the heteroduplexes in each tube with 0.1 volume of 3 M sodium acetate (pH 5.2) and 2–2.5 volumes of absolute ethanol and dry the pellet under vacuum (see Appendix).

5. Dissolve each heteroduplex pellet in 50 μl of H_2O. Store at −80°C until needed for modification of the mismatched bases as described on pp. 357–358.

   Note: These heteroduplexes can be stored at −80°C for a few days, but damage from radioactive decay may begin to cause background problems if the heteroduplexes are stored for too long.

PROTOCOL

Modification of Mismatched Bases with Hydroxylamine or Osmium Tetroxide

1. In a chemical fume hood, set up six reactions for each test and control heteroduplex in separate numbered 1.5-ml microcentrifuge tubes and incubate at room temperature as described below. For the zero time points (tubes 1 and 4), add CCM stop buffer (see step 2a) at the same time that the heteroduplexes are added; for the other tubes, proceed with step 2a immediately after the appropriate incubation period.

	tube 1	tube 2	tube 3	tube 4	tube 5	tube 6
heteroduplex (from p. 356)	5 µl	5 µl	5 µl	5 µl	5 µl	5 µl
hydroxylamine solution	20 µl	20 µl	20 µl	–	–	–
osmium tetroxide buffer	–	–	–	2.5 µl	2.5 µl	2.5 µl
4% **osmium tetroxide** solution	–	–	–	17.5 µl	17.5 µl	17.5 µl
incubation period (minutes)	0	1	4	0	1	4

Hydroxylamine solution
Dissolve 2 g of hydroxylammonium chloride in 2.3 ml of H_2O in a glass test tube. Slowly and gently mix 2.2 ml of **diethylamine** into this solution. Place on ice. Prepare just before use.

Osmium tetroxide buffer

Component and final concentration	Amount to add per 1 ml
100 mM Tris-Cl	100 µl of 1 M (pH 7.8 at room temperature)
10 mM EDTA	20 µl of 0.5 M (pH 8.0)
15% **pyridine** (HPLC grade)	150 µl
H_2O	0.73 ml

Prepare just before use.

Osmium tetroxide solution
A 4% aqueous osmium tetroxide solution can be purchased from Aldrich. Unfortunately, the chemical is unstable and loses its potency after a few months. A greenish/gray discoloration or a decrease in intense yellow color that occurs when osmium tetroxide is mixed with pyridine indicates that the chemical reactivity has diminished. The solution should be stored at 4ºC, but it will still probably need to be replaced with fresh solution every few months.

hydroxylamine, osmium tetroxide, diethylamine, pyridine (see Appendix for Caution)

2. Stop each reaction as follows:

   a. Add 250 μl of CCM stop buffer and then 1 ml of ice-cold absolute ethanol to each tube and mix by vortexing.

      CCM stop buffer

Component and final concentration	Amount to add per 100 ml
300 mM sodium acetate	10 ml of 3 M (pH 5.2)
100 μM EDTA	20 μl of 0.5 M (pH 8.0)
25 μg/ml tRNA (yeast or *E. coli*)	100 μl of 25 mg/ml (in H_2O)
H_2O	90 ml

      Divide into aliquots and store at –20°C indefinitely.

   b. Place each tube at –20°C for 20 minutes.

   c. Centrifuge in a microcentrifuge at 12,000*g* at 4°C for 15 minutes. Discard the supernatant.

   d. Wash each pellet by adding 1 ml of 70% ethanol and recentrifuging for 5 minutes. Decant the supernatant and invert the tube on a paper towel to dry the pellet.

3. Do not dissolve the pellets until needed for cleavage with piperidine on p. 359.

   Note: These pellets can be stored at room temperature for a few days.

PROTOCOL

Cleavage at Modified Bases with Piperidine

1. Dissolve each pellet (from pp. 357–358) in 50 µl of 1 M **piperidine** by vigorous vortexing for 10–20 seconds.

   piperidine (see Appendix for Caution)

2. Place the samples in a heating block and heat at 90°C for 30 minutes.

   Note: The samples treated with osmium tetroxide may turn dark during the reaction with piperidine.

3. Allow the samples to cool on ice and then precipitate each as follows:

   a. Add 50 µl of 0.6 M sodium acetate (pH 5.2), 300 µl of absolute ethanol, and 5 µl of glycogen solution (10 mg/ml in H$_2$O) as a carrier for the precipitation.

   b. Place at –20°C for 30 minutes.

   c. Centrifuge in a microcentrifuge at 12,000g at 4°C for 10 minutes. Discard the supernatant.

   d. Wash each pellet by adding 1 ml of 70% ethanol and recentrifuging for 5 minutes. Decant the supernatant and invert the tube on a paper towel to dry the pellet.

4. Dissolve each pellet in 10 µl of 1x **formamide**/TBE gel-loading buffer (for preparation, see p. 349). Store at –20°C until needed for the sequencing gel on p. 360.

   formamide (see Appendix for Caution)

PROTOCOL

Analysis of Chemically Cleaved Products by Denaturing Gel Electrophoresis

1. Prepare a 6.5% denaturing polyacrylamide sequencing gel (see, e.g., Sambrook et al. 1989) and run at 15–20 W for 30–60 minutes before loading the samples.

2. Heat the samples (from p. 359) at 95°C for 3 minutes and then quickly chill by placing on ice.

3. Turn off the power supply to the gel, disconnect the electrodes, and load 2.5 µl of each sample and appropriate molecular-weight markers (e.g., 100–2000 bp) in separate lanes of the gel.

4. Perform electrophoresis and autoradiography, and interpret the results as described on p. 351, steps 4–7.

   Note: Results obtained with the CCM method are similar to those obtained with the RNase cleavage method.

Direct Sequencing of Uncloned PCR Products

The method of direct sequencing of uncloned PCR products to screen for mutations and polymorphisms (see, e.g., Nigro et al. 1989; Leach et al. 1993; Papadopoulos et al. 1994; Liu et al. 1995) is based on the ddNTP-mediated chain-termination method of Sanger et al. (1977) and the PCR method of Mullis et al. (1986).

The adaptation of PCR methodology to conventional ddNTP-mediated chain-termination sequencing contributes several notable features to cycle sequencing (see Wilson and Mardis 1997). Unlike conventional chain-termination sequencing, in which the annealing, extension, and termination steps require manually manipulated specific reaction times, cycle sequencing exploits the use of automated programmable PCR machines to control the reaction times. Consequently, cycle sequencing requires fewer manipulations and is prone to fewer errors during the reaction. In addition, because of the linear amplification of the labeled products in each cycle, cycle sequencing requires less DNA template. A final feature is that *Taq* DNA polymerase has properties that are beneficial for sequencing (Chien et al. 1976). It has a relatively high degree of processivity, and the temperature at which the enzyme is active markedly reduces both the second-ary structure of DNA and nonspecific priming. These qualities significantly increase the clarity of the sequencing ladder by reducing the number of background bands.

All of these features of cycle sequencing have made it possible to generate high-quality, easily interpretable sequence data by direct sequencing of uncloned PCR products.

The most obvious advantage offered by direct sequencing of PCR products is its expediency. The procedure provided in this chapter eliminates the cloning step and/or extensive preparation of PCR products necessary for conventional sequencing methods. This makes it possible to obtain sequence data within 1 day of generating PCR products. In addition, because the PCR product is generated from a large number of starting template molecules, errors introduced by the DNA polymerase used to amplify the target DNA segment do not confound the interpretation of the sequence. This alleviates the need for sequencing PCR products of multiple amplifications, as is necessary when cloned products that represent only single molecules are sequenced (Nigro et al. 1989).

The most commonly experienced disadvantage of direct sequencing is the production of poor-quality sequence data. Since current methods that use cycle sequencing have vastly improved the quality of sequence data generated by direct sequencing of uncloned PCR products, these may be the methods of choice under certain situations. For example, the method presented here has been extensively used to screen for and characterize mutations in known DNA sequences isolated from many different individuals (Leach et al. 1993; Papadopoulos et al. 1994). It has been applied to PCR products generated from either genomic DNA or first-strand cDNA synthesized from total cellular RNA. The method has worked well with DNA and RNA isolated from several sources, including cell lines, xenografts, and primary human tissues (Liu et al. 1995).

PROTOCOLS FOR DETECTING DNA VARIATION USING DIRECT SEQUENCING OF PCR PRODUCTS

Preparation of PCR Products for Sequencing

Direct sequencing of uncloned PCR products relies on the ability to obtain a relatively pure PCR product free of dNTPs, primers from the amplification, and spurious unwanted PCR products. The first two contaminants can be easily removed by using one of several methods, two of which are provided here—precipitation of PCR products with isopropanol plus sodium perchlorate (pp. 364–365) and purification of PCR products on exclusion spin columns (as described in the protocol for end-labeling on p. 366).

Exclusion spin columns that effectively separate PCR products larger than 100 bp from primers and unincorporated nucleotides are commercially available from many suppliers (e.g., Chromaspin+TE exclusion columns [Clontech Laboratories]). These columns are less time-consuming to use and often give a better yield than the precipitation method on pp. 364–365. However, they are more expensive.

After one of these purification steps, the PCR product can be qualitatively and quantitatively evaluated by gel electrophoresis. If the PCR product contains a significant amount or number of undesired fragments, the desired fragment can be isolated by agarose gel electrophoresis and purified. Although this additional step is generally not necessary, it makes it possible to sequence products that would otherwise yield poor-quality sequence.

End-labeling of Oligonucleotide Primers

The DNA products in cycle sequencing can be labeled with either ^{35}S, ^{33}P, or ^{32}P. The first two require incorporation of an α-labeled nucleotide during the sequencing reaction, but labeling with ^{32}P can be done most readily by phosphorylating the oligonucleotide that is to be used as the sequencing primer. This last method, which is provided on p. 366, has the advantage that a large amount of primer can be labeled and used in multiple sequencing reactions. (The end-labeling protocol on pp. 354–355 could be used as an alternative to the protocol here.)

An exclusion spin column can be used to remove residual labeled ATP that was not incorporated into the oligonucleotide in the kinase reaction. In many cases, however, the results of cycle sequencing are not improved by the use of this step. Therefore, it is optional.

Cycle Sequencing

The cycle sequencing protocol on pp. 367–370 is based on the methods and reagents provided by Epicentre Technologies in the SequiTherm Cycle Sequencing Kit. Similar conditions can be used with other enzymes from other commercial sources. (For further discussion of cycle sequencing, see Wilson and Mardis 1997.)

An optional final incubation of the sequencing reaction with terminal dNTP transferase is provided in the protocol on pp. 367–370. Although this step markedly improves the quality of the sequence data by overcoming the problem of premature terminations, it is also cumbersome, expensive, and time-consuming and may not be necessary for all DNA sequences.

PROTOCOL

Precipitation of PCR Products with Isopropanol Plus Sodium Perchlorate

1. For each wild-type and mutant sample, adjust the volume of the PCR product (~500–1000 ng of DNA) to 200 µl with 3 mM Tris-Cl/200 µM EDTA in a 1.5-ml microcentrifuge tube.

3 mM Tris-Cl/200 µM EDTA

Component and final concentration	Amount to add per 1 liter
3 mM Tris-Cl	3 ml of 1 M (pH 7.5 at room temperature)
200 µM EDTA	400 µl of 0.5 M (pH 8.0)
H₂O	0.997 liter

Combine the Tris-Cl, EDTA, and 0.9 liter of H₂O. Adjust the pH to 7.5 with **HCl** and then adjust the volume to 1 liter with H₂O. Store at room temperature indefinitely.

concentrated acids (see Appendix for Caution)

2. Extract the PCR product with 200 µl of buffered phenol:chloroform (for extraction protocol, see Appendix).

Buffered phenol:chloroform

Component and final concentration	Amount to add per 1.44 liter
3 parts **phenol**	480 ml
2 parts Tris/EDTA/NaCl buffer	320 ml
4 parts **chloroform**	640 ml

Warm the phenol to 65°C. Combine the components in the order listed and shake vigorously. Place at 4°C for 2–3 hours and then shake vigorously again. Place at 4°C for an additional 2–3 hours and then remove the aqueous (upper) phase by using with a pasteur pipette connected to an aspirator. Divide the organic phase into aliquots in three 0.5-liter bottles wrapped in aluminum foil. Store at –20°C for up to 1 year.

Tris/EDTA/NaCl buffer

Component and final concentration	Amount to add per 2 liters
0.5 M Tris base	121 g
200 mM EDTA (free acid anhydrous)	117 g
10 mM NaCl	4 ml of 5 M
H₂O	to make 2 liters

Combine the Tris base, EDTA, and NaCl in approximately 1.5 liters of H_2O. Adjust the pH to 8.9 with HCl and then adjust the volume to 2 liters with H_2O. Store at room temperature indefinitely.

phenol, chloroform (see Appendix for Caution)

3. Precipitate the extracted PCR product from the aqueous (upper) phase in a new 1.5-ml microcentrifuge tube as follows:

   a. Add the components below (in the order listed) to the aqueous phase from the extracted PCR product and mix by vortexing.

glycogen solution (20 mg/ml in H_2O)	2 μl
3 mM Tris-Cl/200 μM EDTA	400 μl
2 M **sodium perchlorate**	200 μl
isopropanol	400 μl

   2 M Sodium perchlorate
   Dissolve 28.1 g of sodium perchlorate (monohydrate) in sufficient H_2O to make a final volume of 100 ml. Divide into 5–10-ml aliquots and store wrapped in foil at room temperature for up to 6 months.

   sodium perchlorate (see Appendix for Caution)

   b. Centrifuge in a microcentrifuge at 12,000*g* at room temperature for 15 minutes. Decant the supernatant.

   c. Wash the pellet by adding 1 ml of 70% ethanol and recentrifuging for 5 minutes. Carefully remove the supernatant with a pipette and discard.

   d. Dry the pellet under vacuum for 10 minutes in a SpeedVac Concentrator.

4. Dissolve the pellet of PCR product in 15 μl of 3 mM Tris-Cl/200 μM EDTA.

5. Quantitate the PCR product by analyzing a 1-μl aliquot and appropriate concentration markers on a 1% agarose gel and then comparing the intensities after staining with **ethidium bromide** (see Appendix).

   Notes: The PCR product can be stored at −20°C until needed for sequencing on pp. 367–370. It should be stable indefinitely.

   The PCR product can also be qualitatively analyzed on the gel at this point. A single PCR product should be seen. If there is a substantial amount of undesired PCR product present in the sample, the entire sample can be loaded on a 1% agarose gel (along with appropriate molecular-weight markers) for the isolation and purification of the desired PCR fragment. The gel purification can be performed by routine protocols. For example, the DNA can be extracted from the agarose as specified by the manufacturer of one of the kits for purification of DNA (e.g., GENECLEAN [BIO 101], Prep-A-Gene [Bio-Rad], QIAEX [Qiagen], or Band-Prep [Pharmacia]).

   ethidium bromide (see Appendix for Caution)

PROTOCOL

End Labeling the Oligonucleotide Primer Using Bacteriophage T4 Polynucleotide Kinase

1. Prepare a 20-µl reaction mixture by combining the components below (in the order listed) in a 1.5-ml microcentrifuge tube:

oligonucleotide primer (350 ng/µl)	2 µl
10x bacteriophage T4 polynucleotide kinase buffer (pH 7.6; for preparation, see p. 297)	2 µl
H$_2$O	13 µl
[γ-^{32}P]ATP (10 µCi/µl, 4000–7000 Ci/mmole)	2 µl
bacteriophage T4 polynucleotide kinase (10 units/µl)	1 µl

   Note: An appropriate primer for this step should be 18–25 nucleotides in length.

   radioactive substances (see Appendix for Caution)

2. Incubate at 37°C for 20–40 minutes.

3. Adjust the reaction volume to 50 µl by adding 30 µl of 3 mM Tris-Cl/200 µM EDTA (for preparation, see p. 364).

4. (*Optional*) Place the sample on an exclusion spin column (Chromaspin+TE column [Clontech Laboratories] or equivalent). Centrifuge at 700*g* at room temperature for 5 minutes.

   Note: This is an optional procedure to remove residual labeled ATP that was not incorporated into the oligonucleotide in the kinase reaction. In many cases, the results of cycle sequencing are not improved by the use of this additional step.

5. Adjust the sample volume with TE (pH 7.8) to make a final primer concentration of approximately 10 ng/µl. Store at –20°C until needed for sequencing on pp. 367–370.

   Notes: The primer can be stored for up to a few days.
   Each of the cycle sequencing reactions in this chapter requires 1–2 µl (10–20 ng) of end-labeled primer.

PROTOCOL

Cycle Sequencing Method

1. For each DNA template and primer set, prepare the following 8.5-µl reaction mixture:

DNA template (prepared wild-type or mutant PCR product; e.g., from pp. 364–365)	50–100 ng
^{32}P-end-labeled primer (from p. 366)	1–2 µl
10x sequencing buffer	1.25 µl
SequiTherm DNA polymerase (5 units/µl; Epicentre Technologies)	0.5 µl
H_2O	to a final volume of 8.5 µl

    Note: The source of the DNA polymerase used in this reaction does affect the quality of the sequence data. The SequiTherm DNA polymerase may be purchased separately or as part of a kit (Epicentre Technologies). Other DNA polymerases are likely to generate data of similar quality.

    10x Sequencing buffer

Component and final concentration	Amount to add per 1 ml
0.5 M Tris-Cl	0.5 ml of 1 M (pH 9.3 at room temperature)
25 mM $MgCl_2$	25 µl of 1 M
H_2O	475 µl

    Store at 4°C for up to 6 months.

    radioactive substances (see Appendix for Caution)

2. For each reaction mixture from step 1, set up four reactions in labeled, thin-walled 0.5-ml PCR tubes. Combine the components below in the order listed.

	A	C	G	T
appropriate ddNTP termination mixture	1 µl	1 µl	1 µl	1 µl
reaction mixture (from step 1)	2 µl	2 µl	2 µl	2 µl

    Note: For thermal cyclers that do not have a top heating block, overlay each reaction mixture with one drop of light mineral oil and then cover the tube.

ddNTP termination mixtures

These solutions are available as part of the SequiTherm Cycle-sequencing Kit (Epicentre Technologies).

For the A tube:
    0.45 mM ddATP
    15 μM dATP
    15 μM dCTP
    15 μM dTTP
    15 μM 7-aza-dGTP
For the C tube:
    0.3 mM ddCTP
    15 μM dATP
    15 μM dCTP
    15 μM dTTP
    15 μM 7-aza-dGTP
For the G tube:
    0.03 mM ddGTP
    15 μM dATP
    15 μM dCTP
    15 μM dTTP
    15 μM 7-aza-dGTP
For the T tube:
    0.9 mM ddTTP
    15 μM dATP
    15 μM dCTP
    15 μM dTTP
    15 μM 7-aza-dGTP

3. Transfer the samples into a thermal cycler and perform thermal cycling as follows for 30 cycles:

    95°C for 30 seconds
    50°C for 30 seconds
    70°C for 60 seconds

Perform an additional extension cycle of 70°C for 5 minutes and then hold the samples at 4°C.

Note: Although the conditions described here typically work well, it may be necessary to adjust the cycling parameters with some primer-template combinations or with some thermal cyclers.

4. (*Optional*) Add 1 μl of the transferase reaction mixture below to each tube and incubate at 37°C for 30 minutes.

10x terminal dNTP transferase buffer	1 μl
10 mM dNTP mixture (see Appendix)	0.5 μl
terminal dNTP transferase (17 units/μl)	0.5 μl
H$_2$O	8 μl

Notes: If a mineral oil overlay was used, be sure to insert a pipette through the oil and place the transferase reaction mixture in the aqueous phase below the oil. To further ensure mixing, gently agitate the tubes and centrifuge briefly in the microcentrifuge.

As in all sequencing reactions, premature terminations can contribute to the poor quality of sequence data produced in this protocol. One method that helps overcome this problem is a final incubation of the sequencing reaction with terminal dNTP transferase. This enzyme adds additional dNTPs to the 3′ end of DNA molecules that lack an incorporated ddNTP, thus extending the molecules that did not terminate because the appropriate ddNTP was incorporated. This step markedly improves the quality of the sequence data; however, it is cumbersome, expensive, and time-consuming and may not be necessary for all DNA sequences.

10x Terminal dNTP transferase buffer

Component and final concentration	Amount to add per 10 ml
200 mM Tris-Cl	2 ml of 1 M (pH 7.5 at room temperature)
100 mM MgCl$_2$	1 ml of 1 M
250 mM NaCl	0.5 ml of 5 M
H$_2$O	6.5 ml

Store at room temperature indefinitely.

5. Add 6 μl of stop/loading solution to each tube and mix by vigorous vortexing.

   Note: If desired, samples mixed with stop/loading solution can be stored at –20°C for several days at this point.

Stop/loading solution

Component and final concentration	Amount to add per ~10 ml
~90% (v/v) deionized **formamide**	9 ml
10 mM EDTA	200 μl of 0.5 M (pH 8.0)
0.05% xylene cyanole FF	0.5 ml of 1%
0.05% bromophenol blue	0.5 ml of 1%

Store at –20°C or at –80°C for up to 1 year.

formamide (see Appendix for Caution)

6. Prepare a 6% denaturing polyacrylamide sequencing gel (see, e.g., Sambrook et al. 1989) and run at 15–20 W for 30–60 minutes before loading the samples.

7. Just before loading the gel, heat the samples at 95°C for 5 minutes.

8. Turn off the power supply to the gel, disconnect the electrodes, and load 6 μl of each sample in separate lanes of the gel.

   Note: Removing 6 μl of sample from the aqueous layer that lies below the mineral oil can be technically difficult. The use of long rounded pipette tips will help if the orifice of the tip is placed under the mineral oil in the middle of the reaction mixture. If the samples are consistently contaminated with mineral oil, extraction with chloroform can be performed. However, this adds a timely and cumbersome step that is usually not necessary. In addition, contamination of the samples with mineral oil does not drastically reduce the quality of the sequencing reaction.

9. Reconnect the electrodes and perform electrophoresis at 15–20 W for the appropriate length of time.

   Note: In most cases, the duration of electrophoresis should be adjusted so that the bromophenol blue dye just runs off the gel.

10. When electrophoresis is completed, fix the gel in 5% **methanol**/5% **acetic acid** solution at room temperature for 30 minutes.

    methanol, glacial acetic acid (see Appendix for Caution)

11. Transfer the gel onto a sheet of filter paper, cover with plastic wrap, and dry in a gel dryer.

12. Place the dried gel on X-ray film. Expose and develop the film.

    Note: Appropriate exposure varies from experiment to experiment. Under the conditions described here, reasonable signals can generally be seen after exposing the film at room temperature for a few hours without an intensifying screen.

13. Compare the nucleotide sequences of the wild-type control sample and the test samples. A difference in sequence is indicative of a polymorphism or a mutation in the test DNA sample.

    Note: Any difference seen can be confirmed by identifying the presence of the same difference in other members of the family or pedigree in expected Mendelian frequencies.

The PTT Method

The PTT method was developed to screen for mutations that produce an alteration in the size of the protein product of a gene (Roest et al. 1993; Hogervorst et al. 1995). Differences in polypeptide migration on gels may be the result of insertions, deletions, frameshifts, and nonsense mutations and may inadvertently result from splice-site mutations. The PTT method uses coupled in vitro transcription and translation to screen for such mutations by detecting alterations in the electrophoretic mobility of polypeptides translated in vitro.

The detection of genetic alterations by the PTT method usually relies on the availability of RNA from cells or tissues to be analyzed. The RNA is used as a template to generate first-strand cDNA with random hexamers. The target gene is amplified from this cDNA in overlapping segments that contain the entire ORF. The 5′ primer of each primer set is designed to contain sequences attached to the 5′ end of a segment that will anneal to the target gene and that will efficiently direct transcription and translation (Hogervorst et al. 1995). The 3′ primer consists only of target gene sequences. The PCR product generated with these oligonucleotide primers is used in a rabbit reticulocyte-lysate-based coupled transcription/translation system that uses RNA polymerase. The PCR product is transcribed and the resulting RNA is translated in the same tube without further experimental manipulation. The reaction is performed in the presence of [^{35}S]methionine, which allows the polypeptides synthesized in the coupled system to be detected by autoradiography after separation on denaturing polyacrylamide gels. An aberrantly migrating band on the autoradiograph suggests that a mutation is present in the target gene. In such cases, the gene is further analyzed to characterize the nature of the mutation that results in the altered polypeptide.

The PTT method offers several advantages over some other mutation detection methods: (1) it is rapid and relatively simple, allowing many samples to be screened simultaneously, (2) it uses routine molecular biological reagents and methods, (3) the results are usually unambiguous, and (4) it can be used to screen for mutations in relatively large target DNA sequences. The PTT method has been used on PCR products up to 2 kb, which is larger than the sizes of target sequences that are typically used with other mutation detection methods (Nicolaides et al. 1994). On the other hand, the PTT method has several limitations that should be considered before it is used for mutation detection: (1) it detects only mutations that lead to an alteration in the size of the encoded protein, such that missense and some small insertions and deletions escape detection, (2) mutations that cause a lack of production of mRNA molecules are not detected, and (3) the assay requires the availability of mRNA from tissues that express the target gene. Exceptions to this last limitation are genes that contain large uninterrupted ORFs (exons) that can be amplified from genomic DNA and used to direct polypeptide synthesis (Powell et al. 1993; Hogervorst et al. 1995). The requirement for mRNA excludes the possibility of using this method for mutation detection in archival tissue specimens; this eliminates a valuable source of material, particularly in studies of tumors, that can be used with many of the other mutation detection methods.

Samples that yield an abnormal sized polypeptide in the PTT method should be characterized further to determine the nature of the mutation responsible for the altered protein. Determination of the estimated size of the aberrantly migrating polypeptide can often provide information about the possible location of the causative mutation. Direct sequencing or other methods can be used to uncover the exact sequence alteration in the gene. The same PCR product that gives rise to an abnormal polypeptide in the PTT method can be used as a template for sequence analysis. In the case of splice-site mutations, evaluation of the genomic DNA may be necessary to find an explanation for the change in the coding sequence of the gene (Chomczynski and Sacchi 1987).

Mutations are detected with the PTT method by testing for the presence of aberrantly migrating polypeptides in test samples versus a normal sample. The size of the normal, nonmutant polypeptide can be estimated from the size of the PCR product from which transcription and translation are directed, and this product is typically the dominant band in the control lane. Degradation or incomplete products appear as smaller, faster-migrating fragments that are usually present in smaller amounts than the full-length wild-type polypeptide. For a given polypeptide, these smaller-molecular-weight products are typically present in a pattern that is reproducible from experiment to experiment. Thus, any new polypeptide band that appears in the lane from a test sample is probably due to the presence of a stop codon mutation. Furthermore, in the case of a mutant sample, the truncated polypeptide band is typically present as the major band in the lane.

PROTOCOLS FOR DETECTING DNA VARIATION USING PTT

Preparation of Tissue

Tissues must be disrupted into single-cell suspensions before the cells are lysed and used as a source of RNA. This is an important step, since the extent to which the tissue is disrupted has a substantial influence on the yield of RNA. Once the tissue is disrupted, the RNA isolation procedure is identical to that used for cell lines grown in culture.

The method of tissue disruption that should be used depends on the specific mutations to be analyzed. In the evaluation of germ-line mutations, the method used to prepare the tissue is not critical. However, detection of somatic mutations in tumors requires a preparation technique that eliminates the majority of normal cells inevitably present in tumor tissue (Fearon et al. 1987). The first method in this chapter (p. 374) is one of many ways to disrupt tissue mechanically; the second (pp. 375–376) is a method that has been developed specifically for enrichment of tumor cells. The second method is more time-consuming, but it is critical for isolating relatively pure tumor cell RNA, a crucial aspect in the detection of many somatic mutations in tumors. This second technique has become a mainstay in preparing tumor tissue for mutation analyses. It requires the availability of a cryostat and knowledge of the histopathology of the tumors being evaluated.

All of the methods below should be performed observing the special considerations for working with RNA (see p. 343 and pp. 67–68). Normal and tumor tissue should be processed separately to avoid contamination.

Preparation of Total Cellular RNA

Total cellular RNA can be prepared for the PTT method by many methods commonly used in the laboratory. Various methods for preparing total cellular RNA can be found in Chapter 2. In addition, a simplified version for making total cellular RNA can be performed using Trizol (Life Technologies), or RNA can be prepared using the RNAgents Total RNA Isolation System (Promega). These protocols have been used successfully to isolate RNA from both cultured cells and primary human tissues. Isolating RNA from cell lines for mutation analysis is straightforward, since cell lines represent a pure population of cells that simply need to be harvested and lysed before isolation. Isolation of RNA from primary tissues is more difficult. Because RNA is susceptible to degradation by ubiquitous RNases, extreme precautions (see p. 343 and pp. 67–68) are necessary to ensure that RNA molecules remain intact during any isolation procedure.

Reverse Transcription

Total cellular RNA is used as a template to generate first-strand cDNA with random hexamers in the protocol on p. 377. The target gene can then be amplified from this cDNA in overlapping segments that contain the entire ORF.

Preparation of the PCR Product

The exact conditions for performing the amplification vary with each primer set, and conditions will need to be optimized to obtain reasonable amounts of specific PCR product. The PCR buffer conditions described on p. 378 work for amplifying fragments up to 2 kb in length with many primer sets and for RNA from tissue-culture cell lines. However, RNA isolated from primary human tissue can be more difficult to amplify by PCR because of contaminants and, in such cases, the conditions may have to be carefully optimized. Most of the primer sets that have been used have T_m values of approximately 56–58°C.

In Vitro Transcription and Translation of the PCR Product

The PCR product is generally used directly, without purification, in the in vitro transcription and translation reaction. If a protein product is not obtained, it may be necessary to purify the PCR products and then repeat the experiment. In such cases, the precipitation protocol on pp. 364–365 is recommended for purification.

The most important precaution in this protocol is to prevent degradation of the polypeptides synthesized in vitro. Degradation of the proteins obscures the ability to observe mutant polypeptides that migrate faster than their normal counterpart. To avoid degradation, add 2x loading buffer immediately after the reaction is completed, heat to denature, and store at –80°C.

The protocol on pp. 379–380 uses the TNT T7-coupled Reticulocyte Lysate System (Promega) (Roest et al. 1993; Hogervorst et al. 1995).

PROTOCOL

Mechanical Disruption of Tissues with Glass Beads

Be sure to observe the special considerations for working with RNA throughout this procedure (see p. 343).

1. Remove tissue from storage at –80ºC. Use a clean single-edged razor blade to cut off a small piece of tissue weighing approximately 0.2–0.5 g.

   Note: Prepare and process normal and tumor tissue samples separately throughout all protocols for the PTT method.

   human blood, blood products, and tissues (see Appendix for Caution)

2. Use a clean razor blade to mince the tissue into small (~1 mm³) fragments. Work fast so that the tissue does not thaw.

   Note: This step should be done on a disposable surface (e.g., a weighing boat placed on ice) to avoid contamination by other samples.

3. Place the tissue in a chilled 2-ml screwcap tube and add 1.5 ml of ice-cold guanidinium thiocyanate (available as part of the RNAgents Total RNA Isolation System [Promega]). Place on ice.

4. Add enough 1-mm glass beads (Biospec Products) to fill the tube. Place the tube in a Mini-BeadBeater (Biospec Products) and beat at room temperature for 1–2 minutes on the high setting (until minced pieces are no longer visible).

   Note: Some tissues, depending on the amount of fibrous tissue, may be difficult to disrupt. In such cases, some small visible pieces of tissue may remain.

5. Use a long rounded pipette tip to transfer the solution from the beads into a clean, chilled 6-ml polypropylene tube. Avoid transferring any visible pieces of intact tissue. Place the tube on ice.

6. Immediately prepare total cellular RNA from the disrupted tissue (e.g., as described in one of the methods in Chapter 2) and then reverse-transcribe the RNA as described on p. 377.

Sectioning Tissue Using a Cryostat

Be sure to observe the special considerations for working with RNA throughout this procedure (see p. 343).

1. Remove tumor tissue from storage at –80ºC. Use a clean single-edged razor blade to cut off a piece of tumor **tissue** approximately 0.5 x 0.5 x 1 cm in size.

   human blood, blood products, and tissues (see Appendix for Caution)

2. Prepare the tissue block for sectioning as follows:

   a. Place a small amount of freezing compound on a cryostat chuck and freeze it in **liquid nitrogen** until it is slightly firm.

      Note: Freezing compound is used to inhibit the formation of ice crystals and reduce the tissue artifacts produced by freezing. Tissue-Tek O.C.T. compound (Miles) contains 10.24% (w/w) polyvinyl alcohol, 4.26% (w/w) PEG, and 85.5% (w/w) nonreactive ingredients.

      liquid nitrogen (see Appendix for Caution)

   b. Embed the tissue fragment in the partially frozen freezing compound and cover the tissue with additional compound. Freeze the tissue again in liquid nitrogen until the compound is just frozen. Do not allow the compound to become brittle by excess freezing.

   c. Mount the prepared chuck in the cryostat with the largest surface area facing toward the blade. Cut into the tissue block until the majority of the surface area of the tissue is visible.

3. Eliminate most of the normal tissue from the tissue block as follows:

   a. Cut one 6-μm tissue section, and place it on a glass slide. Stain the section with hematoxylin and eosin by sequentially placing the slide in Coplin jars containing each of the following solutions:

      - two changes of absolute ethanol
      - one change of 95% ethanol
      - one change of 75% ethanol
      - two changes of H_2O
      - hematoxylin (Sigma) for 1–2 minutes
      - 0.5% lithium carbonate in H_2O for 1 minute
      - two changes of H_2O
      - eosin (Sigma) for 30 seconds
      - two changes of H_2O
      - one change of 95% ethanol
      - two changes of absolute ethanol
      - two changes of **xylene**

      xylene (see Appendix for Caution)

b. Cover the tissue section with mounting medium and a coverslip. Examine it by light microscopy.

c. If there is any normal tissue in the section, use ink to circle it on the slide.

d. Align the slide with the tissue block on the cryostat chuck. Use a clean razor blade to remove the area on the block that corresponds to the area circled on the slide. Discard this normal tissue.

Note: Prepare and process normal tissue sections separately throughout all protocols for the PTT method.

4. Cut 150 12-μm sections into a 50-ml conical polypropylene tube. Add 6 ml of ice-cold guanidinium thiocyanate (available as part of the RNAgents Total RNA Isolation System [Promega]) to the tube and mix by inverting. Immediately prepare total cellular RNA from the sectioned tissue (e.g., as described in one of the methods in Chapter 2) and then reverse-transcribe the RNA as described on p. 377.

Note: Sectioning using a cryostat is effective at disrupting tissue. No further disruption is required before RNA is isolated.

5. Prepare additional sections as follows:

a. Repeat steps 3a–b. Make sure that the tumor is still free of normal tissue.

b. If normal tissue has reappeared on the block and contributes more than 10–20% of the cell nuclei, remove the contaminating normal tissue as described in steps 3c–d.

c. If the tissue is free of normal cells, cut 50 additional 12-μm sections into a 50-ml conical polypropylene tube.

d. Repeat steps a–c until a total of 150 12-μm sections have been collected. Add 6 ml of ice-cold guanidinium thiocyanate to the tube and mix by inverting. Immediately prepare total cellular RNA from the sectioned tissue and reverse-transcribe the RNA as described on p. 377.

PROTOCOL

Reverse Transcription of the Total Cellular RNA

Be sure to observe the special considerations for working with RNA throughout this procedure (see p. 343).

1. For each normal or tumor sample, combine the following in the order listed in a 1.5-ml microcentrifuge tube:

total cellular RNA (from p. 374 or pp. 375–376)	5 µg
random hexamers (1 mg/ml; Pharmacia)	4 µl
single-stranded binding	
protein (1 mg/ml; USB)	2 µl
DEPC-treated H_2O	to a final volume of 25 µl

   DEPC (see Appendix for Caution)

2. Incubate at 70°C for 10 minutes.

3. Immediately place the tube on ice and allow to cool. Centrifuge briefly at room temperature to collect the liquid at the bottom of the tube.

4. Transfer 11.9-µl aliquots of the mixture into two separate 1.5-ml microcentrifuge tubes labeled RT– (minus reverse transcriptase) and RT+ (plus reverse transcriptase). Place on ice.

5. Combine the following in the order listed in a separate tube on ice:

5x reverse transcriptase buffer (supplied	
with the enzyme)	8 µl
100 mM DTT	5 µl
25 mM dNTP mixture (see Appendix)	0.8 µl
RNasin RNase inhibitor (40 units/µl; Promega)	0.4 µl

6. Pipette 6.6 µl of the mixture from step 5 into each of the tubes from step 4.

7. To the RT+ tube, add 1.5 µl of reverse transcriptase (5–10 units/µl) and mix briefly by vortexing. Centrifuge briefly to collect the liquid at the bottom of the tube.

8. Incubate both the RT– and RT+ tubes at 37°C for 1 hour.

9. Store the samples at –20°C until needed for preparing the PCR products on p. 378.

   Note: The samples are stable at –20°C for up to 6 months.

PROTOCOL

Preparation of PCR Products for the PTT Method

Be sure to observe the special considerations for working with RNA throughout this procedure (see p. 343).

1. For each normal or tumor sample, combine the components below (in the order listed) in each of two separate 1.5-ml microcentrifuge tubes and mix by vigorous vortexing.

10x amplification buffer (see p. 298)	5 μl
DEPC-treated H$_2$O	30.5 μl
25 mM dNTP mixture (see Appendix)	3 μl
DMSO	5 μl

   DEPC (see Appendix for Caution)

2. Add the following to each tube:

78 mM MgCl$_2$	3 μl
primer 1 (350 ng/μl)	1 μl
primer 2 (350 ng/μl)	1 μl
Taq DNA polymerase	0.5 μl

3. Add 1 μl of first-strand cDNA (the sample in the RT+ tube from p. 377) to one of the tubes, and add 1 μl of the negative control (the sample in the RT– tube from p. 377) to the other to make each reaction mixture 50 μl.

   Note: For thermal cyclers that do not have a top heating block, overlay each reaction mixture with 10–20 μl of light mineral oil and then cover the tube.

4. Perform thermal cycling as follows for 35 cycles:

   95°C for 30 seconds
   52°C for 2 minutes
   70°C for 2–3 minutes

   Perform an additional extension cycle of 70°C for 5 minutes and then hold the samples at 4°C.

   Note: Although the conditions described here typically work well for primers with a T_m of 56–58°C and expected products of 1–2 kb, it may be necessary to adjust the cycling parameters with some primer-template combinations or with some thermal cyclers.

5. Analyze 5 μl of each PCR product and appropriate molecular-weight markers (100–5000 bp) on a 1% agarose gel to evaluate yield and quality.

   Notes: A single band should be seen on the gel.
   If a mineral oil overlay was used, remove the oil.

6. Store the PCR products at –20°C until needed for in vitro transcription and translation on pp. 379–380.

   Note: The PCR products should be stable for months to years.

PROTOCOL

In Vitro Transcription and Translation

Be sure to observe the special considerations for working with RNA throughout this procedure (see p. 343).

1. Prepare a 12% SDS-polyacrylamide gel (e.g., as described in Sambrook et al. 1989, pp. 18.47–18.54).

2. For each PCR product, prepare the following 25-μl reaction mixture:

TNT rabbit reticulocyte lysate	12.5 μl
TNT reaction buffer	1 μl
TNT T7 RNA polymerase	0.5 μl
amino acid mixture minus methionine	0.5 μl
[^{35}S]methionine (10 mCi/ml, 1000 Ci/mmole)	2 μl
RNasin RNase inhibitor (40 units/μl)	0.5 μl
PCR product (from p. 378; ~10 ng/μl)	8 μl

radioactive substances (see Appendix for Caution)

Note: This protocol uses reagents in the TNT T7-coupled Reticulocyte Lysate System (Promega).

3. Incubate at 30°C for 30–60 minutes.

4. Immediately add 25 μl of 2x loading buffer to each tube, and heat the samples at 95°C for 10 minutes.

Note: To avoid degradation of the polypeptides synthesized in vitro, it is important to add the 2x loading buffer immediately after the reaction is completed and then to heat the samples to denature them. Samples can then be stored at –80°C for a few weeks.

2x Loading buffer for SDS-PAGE

Component and final concentration	Amount to add per 5 ml
125 mM Tris-Cl	0.5 ml of 1.25 M (pH 6.8 at room temperature)
4% **SDS**	2 ml of 10% (ultrapure)
1.43 M β-**mercaptoethanol**	0.5 ml of 14.3 M
20% glycerol	1 ml of 100%
0.05% bromophenol blue	250 μl of 1%
H_2O	0.75 ml

Store at 4°C for up to 2 weeks.

1.25 M Tris-Cl (pH 6.8)
Dissolve 75.7 g of Tris base in approximately 400 ml of H_2O. Adjust the pH to 6.8 with **concentrated HCl** and then adjust the final volume to 0.5 liter with H_2O. Store at room temperature for at least 6 months.

SDS, β-mercaptoethanol, concentrated HCl (see Appendix for Caution)

5. Load 20 μl of each sample and appropriate molecular-weight markers (Rainbow markers; Amersham) in separate wells of the gel.

6. Perform electrophoresis at 50 mA for 2 hours (or until the bromophenol blue dye reaches the bottom of the gel).

7. When electrophoresis is completed, place the gel in a suitable container and cover the gel with 3–4 volumes of fixative (30% **methanol**/10% **acetic acid**). Incubate at room temperature for 1 hour.

methanol, glacial acetic acid (see Appendix for Caution)

8. Remove the fixative. Cover the gel with EnHance (NEN Life Science Products) and incubate at room temperature for 45–60 minutes.

9. Decant the EnHance. Rinse the gel with H_2O. Add enough H_2O to cover the gel and incubate at room temperature for 30 minutes.

Note: The incubation period for the gel in H_2O is critical and should be as close to 30 minutes as possible. However, the incubation period in fixative is not critical, and the gel can be left in fixative for long periods.

10. Transfer the gel onto a sheet of filter paper, cover with plastic wrap, and dry on a gel dryer at 60ºC.

11. Place the dried gel on X-ray film. Expose at −80ºC, typically for 12–24 hours, and develop the film.

12. Interpret the pattern on the autoradiograph to determine whether or not aberrantly migrating polypeptides are present in test samples (for further description, see p. 371) (see Figure 11).

Figure 11 Mock autoradiograph of the results of the PTT analysis. The tumor tissue (lane T) produces one smaller polypeptide not present in cDNA extracted from matched normal tissue (lane N). The polypeptide seen in lane N is approximately 46 kD. The arrow represents the truncated polypeptide resulting from the introduction of a premature stop codon. In addition, the normal-size polypeptide is present in the tumor because of a small amount of normal DNA from normal contaminating tissue.

REFERENCES

Ainsworth, P.J., L.C. Surh, and M.B. Coulter-Mackie. 1991. Diagnostic single strand conformational polymorphism (SSCP): A simple non-radioisotopic method as applied to a Tay-Sachs B1 variant. *Nucleic Acids Res.* **19:** 405–406.

Ballabio, A. 1993. The rise and fall of positional cloning? *Nat. Genet.* **3:** 277–279.

Børresen, A.L., E. Hovig, and A. Brogger. 1988. Detection of base mutations in genomic DNA using denaturing gradient gel electrophoresis (DGGE) followed by transfer and hybridization with gene-specific probes. *Mutat. Res.* **202:** 77–83.

Børresen, A.-L., E. Hovig, B. Smith-Sørensen, D. Malkin, S. Lystad, T.I. Andersen, J.M. Nesland, K.H. Isselbacher, and S.H. Friend. 1991. Constant denaturant gel electrophoresis as a rapid screening technique for p53 mutations. *Proc. Natl. Acad. Sci.* **88:** 8405–8409.

Botstein, D., R.L. White, M. Skolnick, and R.W. Davis. 1980. Construction of a genetic linkage map in man using restriction fragment length polymorphisms. *Am. J. Hum. Genet.* **32:** 314–331.

Burmeister, M., G. diSibio, D.R. Cox, and R.M. Myers. 1991. Identification of polymorphisms by genomic denaturing gradient gel electrophoresis: Application to the proximal region of human chromosome 21. *Nucleic Acids Res.* **19:** 1475–1481.

Cai, Q-Q. and I. Touitou. 1993. Excess PCR primers may dramatically affect SSCP efficiency. *Nucleic Acids Res.* **21:** 3909–3910.

Chakravarti, A. 1998. Meiotic mapping of human chromosomes. In *Genome analysis: A laboratory manual*. Vol. 4 *Mapping genomes* (ed. B. Birren et al.). Cold Spring Harbor Laboratory Press, Cold Spring Harborm New York. (In press.)

Chee, M., R. Yang, E. Hubbell, A. Berno, X.C. Huang, D. Stern, J. Winkler, D.J. Lockhart, M.S. Morris, and S.P. Fodor. 1996. Accessing genetic information with high-density DNA arrays. *Science* **274:** 610– 614.

Chen, X. and P.-Y. Kwok. 1997. Template-directed dye-terminator incorporation (TDI) assay: A homogeneous DNA diagnostic method based on fluorescence resonance energy transfer. *Nucleic Acids Res.* **25:** 347–353.

Chien, A., D.B. Edgar, and J.M. Trela. 1976. Deoxyribonucleic acid polymerase from the extreme thermophile *Thermus aquaticus*. *J. Bacteriol.* **127:** 1550–1557.

Chomczynski, P. and N. Sacchi. 1987. Single step method of RNA isolation by acid guanidinium thiocyanate-phenol-chloroform extraction. *Anal. Biochem.* **162:** 156–159.

Collins, F.S. 1992. Positional cloning: Let's not call it reverse anymore. *Nat. Genet.* **1:** 3–6.

Conner, B.J., A.A. Reyes, C. Morin, K. Itakura, R.L. Teplitz, and R.B. Wallace. 1983. Detection of sickle cell β^S-globin allele by hybridization with synthetic oligonucleotides. *Proc. Natl. Acad. Sci.* **80:** 278–282.

Cooper, D.N., B.A. Smith, H.J. Cooke, S. Niemann, and J. Schmidtke. 1985. An estimate of unique DNA sequence heterozygosity in the human genome. *Hum. Genet.* **69:** 201–205.

Costes, B., E. Girodon, N. Ghanem, M. Chassignol, N.T. Thuong, D. Dupret, and M. Goossens. 1993. Psoralen-modified oligonucleotide primers improve detection of mutations by denaturing gradient gel electrophoresis and provide an alternative to GC-clamping. *Hum. Mol. Genet.* **2:** 393–397.

Cotton, R.G., N.R. Rodrigues, and Campbell. 1988. Reactivity of cytosine and thymine in single-base-pair mismatches with hydroxylamine and osmium tetroxide and its application to the study of mutations. *Proc. Natl. Acad. Sci.* **85:** 4397–4401.

Dietrich, W., J. Weber, P. Kwok, and D. Nickerson. 1998. Isolation and analysis of DNA polymorphisms. In *Genome analysis: A laboratory manual*. Vol. 4 *Mapping genomes* (ed. B. Birren et al.). Cold Spring Harbor Laboratory Press, Cold Spring Harbor, New York. (In press.)

Donis-Keller, H., D.F. Barker, R.G. Knowlton, J.W. Schumm, J.C. Braman, and P. Green. 1986. Highly polymorphic RFLP probes as diagnostic tools. *Cold Spring Harbor Symp. Quant. Biol.* **51:** 317–324.

Faham, M. and D.R. Cox. 1996. A novel in vivo method to detect DNA sequence variation. *Genome Res.* **5:** 474–482.

Fearon, E.R, S.R. Hamilton, and B. Vogelstein. 1987. Clonal analysis of human colorectal tumors. *Science* **238:** 193– 197.

Fischer, S.G. and L.S. Lerman. 1983. DNA fragments differing by single base-pair substitutions are separated in denaturing gradient gels: Correspondence with melting theory. *Proc. Natl. Acad. Sci.* **80:** 1579–1583.

Ganguly, A. and D.J. Prockop. 1990. Detection of single-base mutations by reaction of DNA heteroduplexes with a water-soluble carbodiimide followed by primer extension: Application to products from the polymerase chain reaction. *Nucleic Acids Res.* **18:** 3933–3939.

Gibbs, R.A., P.-N. Nguyen, and C.T. Caskey. 1989. Detection of single DNA base differences by competitive oligonucleotide priming. *Nucleic Acids Res.* **17:** 2437–2448.

Glavac, D. and M. Dean. 1993. Optimization of the single-strand conformation polymorphism (SSCP) technique for detection of point mutations. *Hum. Mutat.* **2:** 404–414.

Gotoh, O. 1983. Prediction of melting profiles and local helix stability for sequenced DNA. *Adv. Biophys.* **16:** 1–52.

Gotoh, O. and Y. Tagashira. 1981. Stabilities of nearest-neighbor doublets in double-helical DNA determined by fitting calculated melting profiles to observed profiles. *Biopolymers* 20: 1033–1042.

Gray, M., A. Charpentier, K. Walsh, P. Wu, and W. Bender. 1991. Mapping point mutations in the *Drosophila rosy* locus using denaturing gradient gel blots. *Genetics* 127: 139–149.

Hacia, J.G., L.C. Brody, M.S. Chee, S.P. Fodor, and F.S. Collins. 1996. Detection of heterozygous mutations in BRCA1 using high density oligonucleotide arrays and two-colour fluorescence analysis [see comments]. *Nat. Genet.* 14: 441–447.

Hayashi, K. 1991. PCR-SSCP: A simple and sensitive method for detection of mutations in the genomic DNA. *PCR Methods Appl.* 1: 34–38.

———. 1992. PCR-SSCP: A method for detection of mutations. *Genet. Anal.Tech. Appl.* 9: 73–79.

Hayashi, K. and D.W. Yandell. 1993. How sensitive is PCR-SSCP? *Hum. Mutat.* 2: 338–346.

Hogervorst, F.B.L., R.S. Cornelis, M. Bout, M. van Vliet, J.C. Oosterwijk, R. Olmer, B. Bakker, J.G.M. Klijn, H.F.A. Vasen, H. Meijers-Heijboer, F.H. Menko, C.J. Cornelisse, J.T. den Dunnen, P. Devilee, and G.B. van Ommen. 1995. Rapid detection of *BRCA1* mutations by the protein truncation test. *Nat. Genet.* 10: 208–212.

Holland, P.M., R.D. Abramson, R. Watson, and D.H. Gelfand. 1991. Detection of specific polymerase chain reaction product by utilizing the $5' \rightarrow 3'$ exonuclease activity of *Thermus aquaticus* DNA polymerase. *Proc. Natl. Acad. Sci.* 88: 7276–7280.

Keen, J., D. Lester, C. Inglehearn, A. Curtis, and S. Bhattacharya. 1991. Rapid detection of single base mismatches as heteroduplexes on hydrolink gel. *Trends Genet.* 7: 5.

Krieg, P.A. and D. Melton. 1987. In vitro RNA synthesis with SP6 RNA polymerase. *Methods Enzymol.* 155: 397–415.

Kukita, Y., T. Thaira, S.S. Sommer, and K. Hayashi. 1997. SSCP analysis of long DNA fragments in low pH gels. *Hum. Mutat.* 10: 400–407.

Kwok, P.-Y., C. Carlson, T.D. Yager, W. Ankener, and D.A. Nickerson. 1994. Comparative analysis of human DNA variations by fluorescence-based sequencing of PCR products. *Genomics* 23: 138–144.

Kwok, P.-Y., Q. Deng, H. Zakeri, S.L. Taylor, and D.A. Nickerson. 1996. Increasing the information content of STS-based genome maps: Identifying polymorphisms in mapped STSs. *Genomics* 31: 123–126.

Landegren, U., R. Kaiser, J. Sanders, and L. Hood. 1988. A ligase-mediated gene detection technique. *Science* 241: 1077–1080.

Leach, F.S., N.C. Nicolaides, N. Papadopoulos, B. Liu, J. Jen, R. Parsons, P. Peltomaki, P. Sistonen, L.A. Aaltonen, M. Nystrom-Lahti, X.-Y. Guan, J. Zhang, P.S. Meltzer, J.-W. Yu, F.-T. Kao, D.J. Chen, K.M. Cerosaletti, R.E.K. Fournier, S. Todd, T. Lewis, R.J. Leach, S.L. Naylor, J. Weissenbach, J.-P. Mecklin, H.

Jarvinen, G.M. Petersen, S.R. Hamilton, J. Green, J. Jass, P. Watson, H.T. Lynch, J.M. Trent, A. de la Chapelle, K.W. Kinzler, and B. Vogelstein. 1993. Mutations of a *mutS* homolog in hereditary non-polyoposis colorectal cancer. *Cell* 75: 1215–1225.

Lee, L.G., C.R. Connell, and W. Bloch. 1993. Allelic discrimination by nick-translation PCR with fluorogenic probes. *Nucleic Acids Res.* 21: 3761–3766.

Lerman, L.S., K. Silverstein, and E. Grinfeld. 1986. Searching for gene defects by denaturing gradient gel electrophoresis. *Cold Spring Harbor Symp. Quant. Biol.* 51: 285–297.

Lerman, L.S. and K. Silverstein. 1987. Computational simulation of DNA melting and its application to denaturing gradient gel electrophoresis. *Methods Enzymol.* 155: 482–501.

Liu, B., R.E. Parsons, S.R. Hamilton, G.M. Petersen, H.T. Lynch, P. Watson, S. Markowitz, J.K.V. Willson, J. Green, A. de la Chapelle, K.W. Kinzler, and B. Vogelstein. 1994. hMSH2 Mutations in hereditary nonpolyposis colorectal cancer kindreds. *Cancer Res.* 54: 4590–4594.

Liu, B., N.C. Nicolaides, S. Markowitz, J.K.V. Willson, R.E. Parsons, J. Jen, N. Papadopolous, P. Peltomaki, A. de la Chapelle, S.R. Hamilton, K.W. Kinzler, and B. Vogelstein. 1995. Mismatch repair gene defects in sporadic colorectal cancers with microsatellite instability. *Nat. Genet.* 9: 48–55.

Liu, Q. and S.S. Sommer. 1995. Restriction endonuclease fingerprinting (REF): A sensitive method for screening mutations in long, contiguous segments of DNA. *BioTechniques* 18: 470–477.

Livak, K.J., J. Marmaro, and J.A. Todd. 1995. Towards fully automated genome-wide polymorphism screening. *Nat. Genet.* 9: 341–342.

Lu, A.-L. and I.-C. Hsu. 1991. Detection of single DNA base mutations with mismatch repair enzymes. *Genomics* 14: 249–255.

Makino, R., H. Yazyu, Y. Kishimoto, T. Sekiya, and K. Hayashi. 1992. F-SSCP: A fluorescent polymerase chain reaction-single strand conformation polymorphism (PCR-SSCP) analysis. *PCR Methods Appl.* 2: 10–13.

Mashiyama, S., T. Sekiya, and K. Hayashi. 1990. Screening of multiple DNA samples for detection of sequence changes. *Technique* 2: 304–306.

Maxam, A.M. and W. Gilbert. 1980. Sequencing end-labeled DNA with base-specific chemical cleavages. *Methods Enzymol.* 65: 499–540.

Mullis, K., F. Faloona, S. Scharf, R. Saiki, G. Horn, and H. Erlich. 1986. Specific enzymatic amplification of DNA *in vitro*: The polymerase chain reaction. *Cold Spring Harbor Symp. Quant. Biol.* 51: 263–273.

Myers, R.M., Z. Larin, and T. Maniatis. 1985a. Detection of single base substitutions by ribonuclease cleavage of mismatches in RNA:DNA duplexes. *Science* 230: 1242–1246.

Myers, R.M., L.S. Lerman, and T. Maniatis. 1995b. A general method for saturation mutagenesis of cloned DNA fragments. *Science* **229**: 242–247.

Myers, R.M., T. Maniatis, and L.S. Lerman. 1987. Detection and localization of single base changes by denaturing gradient gel electrophoresis. *Methods Enzymol.* **155**: 501–527.

Myers, R.M., V. Sheffield, and D.R. Cox. 1988. Detection of single base changes in DNA: Ribonuclease cleavage and denaturing gradient gel electrophoresis. In *Genomic analysis: A practical approach* (ed. K. Davies), pp. 95–139. IRL Press, Oxford.

———. 1989a. PCR and denaturing gradient gel electrophoresis. In *The polymerase chain reaction* (ed. H. Erlich et al.), pp. 177–181. Cold Spring Harbor Laboratory Press, Cold Spring Harbor, New York.

———. 1989b. Mutation detection by PCR, GC-clamps and denaturing gradient gel electrophoresis. In *PCR technology: Principles and applications for DNA amplification* (ed. H. Erlich), pp. 71–88. Stockton Press, New York.

Myers, R.M., S.G. Fischer, L.S. Lerman, and T. Maniatis. 1985c. Nearly all single base substitutions in DNA fragments joined to a GC-clamp can be detected by denaturing gradient gel electrophoresis. *Nucleic Acids Res.* **13**: 3131–3146.

Myers, R.M., S.G. Fischer, T. Maniatis, and L.S. Lerman. 1985d. Modification of the melting properties of duplex DNA by attachment of a GC-rich DNA sequence as determined by denaturing gradient gel electrophoresis. *Nucleic Acids Res.* **13**: 3111–3130.

Myers, R.M., N. Lumelsky, L.S. Lerman, and T. Maniatis. 1985e. Detection of single base substitutions in total genomic DNA. *Nature* **313**: 495–498.

Nickerson, D.A., C. Whitehurst, C. Boysen, P. Charmley, R. Kaiser, and L. Hood. 1992. Identification of clusters of biallelic polymorphic sequence-tagged sites (pSTSs) that generate highly informative and automatable markers for genetic linkage mapping. *Genomics* **12**: 377-387.

Nicolaides, N.C., N. Papadopoulos, B. Liu, Y. Wei, K.C. Carter, S.M. Ruben, C.A. Rosen, W.A. Haseltine, R.D. Fleischmann, C.M. Fraser, M.D. Adams, J.C. Venter, M.G. Dunlop, S.R. Hamilton, G.M. Petersen, A. de la Chapelle, B. Vogelstein, and K.W. Kinzler. 1994. Mutations of two *PMS* homologues in the hereditary nonpolyposis colon cancer. *Nature* **371**: 75–80.

Nigro, J.M., S.J. Baker, A.C. Preisinger, J.M. Jessup, R. Hostetter, K. Clearly, S.H. Bigner, N. Davidson, S. Baylin, P. Devilee, T. Glover, F.S. Collins, A. Weston, R. Modali, C.C. Harris, and B. Vogelstein. 1989. Mutations in the *p53* gene occur in diverse human tumour types. *Nature* **342**: 705–709.

Novack, D.F., N.J. Casna, S.G. Fischer, and J.P. Ford. 1986. Detection of single base-pair mismatches in DNA by chemical modification followed by electrophoresis in 15% polyacrylamide gel. *Proc. Natl. Acad. Sci.* **83**: 586–590.

Orita, M., Y. Suzuki, T. Sekiya, and K. Hayashi. 1989. Rapid and sensitive detection of point mutations and DNA polymorphisms using the polymerase chain reaction. *Genomics* **5**: 874–879.

Papadopoulos, N., N.C. Nicolaides, Y. Wei, S.M. Ruben, K.C. Carter, C.A. Rosen, W.A. Haseltine, R.D. Fleischmann, C.M. Fraser, M.D. Adams, J.C. Venter, S.R. Hamilton, G.M. Petersen, P. Watson, H.T. Lynch, P. Peltomaki, J. Mecklin, A. de la Chapelle, K.W. Kinzler, and B. Vogelstein. 1994. Mutations of a *mutL* homolog in hereditary colon cancer. *Science* **263**: 1625–1629.

Powell, S.M., G.M. Petersen, A.J. Krush, S. Booker, J. Jen, F.M. Giardiello, S.R. Hamilton, B. Vogelstein, and K.W. Kinzler. 1993. Molecular diagnosis of familial adenomatous polyposis. *N. Engl. J. Med.* **329**: 1982–1987.

Reindollar, R.H., B.C. Su, S.R. Bayer, and M.R. Gray. 1992. Rapid identification of deoxyribonucleic acid sequence differences in cytochrome P-450 21-hydroxylase (CYP21) genes with denaturing gradient gel blots. *Am. J. Obstet. Gynecol.* **166**: 184–190.

Roest, P.A.M., R.G. Roberts, S. Sugino, G.-J.B. van Ommen, and J.T. den Dunnen. 1993. Protein truncation test (PTT) for rapid detection of translation-terminating mutations. *Hum. Mol. Genet.* **2**: 1719–1721.

Saiki, R.K., S. Scharf, F. Faloona, K.B. Mullis, G.T. Horn, H.A. Erlich, and N. Arnheim. 1985. Enzymatic amplification of β-globin genomic sequences and restriction site analysis for diagnosis of sickle cell anemia. *Science* **230**: 1350–1354.

Saleeba, J.A. and G.H. Cotton. 1993. Chemical cleavage of mismatch to detect mutations. *Methods Enzymol.* **217**: 286–295.

Sambrook, J., E.F. Fritch, and T. Maniatis. 1989. *Molecular cloning: A laboratory manual*, 2nd edition. Cold Spring Harbor Laboratory Press, Cold Spring Harbor, New York.

Sanger, F., S. Nicklen, and A.R. Carlson. 1977. DNA sequencing with chain-terminating inhibitors. *Proc. Natl. Acad. Sci.* **74**: 5463–5467.

Sarkar, G., H.-S. Yoon, and S.S. Sommer. 1992a. Screening for mutations by RNA single-strand conformation polymorphism (rSSCP): comparison with DNA-SSCP. *Nucleic Acids Res.* **20**: 871–878.

Sarkar, G., H.-S. Yoon, and S.S. Sommer. 1992b. Dideoxy fingerprinting (ddF): A rapid and efficient screen for the presence of mutations. *Genomics* **13**: 441–443.

Savov, A., D. Angelicheva, A. Jordanova, A. Eigel, and L. Kalaydjieva. 1992. High percentage acrylamide gels improve resolution in SSCP analysis. *Nucleic Acids Res.* **20**: 6741–6742.

Sheffield, V.C., D.R. Cox, and R.M. Myers. 1990. Strategies for identifying DNA polymorphisms in

PCR-amplified genomic DNA by denaturing gradient gel electrophoresis. In *PCR protocols and applications: A laboratory manual* (ed. M. Innis et al.), pp. 206–218. Academic Press, San Diego.

Sheffield, V.C., J.S. Beck, E.M. Stone, and R.M. Myers. 1992. A simple and efficient method for attachment of a 40-base pair, GC-rich sequence to PCR-amplified DNA. *BioTechniques* **12:** 386–387.

Sheffield, V.C., D.R. Cox, L.S. Lerman, and R.M. Myers. 1989. Attachment of a 40-base-pair G+C-rich sequence (GC-clamp) to genomic DNA fragments by the polymerase chain reaction results in improved detection of single-base changes. *Proc. Natl. Acad. Sci.* **86:** 232–236.

Suzuki, Y., T. Sekiya, and K. Hayashi. 1991. Allele-specific polymerase chain reaction: A method for amplification and sequence determination of a single component among a mixture of sequence variants. *Anal. Biochem.* **192:** 82–84.

Suzuki, Y., M. Orita, M. Shiraishi, K. Hayashi, and T. Sekiya. 1990. Detection of ras gene mutations in human lung cancers by single-strand conformation polymorphism analysis of polymerase chain reaction products. *Oncogene* **5:** 1037–1043.

Syvänen, A.-C., K. Aalto-Setälä, L. Harju, K. Kontula, and H. Söderlund. 1990. A primer-guided nucleotide incorporation assay in the genotyping of apolipoprotein E. *Genomics* **8:** 684–692.

Syvänen, A.-C., E. Ikonen, T. Manninen, M. Bengtström, H. Söderlund, P. Aula, and L. Peltonen. 1992. Convenient and quantitative determination of the frequency of a mutant allele using solid-phase minisequencing: Application to aspartylglucosaminuria in Finland. *Genomics* **12:** 684–692.

White, M., M. Carvalho, D. Derse, S.J. O'Brien, and M. Dean. 1992. Detecting single base substitutions as heteroduplex polymorphisms. *Genomics* **12:** 301–306.

White, R., M. Leppart, D.T. Bishop, D. Barker, J. Berkowitz, C. Brown, P. Callahan, T. Holm, and L. Jerominski. 1985. Construction of linkage maps with DNA markers for human chromosomes. *Nature* **313:** 101–105.

Wilson, R.K. and E.R. Mardis. 1997. Fluorescence-based DNA sequencing. In *Genome analysis: A laboratory manual.* Vol. 1 *Analyzing DNA* (ed. B. Birren et al.), pp. 301–395. Cold Spring Harbor Laboratory Press, Cold Spring Harbor, New York.

Winter, E., F. Yamamoto, C. Almoguera, and M. Perucho. 1985. A method to detect and characterize point mutations in transcribed genes: Amplification and overexpression of the mutant c-Ki-ras allele in human tumor cells. *Proc. Natl. Acad. Sci.* **82:** 7575–7579.

Wolff, R. and R. Gemmill. 1997. Purifying and analyzing genomic DNA. In *Genome analysis: A laboratory manual.* Vol. 1 *Analyzing DNA* (ed. B. Birren et al.), pp. 1–81. Cold Spring Harbor Laboratory Press, Cold Spring Harbor, New York.

Wu, D.Y. and R.B. Wallace. 1989. The ligation amplification reaction (LAR): Amplification of specific DNA sequences using sequential rounds of template-dependent ligation. *Genomics* **4:** 560–569.

Wu, D.Y., L. Ugozzoli, B.K. Pal, and R.B. Wallace. 1989. Allele-specific enzymatic amplification of β-globin genomic DNA for diagnosis of sickle cell anemia. *Proc. Natl. Acad. Sci.* **86:** 2757–2760.

Youil, R., J.W. Kemper, and R.G.H. Cotton. 1995. Screening for mutations by enzyme mismatch cleavage with T4 endonuclease. *Proc. Natl. Acad. Sci.* **92:** 87–91.

Common Reagents

COMPILED BY JOAN KOBORI

All chemicals must be reagent grade or molecular biology grade. All H_2O used in the preparation of solutions must be of the highest quality available in the standard molecular biology laboratory. Use sterile, glass-distilled, deionized H_2O (purified through a Milli-Q filter or similar type of system) whenever possible. Unless otherwise stated, most prepared solutions require sterilization either by filtration through a 0.22-μm filter or by autoclaving at 15 psi on liquid cycle at 121ºC for 20–30 minutes. Use autoclaved H_2O and sterile measuring devices for the preparation and use of solutions from sterile stock solutions and reagents. These measures ensure that the storage life of the reagents is as long as possible. In general, solutions prepared from dry chemicals and sterile H_2O do not need additional sterilization. For some solutions, sterilization is not required (e.g., acids, bases, and some organic compounds) because microbial growth cannot occur.

Unless specifically stated otherwise, all stock solutions and buffers can be stored at room temperature for at least 6 months. Divide large volumes into smaller aliquots for storage. Allow stock chemicals stored at 4ºC or –20ºC to reach room temperature before opening to prevent the accumulation of condensation within the reagent and therefore ensure more accurate measurement.

Solutions designated by percentage w/v are defined as the solute weight in grams per 100 ml and solutions designated by percentage v/v are defined as the constituent volume in milliliters per 100 ml of total volume. Buffer pH is given for a solution at 25ºC. Information on molarities of standard concentrated acids and bases is also provided.

Important safety precautions for the reagents used here are provided in Appendix 3. In general, the use of protective clothing, gloves, and face protection are required when exposure to the skin can result in burns or absorption of a toxic chemical. Face masks (disposable covers for the mouth and nose) are needed when inhalation of a material is to be avoided.

STANDARD STOCK SOLUTIONS

10 M Ammonium Acetate

Dissolve 771 g of ammonium acetate (m.w. = 77.1 g/mole) in sufficient H_2O to make a final volume of 1 liter. Sterilize the solution by passing it through a 0.22-μm filter.

10 mg/ml BSA

Add 100 mg of BSA (Fraction V or molecular biology grade, DNase-free) to a 15-ml polypropylene tube containing 9.5 ml of H_2O. To reduce denaturation, always add protein to an aqueous solution instead of adding an aqueous solution to a protein. Gently rock the capped tube until the BSA is completely dissolved. Do not mix by vortexing (this causes foaming, which indicates protein denaturation). Adjust the final volume to 10 ml with H_2O. Divide into aliquots and store at –20°C. In general, this solution is not sterilized.

100x Denhardt's Reagent

Component and final concentration	Amount to add per 100 ml
2% Ficoll (Type 400)	2 g
2% polyvinylpyrrolidone (PVP-40)	2 g
2% BSA (Fraction V)	2 g
H_2O	to make 100 ml

Dissolve the components in the H_2O. Filter to sterilize and remove particulate matter. Divide into aliquots and store at –20°C.

10x Standard DNA Ligase Buffer

Different conditions for using bacteriophage T4 DNA ligase have been reported. A standard buffer for sticky-end and blunt-end ligation is given here. The optional spermidine is recommended for blunt-end ligations.

Component and final concentration	Amount to add per 10 ml
0.5 M Tris-Cl	5 ml of 1 M (pH 7.6 at 25°C)
100 mM $MgCl_2$	1 ml of 1 M
100 mM DTT	1 ml of 1 M
2 mM ATP	200 μl of 100 mM
5 mM spermidine HCl (*optional*)	50 μl of 1 M
0.5 mg/ml BSA (Fraction V) (*optional*)	0.5 ml of 10 mg/ml
H_2O	2.25 ml

Divide into small aliquots and store at –20°C. In general, this buffer is
not sterilized.

100 mM dNTP Solutions

Stock solutions of purified dNTPs can be purchased as 100 mM solutions. These
solutions can be stored at –80°C for at least 6 months.

To prepare 100 mM stock solutions, dissolve the appropriate amount of dNTP
in H_2O, adjust the pH to approximately 7 with 1 M Tris base, and then
determine the concentration precisely. These 100 mM stock preparations are
typically found to be 85–95 mM. Therefore, adding less than the calculated
volume of H_2O is recommended. In general, these stock solutions are not sterilized.

To determine the concentrations, serially dilute the 100 mM stock solutions
with 10 mM Tris-Cl or phosphate buffer (pH 7.0) to approximately 10 μM. Adjust the spectrophotometer to zero with the dilution buffer. Use a quartz cuvette
with a 1-cm path length and read the OD of each solution at the wavelength
designated in the table below. Using the extinction coefficients (ε) listed in the
table below, calculate the concentration of each dNTP solution as follows:

$$\text{molar concentration} = \frac{\text{measured OD x dilution factor}}{\varepsilon}$$

dNTP	Wavelength (nm)	ε ($M^{-1}cm^{-1}$)
dATP	259	1.54×10^4
dCTP	271	9.10×10^3
dGTP	253	1.37×10^4
dTTP	260	7.40×10^3

Note: Many protocols call for mixtures of dNTPs at a set molarity
(generally 0.5–10 mM). For example, a 10 mM dNTP mixture denotes a
solution containing all four dNTPs, each at a final concentration of 10
mM, prepared by diluting concentrated dNTP stock solutions with H_2O
as follows:

10 mM dNTP mixture

Component and final concentration	Amount to add per 20 μl
10 mM dATP	2 μl of 100 mM
10 mM dCTP	2 μl of 100 mM
10 mM dGTP	2 μl of 100 mM
10 mM dTTP	2 μl of 100 mM
H_2O	12 μl

Mix the components in a 1-ml microcentrifuge tube. This mixture can
be stored at –20°C for at least 6 months.

1 M DTT

The simplest approach for preparing a 1 M stock solution is to add 32.4 ml of H_2O to a 5-g bottle of DTT. Divide into aliquots and store at −20°C. This procedure avoids the weighing of this foul-smelling chemical.

Alternatively, transfer 100 mg of DTT (m.w. = 154.25 g/mole) into a microcentrifuge tube and add 0.65 ml of H_2O to make a 1 M solution. Sterilization is not required.

0.5 M EDTA

The easiest method for preparing this common stock solution is to generate the trisodium salt of EDTA by preparing an equimolar solution of Na_2 EDTA and NaOH (i.e., each is 0.5 M). The trisodium salt is more soluble than the disodium salt. The pH of this solution should be approximately 8. This stock solution is adequate for all molecular biology protocols.

To prepare 0.5 M EDTA, combine 186.1 g of Na_2 EDTA · $2H_2O$ (m.w. = 372.2 g/mole), 20 g of **NaOH** (m.w. = 40 g/mole), and H_2O to make a final volume of 1 liter.

NaOH (see Appendix for Caution)

1 M HEPES

Dissolve 23.8 g of HEPES free acid (m.w. = 238.3 g/mole) in approximately 90 ml of H_2O. Adjust the pH with **NaOH** (the useful pH range is 6.8 to 8.2) and then adjust the final volume to 100 ml with H_2O.

NaOH (see Appendix for Caution)

1 N HCl

Standard concentrated HCl is 36% (w/w) or 11.6 N. To avoid severe burns caused by splattering, *always add acid to H_2O*; do not add H_2O to acid. Sterilization is not required.

For 100 ml of solution, add 8.6 ml of **concentrated HCl** to 91.4 ml of H_2O.

concentrated HCl (see Appendix for Caution)

25 mg/ml IPTG

Dissolve 250 mg of IPTG (m.w. = 238.3 g/mole) in 10 ml of H_2O. Divide into aliquots and store at −20°C. In general, this solution is not sterilized.

1 M $MgCl_2$

Dissolve 20.3 g of $MgCl_2$ · $6H_2O$ (m.w. = 203.3 g/mole) in sufficient H_2O to make a final volume of 100 ml.

20% PEG 8000/2.5 M NaCl

Component and final concentration	Amount to add per 100 ml
20% (w/v) PEG 8000	20 g
2.5 M NaCl	50 ml of 5 M NaCl or 14.6 g of solid NaCl
H_2O	to make 100 ml

Add the PEG 8000 to a beaker containing the NaCl (m.w. = 58.44 g/mole) and sufficient H_2O to make a final volume of 100 ml. Stir with a magnetic stirring bar.

100 mM PMSF

Dissolve 174 mg of **PMSF** (m.w. = 174.2 g/mole) in sufficient isopropanol to make a final volume of 10 ml. Divide into aliquots and store in foil-wrapped tubes at –20°C. Sterilization is not required.

PMSF (see Appendix for Caution)

8 M Potassium Acetate

Dissolve 78.5 g of potassium acetate (m.w. = 98.14 g/mole) in sufficient H_2O to make a final volume of 100 ml.

1 M KCl

Dissolve 7.46 g of KCl (m.w. = 74.55 g/mole) in sufficient H_2O to make a final volume of 100 ml.

20 mg/ml Proteinase K

Add 200 mg of proteinase K to a 15-ml polypropylene tube containing 9.5 ml of H_2O. To reduce denaturation, always add protein to an aqueous solution instead of adding an aqueous solution to a protein. Gently rock the capped tube until the proteinase K is completely dissolved. Do not mix by vortexing (this causes foaming, which indicates protein denaturation). Adjust the final volume to 10 ml. Divide into aliquots and store at –20°C. In general, this solution is not sterilized.

10 mg/ml RNase A (DNase-free)

To avoid contamination of RNase A, wear gloves and do not allow RNase A to come in contact with any laboratory surfaces or equipment used for RNA work. Dissolve 10 mg of pancreatic RNase A in 1 ml of 10 mM sodium acetate (pH 5.0). Place in a boiling-water bath for 15 minutes to inactivate any contaminating DNase. Adjust the pH to 7.5 with 1 M Tris-Cl. Store at –20°C. In general, this solution is not sterilized.

10 mg/ml Salmon Sperm DNA

Sonicated, denatured salmon sperm DNA is commercially available at a concentration of 10 mg/ml but is fairly expensive. A large economical supply of salmon sperm DNA stock solution can be prepared in the laboratory, but the process is lengthy.

To prepare the stock solution, dissolve 1 g of desiccated salmon sperm DNA in 100 ml of H_2O by stirring for at least 1 day. Add NaCl to a final concentration of 100 mM and extract with **phenol** (see Appendix 2). Shear the extracted DNA by sonication or by repeatedly passing (10–20 times) the DNA through a 16–18-gauge needle. Analyze on an agarose gel along with the appropriate molecular-weight markers to determine the approximate size. (For use in hybridizations, the desired size range is approximately 500–1000 bp. For lithium acetate transformations of yeast, much larger salmon sperm DNA [5–10 kb] is preferred for use as a carrier.) Precipitate the DNA with ethanol (see Appendix 2) and dissolve in H_2O at a final concentration of 10 mg/ml. Divide into aliquots (e.g., 10 ml) and place in a boiling-water bath for 10 minutes to denature the DNA. Rapidly chill the denatured DNA on ice. Store at –20°C.

Some investigators recommend that salmon sperm DNA be boiled and chilled before each use. In general, this solution is not sterilized.

phenol (see Appendix for Caution)

3 M Sodium Acetate

Dissolve 40.8 g of sodium acetate trihydrate (m.w. = 136.1 g/mole) in approximately 90 ml of H_2O. Adjust the pH of the solution to 5.2 with **glacial acetic acid** and then adjust the final volume to 100 ml with H_2O.

glacial acetic acid (see Appendix for Caution)

5 M NaCl

Dissolve 29.2 g of NaCl (m.w. = 58.44 g/mole) in sufficient H_2O to make a final volume of 100 ml.

10 N NaOH

The preparation of a concentrated NaOH (10 N) solution entails an exothermic reaction. Extreme caution must be taken to avoid chemical burns and breakage of glass containers. If possible, use heavy plastic beakers.

To prepare 10 N NaOH, add 400 g of **NaOH** pellets (m.w. = 40 g/mole) to a beaker containing approximately 0.9 liter of H_2O that is being stirred with a magnetic stirring bar. *Do not add H_2O to the NaOH pellets.* The beaker can be placed in a container of ice. After the pellets have completely dissolved, adjust the final volume to 1 liter with H_2O. Sterilization is not required.

To avoid the use of NaOH pellets in preparing 10 N NaOH, use the commercially available concentrated NaOH solution. Add 524 ml of 50% **NaOH** solution (19.1 N) to 476 ml of H_2O while stirring with a magnetic stirring bar.

NaOH (see Appendix for Caution)

10% (w/v) SDS

Carefully weigh 100 g of **SDS** and slowly transfer it into a beaker containing approximately 0.9 liter of H_2O. Stir with a magnetic stirring bar until completely dissolved. Adjust the final volume to 1 liter. A 20% stock solution of SDS (200 g in 1 liter) can also be prepared if desired. Sterilization is not required.

SDS (see Appendix for Caution)

2 M Sorbitol

Dissolve 36.4 g of sorbitol (m.w. = 182.2 g/mole) in sufficient H_2O to make a final volume of 100 ml.

1 M Spermidine

Dissolve 2.55 g of spermidine trihydrochloride (m.w. = 254.6 g/mole) in sufficient H_2O to make a final volume of 10 ml. Divide into aliquots and store at –20°C. In general, this solution is not sterilized.

1 M Spermine

Dissolve 3.48 g of spermine tetrahydrochloride (m.w. = 348.2 g/mole) in sufficient H_2O to make a final volume of 10 ml. Divide into aliquots and store at –20°C. In general, this solution is not sterilized.

20x SSC

Component and final concentration	Amount to add per 1 liter
300 mM trisodium citrate (dihydrate)	88.2 g
3 M NaCl	175.3 g
H_2O	to make 1 liter

Dissolve the trisodium citrate dihydrate (m.w. = 294.1 g/mole) and the NaCl (m.w. = 58.44 g/mole) in approximately 0.9 liter of H_2O. Adjust the pH to 7.0 by adding a few drops of 10 N **NaOH**. Adjust the final volume to 1 liter with H_2O.

NaOH (see Appendix for Caution)

100% (w/v) TCA

The safest method for preparing a TCA stock solution is to avoid weighing out the chemical as follows: Add 100 ml of H_2O to a bottle containing 500 g of **TCA**. (This chemical is very soluble in H_2O.) Stir with a magnetic stirring bar until completely dissolved. Add more H_2O as needed. Adjust the final volume to 500 ml with H_2O. Store in a dark glass bottle. Sterilization is not required.

TCA undergoes decomposition at concentrations below 30%. Dilutions should be prepared just before use.

TCA (see Appendix for Caution)

2.5% X-gal

Dissolve 25 mg of X-gal in 1 ml of **DMF**. Store in a foil-wrapped polypropylene tube at −20ºC. Sterilization is not required.

DMF (see Appendix for Caution)

MOLARITIES OF CONCENTRATED ACIDS AND BASES

	Percentage solution (w/w)	Molarity (M)
Acids		
glacial acetic acid	99–100	17.4
formic acid	90	23.4
HCl	36	11.6
nitric acid	70	15.7
phosphoric acid	85	14.6
sulfuric acid	95	18
Bases		
ammonium hydroxide	28 (NH_3)	14.8
KOH	50	13.5
NaOH	50	19.1

COMMON LABORATORY SOLUTIONS

DEPC-treated H$_2$O

Add 100 µl of fresh **DEPC** to 100 ml of H$_2$O to make a final concentration of 0.1% (v/v). Incubate at 37°C for at least 12 hours and then autoclave at 15 psi on liquid cycle for 20 minutes to inactivate the remaining DEPC.

Note: DEPC reacts with amines. Do not treat Tris buffers with DEPC.

DEPC (see Appendix for Caution)

70% (v/v) Ethanol

To prepare a solution of approximately 70% (v/v), mix 70 ml of absolute ethanol with 30 ml of sterile H$_2$O. Do not autoclave. Prepare as needed or store at −20°C. Sterilization is not required.

Formamide (Deionized)

Reagent-grade formamide can often be used directly, but it decomposes to formic acid and ammonia during storage. Deionized formamide (sometimes called molecular biology grade) can be purchased or prepared as needed. Do not use batches of formamide that have a yellow color.

To deionize formamide, add Dowex XG8 mixed-bed resin to **formamide** in a glass beaker and stir gently with a magnetic stirring bar for 1 hour. Filter through Whatman No. 1 paper to remove the resin. (Dowex XG8 resin can be purchased with a pH-sensitive blue indicator that becomes yellow upon removal of the acid from the formamide.) Divide into small aliquots and store under nitrogen (to prevent oxidation) at −80°C. Sterilization is not required.

formamide (see Appendix for Caution)

80% (v/v) Glycerol

Carefully measure 80 ml of glycerol in a nonwetting plastic graduated cylinder and transfer it into a 250-ml glass bottle or flask. Add 20 ml of H$_2$O and stir with a magnetic stirring bar until homogeneous. Sterilize by autoclaving.

PBS

Historically, this commonly used reagent was sodium phosphate-buffered saline (150 mM NaCl plus 10 mM sodium phosphate). PBS has been modified to suit different applications. D-PBS (8 mM sodium phosphate, 2 mM potassium phosphate, 140 mM NaCl, 2.7 mM KCl [pH 7.4] with or without 0.5 mM MgCl$_2$ and 0.9 mM CaCl$_2$) can be purchased as 1x and 10x stock buffers. One preparation of PBS is provided here.

Component and final concentration	Amount to add per 1 liter
137 mM NaCl	8 g
2.7 mM KCl	200 mg
10 mM Na_2HPO_4 (dibasic, anhydrous)	1.44 g
2 mM KH_2PO_4 (monobasic, anhydrous)	240 mg
H_2O	to make 1 liter

Dissolve the components in approximately 0.9 liter of H_2O. Adjust the pH to 7.4 with **HCl** and then adjust the final volume to 1 liter with H_2O.

concentrated HCl (see Appendix for Caution)

Phosphate Buffers

Mixing 1 M NaH_2PO_4 (monobasic) and 1 M Na_2HPO_4 (dibasic) stock solutions in the volumes designated in the table below results in 1 liter of 1 M sodium phosphate buffer of the desired pH. To prepare the 1 M stock solutions, dissolve 138 g of $NaH_2PO_4 \cdot H_2O$ (monobasic; m.w. = 138 g/mole) in sufficient H_2O to make a final volume of 1 liter and dissolve 142 g of Na_2HPO_4 (dibasic, anhydrous; m.w. = 142 g/mole) in sufficient H_2O to make a final volume of 1 liter.

Volume of 1 M NaH_2PO_4 (ml)	Volume of 1 M Na_2HPO_4 (ml)	Final pH
877	123	6.0
850	150	6.1
815	185	6.2
775	225	6.3
735	265	6.4
685	315	6.5
625	375	6.6
565	435	6.7
510	490	6.8
450	550	6.9
390	610	7.0
330	670	7.1
280	720	7.2

TE

Component and final concentration	Amount to add per 100 ml
10 mM Tris-Cl	1 ml of 1 M (pH 7.4–8.0 at 25°C)
1 mM EDTA	200 μl of 0.5 M (pH 8.0)
H_2O	98.8 ml

This standard buffer is used to resuspend and store DNA. It can be prepared by using 1 M Tris-Cl at pHs ranging from 7.4 to 8.0.

Tris-Cl Buffers

Mixing the volumes of concentrated HCl (11.6 N) designated in the table below with 121 g of Tris base (m.w. = 121 g/mole) results in 1 liter of 1 M Tris-Cl buffer of the desired pH at 25°C. To prepare the 1 M buffer, dissolve 121 g of Tris base in approximately 0.9 liter of H_2O. Add the appropriate volume of **concentrated HCl** and adjust the final volume to 1 liter with H_2O.

Volume of concentrated HCl (ml)	pH
8.6	9.0
14	8.8
21	8.6
28.5	8.4
38	8.2
46	8.0
56	7.8
66	7.6
71.3	7.4
76	7.2

Notes: Some pH electrodes cannot accurately measure the pH of Tris solutions. Be sure to check the information provided by the electrode manufacturer.

Tris has a significant temperature coefficient. As the temperature of the solution decreases from 25°C to 5°C, the pH increases an average of 0.03 pH units per degree centigrade. Conversely, as the temperature increases from 25°C to 37°C, the pH decreases an average of 0.025 pH units per degree centigrade. Slight effects on pH based on the total Tris concentration have also been noted. Most molecular biology protocols do not account for these effects.

concentrated HCl (see Appendix for Caution)

ELECTROPHORESIS BUFFERS, DYES, AND GEL-LOADING SOLUTIONS

50x TAE buffer

Component and final concentration	Amount to add per 1 liter
2 M Tris base	242 g
1 M acetate	57.1 ml of **glacial acetic acid** (17.4 M)
100 mM EDTA	200 ml of 0.5 M (pH 8.0)
H_2O	to make 1 liter

This buffer does not have the buffering capacity of TBE buffer. The 1x TAE buffer (pH 8.1) is 40 mM Tris, 20 mM acetate, and 2 mM EDTA.

glacial acetic acid (see Appendix for Caution)

5x TBE buffer

Component and final concentration	Amount to add per 1 liter
445 mM Tris base	54 g
445 mM borate	27.5 g of boric acid
10 mM EDTA	20 ml of 0.5 M (pH 8.0)
H_2O	to make 1 liter

TBE can be prepared as a 5x or 10x stock buffer, but the 10x stock buffer will precipitate during storage. The 1x buffer (pH 8.3) is 89 mM Tris, 89 mM borate, and 2 mM EDTA.

1% Bromophenol Blue

Add 1 g of the H_2O-soluble sodium form of bromophenol blue to 100 ml of H_2O. Stir or mix by vortexing until fully dissolved. In general, this solution is not sterilized.

10 mg/ml Ethidium Bromide

Since this chemical is commonly used in a molecular biology laboratory, a 100-ml stock solution can be prepared. Carefully weigh 1 g of **ethidium bromide**, avoiding dispersal of the powder. Transfer into a wide-mouth bottle and add 100 ml of H_2O and a magnetic stirring bar. Stir until dissolved. Wrap with foil and store at 4°C.

Use of ethidium bromide powder can be completely avoided by purchasing commercially available solutions or by dissolving one 100-mg tablet of ethidium bromide (Sigma E 2515) in 10 ml of H_2O. Sterilization is not required.

ethidium bromide (see Appendix for Caution)

1% Xylene Cyanole FF

Dissolve 1 g of xylene cyanole FF in sufficient H_2O to make a final volume of 100 ml. In general, this solution is not sterilized.

Gel-loading Solutions

Gel-loading solutions added to DNA samples for analysis on agarose or acrylamide gels can contain sucrose, glycerol, or Ficoll as the agent to increase the density of the sample. A dense sample drops evenly to the bottom of a gel well. Individual preference dictates which to use.

Tracking dyes indicate the extent of electrophoresis. Bromophenol blue migrates at the position of an approximately 300-bp linear double-stranded DNA and xylene cyanole FF migrates at the position of a 4-kb linear double-stranded DNA in 0.5x TBE buffer. The preparations below use dye markers at a final concentration of 0.15–0.25%. In general, these gel-loading solutions are not sterilized. Add 2 μl of 6x buffer per 10 μl of total sample or 1 μl of 10x buffer to 9 μl of total sample.

6x Alkaline Gel-loading Solution

Component and final concentration	Amount to add per 10 ml
0.3 N **NaOH**	300 μl of 10 N
6 mM EDTA	120 μl of 0.5 M (pH 8.0)
18% Ficoll (Type 400)	1.8 g
0.15% bromocresol green	15 mg
0.25% xylene cyanole FF	25 mg
H_2O	to make 10 ml

Store at room temperature.

NaOH (see Appendix for Caution)

6x Ficoll Gel-loading Solution

Component and final concentration	Amount to add per 10 ml
0.15% bromophenol blue	1.5 ml of 1%
0.15% xylene cyanole FF	1.5 ml of 1%
5 mM EDTA	100 μl of 0.5 M (pH 8.0)
15% Ficoll (Type 400)	1.5 g
H_2O	to make 10 ml

Store at room temperature.

6x BP/XC/Ficoll Gel-loading Solution

Component and final concentration	Amount to add per 10 ml
0.25% bromophenol blue	2.5 ml of 1%
0.25% xylene cyanole FF	2.5 ml of 1%
15% Ficoll (Type 400)	1.5 g
H_2O	to make 10 ml

Store at room temperature.

Formamide/EDTA Gel-loading Solution

Component and final concentration	Amount to add per 10 ml
98% deionized **formamide**	9.8 ml
10 mM EDTA	200 µl of 0.5 M (pH 8.0)

Divide into 1-ml aliquots and store at −20°C.

Notes: For fluorescent automated DNA sequence analysis, it is not necessary to add tracking dyes to the gel-loading solution.

For radioactive DNA sequence analysis, the presence of 0.025% bromophenol blue and 0.025% xylene cyanole FF is desirable. Add 2.5 mg of each dye per 10 ml of gel-loading solution.

formamide (see Appendix for Caution)

6x Glycerol Gel-loading Solution

Component and final concentration	Amount to add per 10 ml
0.15% bromophenol blue	1.5 ml of 1%
0.15% xylene cyanole FF	1.5 ml of 1%
5 mM EDTA	100 µl of 0.5 M (pH 8.0)
30% glycerol	3 ml
H_2O	3.9 ml

Store at 4°C.

BP/XC/Glycerol Gel-loading Solution

Component and final concentration	Amount to add per 10 ml
0.25% bromophenol blue	2.5 ml of 1%
0.25% xylene cyanole FF	2.5 ml of 1%
30% glycerol	3 ml
H_2O	2 ml

Store at room temperature.

10x SDS/Glycerol Gel-loading Solution

Component and final concentration	Amount to add per 10 ml
200 mM EDTA	4 ml of 0.5 M (pH 8.0)
0.1% **SDS**	100 µl of 10%
50% glycerol	5 ml
0.2% bromophenol blue	20 mg
0.2% xylene cyanole FF	20 mg
H_2O	to make 10 ml

Store this denaturing gel-loading solution at room temperature.

SDS (see Appendix for Caution)

6x Sucrose Gel-loading Solution

Component and final concentration	Amount to add per 10 ml
0.15% bromophenol blue	1.5 ml of 1%
0.15% xylene cyanole FF	1.5 ml of 1%
5 mM EDTA	100 µl of 0.5 M (pH 8.0)
40% sucrose	4 g
H_2O	to make 10 ml

Store at 4°C.

MEDIA

All preparations are for 1 liter of medium. Autoclave prepared liquid medium in the final flask in which cultures will be grown (ideally, a flask that is at least five times the volume of the medium to allow adequate aeration during growth) or divide into aliquots in conveniently sized bottles.

Wear a face mask to avoid inhalation of the fine powders. Use extreme caution in handling autoclaved medium. Wear thermal gloves and do not swirl hot solutions. Overheating can cause boiling medium to bubble out of containers.

Media Preparation

In general, add reagents to 0.9 liter of H_2O. Shake in a flask (at least a 2-liter flask) or stir with a magnetic stirring bar until dissolved. Adjust the pH if required. Adjust the final volume to 1 liter with H_2O.

Plate Preparation

Add reagents to 0.9 liter of H_2O. Shake in a flask (at least a 2-liter flask) or stir with a magnetic stirring bar until dissolved (agar will not completely dissolve until it is autoclaved). Adjust the pH if required. Adjust the final volume to 1 liter with H_2O. Cover the flask loosely with aluminum foil or a suitable cap. Sterilize by autoclaving. Wear thermal gloves and *carefully* swirl the solution to mix thoroughly. Allow the solution to cool in a water bath set at 55°C. Pour approximately 25–30 ml of the cooled solution into each 10-cm plastic petri dish or approximately 100 ml into each 15-cm plastic petri dish. Use a Bunsen burner to flame the surface of the medium in the plate to remove bubbles. Allow the plates to solidify at room temperature. Store plates upside down at 4°C.

A convenient alternative is to prepare medium in glass bottles and allow it to solidify. Store the solidified medium at room temperature. When needed, melt the solid medium in a microwave oven and pour plates.

Top Agar Preparation

Add reagents to 0.9 liter of H_2O. Shake in a flask (at least a 2-liter flask) or stir with a magnetic stirring bar until dissolved. Adjust the pH if required. Adjust the final volume to 1 liter with H_2O. If desired, divide the medium into 100-ml aliquots in autoclavable bottles before adding the agar and autoclaving. Add the appropriate amount of agar to the medium (for bacterial top agar, add 0.7 g per 100 ml) and then sterilize by autoclaving. Allow the top agar to solidify at room temperature.

Carefully melt the top agar in a microwave oven or in a boiling-water bath. Be sure to loosen caps on bottles before heating them. Do not leave top agar unattended in a microwave oven; the medium can easily bubble over with overheating or swirling of the hot solution. One approach to dealing with boiling over during melting is to break up solidified top agar with a sterile pipette before heating it in a microwave oven. Allow the top agar to cool to 45–48°C before use.

Bacterial/Bacteriophage Media

LB MEDIUM

Combine the following in 0.9 liter of H_2O:

Bacto tryptone	10 g
Bacto yeast extract	5 g
NaCl	10 g

Adjust the pH to 7.0 with 1 N **NaOH** (~1 ml) if desired. Adjust the final volume to 1 liter with H_2O.

Note: For agar plates, include 12 g of Bacto agar per liter. For top agar, include 7 g of Bacto agar per liter.

NaOH (see Appendix for Caution)

SOB MEDIUM

Combine the following in 0.9 liter of H_2O:

Bacto tryptone	20 g
Bacto yeast extract	5 g
NaCl	0.5 g
1 M KCl	2.5 ml

Adjust the final volume to 1 liter with H_2O. Divide the medium into 100-ml aliquots and then sterilize by autoclaving. Allow the medium to cool to room temperature, and then add 1 ml of sterile 1 M $MgCl_2$ to each 100-ml aliquot.

SOC MEDIUM

Prepare SOC medium as described for SOB medium but add 2 ml of sterile 1 M glucose to each 100-ml aliquot (18 g of glucose dissolved in sufficient H_2O to make a final volume of 100 ml; sterilize the solution by passing it through a 0.22-μm filter) in addition to the 1 ml of 1 M $MgCl_2$ after the medium has cooled to room temperature.

TB MEDIUM

Combine the following in sufficient H_2O to make 0.9 liter:

Bacto tryptone	12 g
Bacto yeast extract	24 g
glycerol	4 ml

Dissolve the components and then sterilize by autoclaving. Allow to cool to at least 60°C. Add 100 ml of a sterile solution of 170 mM KH_2PO_4/0.72 M K_2HPO_4

(2.31 g of KH$_2$PO$_4$ [monobasic, anhydrous] and 12.54 g of K$_2$HPO$_4$ [dibasic, anhydrous] in sufficient H$_2$O to make a final volume of 100 ml; sterilize the solution by passing it through a 0.22-μm filter or by autoclaving).

2x YT MEDIUM

Combine the following in 0.9 liter of H$_2$O:

Bacto tryptone	16 g
Bacto yeast extract	10 g
NaCl	5 g

Adjust the pH to 7.0 with 1 N **NaOH** (~1 ml) if desired. Adjust the final volume to 1 liter with H$_2$O.

Note: For agar plates, include 12 g of Bacto agar per liter. For top agar, include 7 g of Bacto agar per liter.

NaOH (see Appendix for Caution)

Yeast Media

AHC MEDIUM

Combine the following in 0.8 liter of H$_2$O:

YNB without amino acids (Difco 0919-15)	6.7 g
casein acid hydrolysate, low salt (USB 12852)	10 g
adenine hemisulfate monohydrate (Sigma A 9126)	20 mg

Adjust the pH to 5.8 with **HCl**. Adjust the final volume to 0.95 liter with H$_2$O. Sterilize by autoclaving. Allow to cool. Add 50 ml of sterile 40% glucose (40 g of glucose dissolved in sufficient H$_2$O to make a final volume of 100 ml; sterilize the solution by passing it through a 0.22-μm filter or by autoclaving) per liter and mix well.

Notes: To prepare plates, add 20 g of Bacto agar before autoclaving. Agar from Difco is highly recommended, otherwise a precipitate may form upon autoclaving.

Instead of the YNB without amino acids used above, 1.7 g of YNB without ammonium sulfate or amino acids (Difco 0335-15-9) plus 5 g of ammonium sulfate can be substituted in each liter.

For high-adenine AHC medium, prepare liquid or solid medium as described above but add 50–100 mg (instead of 20 mg) of adenine hemisulfate to each liter. The higher adenine concentration in high-adenine AHC medium allows Ade$^-$ strains (which includes many YAC-containing strains) to grow faster and to reach a higher cell density (but note that certain Ade$^-$ strains, e.g., *ade2*, will not develop the characteristic red color).

For AHC + Trp medium, add 4 ml of sterile 1% tryptophan per liter after autoclaving.

concentrated HCl (see Appendix for Caution)

SC MEDIUM AND "DROP OUT" MEDIA

Combine the following in sufficient H_2O to make 0.9 liter:

YNB without amino acids (Difco 0919-15)	6.7 g
powdered supplement mixture	2 g

Sterilize by autoclaving. Allow to cool. Add 100 ml of 20% glucose (20 g of glucose dissolved in sufficient H_2O to make a final volume of 100 ml; sterilize the solution by passing it through a 0.22-μm filter or by autoclaving) and mix well.

Notes: To prepare plates, add 20 g of Bacto agar before autoclaving.

Instead of the YNB without amino acids used above, 1.7 g of YNB without ammonium sulfate or amino acids (Difco 0335-15-9) plus 5 g of ammonium sulfate can be substituted in each liter.

Solid or liquid SC medium contains all supplements listed in the table below. These supplements are added in the form of a well-mixed powder (2 g of powder per 1 liter of medium). SC medium can be made selective for the growth of the desired prototrophs by dropping out ingredients. For example, "drop out" medium referred to as SC – Ura,Trp contains all powdered supplements except uracil and tryptophan.

Powdered supplement mixture for SC medium or "drop out" media

Supplement	Amount to add
adenine (hemisulfate salt)	0.5 g
L-alanine	2 g
L-arginine HCl	2 g
L-asparagine (monohydrate)	2 g
L-aspartic acid	2 g
L-cysteine HCl	2 g
glutamine	2 g
L-glutamic acid (monosodium salt)	2 g
glycine (sodium salt)	2 g
L-histidine HCl	2 g
myo-inositol	2 g
L-isoleucine	2 g
L-leucine	4 g
L-lysine HCl	2 g
L-methionine	2 g
p-aminobenzoic acid	0.2 g
L-phenylalanine	2 g
L-proline	2 g
L-serine	2 g
L-threonine	2 g
L-tryptophan	2 g
L-tyrosine	2 g
uracil	2 g
L-valine	2 g

Before adding supplements to the medium, thoroughly mix the powdered supplements in a plastic bottle by adding several clean glass marbles and shaking vigorously.

YPD MEDIUM

Combine the following in 0.9 liter of H_2O:

Bacto peptone	20 g
Bacto yeast extract	10 g
glucose	20 g

Adjust the final volume to 1 liter with H_2O. Sterilize by autoclaving.

Notes: Adding supplemental tryptophan (1.6 g of tryptophan per liter) before autoclaving is recommended for Trp⁻ auxotrophs since YPD medium is limiting for tryptophan.

Adding supplemental adenine (50 mg of adenine hemisulfate monohydrate per liter) before autoclaving is recommended for Ade⁻ auxotrophs.

To prepare plates, add 20 g of Bacto agar before autoclaving.

ANTIMICROBIAL AGENTS

Whenever possible, purchase the H_2O-soluble salt form (sodium, hydrochloride, or sulfate) of antimicrobial agents. All stock solutions are prepared with either sterile H_2O or absolute ethanol. No additional sterilization is needed because of the nature of the chemicals. The small volume of ethanol added to media or plates is of no consequence.

Ampicillin (100 mg/ml)

Dissolve 1 g of sodium ampicillin in sufficient H_2O to make a final volume of 10 ml. Divide into aliquots and store at -20^oC.

Ampicillin, a penicillin derivative, is bactericidal only to growing cells. It inhibits cell wall biosynthesis by preventing cross-linking of the peptidoglycan. β-Lactamase, which is encoded by the *bla* gene, confers resistance by cleaving ampicillin's β-lactam ring. Ampicillin is typically used in growth medium at a final concentration of 25–50 µg/ml.

Carbenicillin (50 mg/ml)

Dissolve 0.5 g of disodium carbenicillin in sufficient H_2O to make a final volume of 10 ml. Divide into aliquots and store at -20^oC.

Like ampicillin above, carbenicillin is a penicillin derivative and is typically used in growth medium at a final concentration of 25–50 µg/ml.

Chloramphenicol (25 mg/ml)

Dissolve 250 mg of chloramphenicol in sufficient absolute ethanol to make a final volume of 10 ml. Divide into aliquots and store at -20^oC.

Chloramphenicol is bacteriostatic because it inhibits protein synthesis. Chloramphenicol transacetylase, which is encoded by the *cam* gene, confers resistance by acetylating chloramphenicol and thus preventing its inhibitory activity. Chloramphenicol is typically used in growth medium at a final concentration of 12.5–25 µg/ml. For complete inhibition of host protein synthesis, use a final concentration of 170 µg/ml.

Kanamycin (10 mg/ml)

Dissolve 100 mg of kanamycin monosulfate in sufficient H_2O to make a final volume of 10 ml. Divide into aliquots and store at -20^oC.

Kanamycin is bactericidal because it inhibits 70S ribosomal subunit translocation during protein synthesis. Aminoglycoside-modifying enzymes confer resistance by modifying kanamycin and thus preventing its inhibitory activity. Kanamycin is typically used in growth medium at a final concentration of 10–50 µg/ml.

Methicillin (100 mg/ml)

Dissolve 1 g of sodium methicillin in sufficient H$_2$O to make a final volume of 10 ml. Divide into aliquots and store at −20°C.

Methicillin is a penicillin derivative that is used to prevent the formation of satellite colonies during selection with ampicillin. Methicillin can be used in growth medium at a final concentration of 37.5 μg/ml in combination with ampicillin at a final concentration of 100 μg/ml.

Nalidixic Acid (5 mg/ml)

Dissolve 50 mg of the sodium salt of nalidixic acid in sufficient H$_2$O to make a final volume of 10 ml. Divide into aliquots and store at −20°C.

Nalidixic acid is bacteriostatic because it inhibits DNA synthesis through its action on DNA gyrase. Mutations in the gyrase gene confer resistance. Nalidixic acid is typically used in growth medium at a final concentration of 15 μg/ml.

Streptomycin (50 mg/ml)

Dissolve 0.5 g of streptomycin sulfate in sufficient absolute ethanol to make a final volume of 10 ml. Divide into aliquots and store at −20°C.

Streptomycin is bactericidal because it acts on the S12 protein of the 30S ribosomal subunit, thereby inhibiting protein synthesis. Mutations in the gene encoding the S12 protein (*rpsL*) confer resistance by preventing binding of streptomycin. Inactivation by aminoglycoside phosphotransferase can also occur. Streptomycin is typically used in growth medium at a final concentration of 10–50 μg/ml.

Tetracycline (10 mg/ml)

Dissolve 100 mg of tetracycline hydrochloride in sufficient H$_2$O to make a final volume of 10 ml. Alternatively, dissolve the free base form of tetracycline in absolute ethanol. Divide into aliquots and store in foil-wrapped tubes at −20°C to protect the solution from light.

Tetracycline is bacteriostatic because it inhibits elongation during protein synthesis. It also prevents binding of aminoacyl tRNA to the ribosome. Loss of cell wall permeability confers resistance to tetracycline. Tetracycline is typically used in growth medium at a final concentration of 10–50 μg/ml.

APPENDIX 2

Basic Procedures

COMPILED BY JOAN KOBORI

Appendix 2 details commonly used molecular biology methods that are frequently referred to in this volume. These methods are intended to aid investigators unfamiliar with these basic procedures. Many variations and modifications of these procedures exist (e.g., see Sambrook et al. [1989], *Molecular cloning: A laboratory manual*), but the generic procedures provided here are applicable to most situations.

QUANTITATION OF CELL CONCENTRATION

Measuring the OD$_{600}$ of a Culture

OD$_{600}$ readings can be used to estimate the number of bacteria or yeast in a culture and to monitor growth. The relationship of light scattering of a culture to the actual cell number is dependent on the growth conditions (e.g., medium, temperature, strain) of the culture. However, approximations are generally adequate.

BACTERIAL CULTURES

For bacteria grown in LB (or equivalent rich) medium:

   OD$_{600}$ of $1 \cong 5 \times 10^8$ (to 1×10^9) cells/ml

Precise correlations can be determined for each strain and growth condition by plating dilutions of a culture on solid medium and counting colonies.

   To determine the bacterial cell concentration, set the spectrophotometer to a wavelength of 600 nm and adjust the spectrophotometer to zero with a sample of fresh medium (the same medium that was used for growing the culture) in a cuvette with a 1-cm path length. The OD$_{600}$ of an undiluted actively growing *E. coli* culture ranges from 0.1 to approximately 1 during log growth phase. Typically, a saturated overnight bacterial culture grown in rich medium (e.g., LB medium) has an OD$_{600}$ of approximately 5. Cells grown in TB medium or Super Broth reach different densities.

YEAST CULTURES

For yeast grown in rich medium:

   OD$_{600}$ of $1 \cong 1 \times 10^7$ cells/ml

This conversion factor for yeast cell concentration varies with haploid versus diploid cells and among different genetic backgrounds. The precise conversion factor should be determined empirically by comparing the actual cell count obtained with a hemocytometer and the corresponding OD$_{600}$ measurement for any previously uncharacterized strain.

   To determine the yeast cell concentration, follow the procedure above for bacterial cells but use the conversion factor for yeast cell concentration.

   A saturated diploid yeast culture grown in YPD medium contains approximately 10^8 cells/ml and has an OD$_{600}$ of approximately 10–15. Saturated cultures grown in minimal medium reach an OD$_{600}$ of approximately 3–5. A medium-sized yeast colony contains approximately 10^6 cells. A "matchhead full" of yeast contains 1×10^7 to 5×10^7 cells.

Counting the Cells in a Hemocytometer

A hemocytometer consists of two chambers, each of which is divided into nine 1-mm squares. A glass coverslip is supported 0.1 mm above the squares, providing a 1 mm x 1 mm x 0.1 mm (10^{-4} cm^3 or 10^{-4} ml) volume per square. The cell concentration per milliliter is the average count per square x 10^4. This method is used to determine accurately cell concentrations for yeast and mammalian cells. Cell counts for viable mammalian cells can be determined by exclusion of the dye trypan blue.

1. Clean the hemocytometer under running H_2O. Rinse with 70% ethanol. Wipe dry with a Kimwipe.

2. Dilute the cells in medium at a concentration of approximately 20–50 cells per 1-mm square on the hemocytometer and suspend them uniformly (i.e., there should be no clumps of cells).

    Note: Viable mammalian tissue-culture cells can be quantitated by using trypan blue, a dye excluded from viable cells. If viable cells must be counted, add 20 μl of 0.4% trypan blue solution to 20 μl of cell suspension and mix thoroughly.

    0.4% Trypan blue solution

    Dissolve 40 mg in sufficient PBS (pH 7.0) to make a final volume of 10 ml. (This stock solution can also be purchased.) This solution can be stored at 4°C for at least 6 months.

3. Place a clean hemocytometer coverslip on the hemocytometer. Load each chamber of the hemocytometer with cell suspension, allowing capillary action to draw the solution into the chamber.

4. Using 100x magnification on a microscope, count the cells in the four large corner squares and the central large square for each chamber (see Figure 1). Count the cells touching the top and left central lines but not those touching the bottom and right central lines.

    Notes: If more than 200 cells are counted per square, the cell suspension is too concentrated and must be diluted for more reliable cell counting. Reliable cell counts are typically 20–50 cells per square.

    If desired, keep a separate count of nonviable and viable mammalian cells stained with trypan blue. Nonviable cells stain blue, but viable cells appear clear with a distinct, sharp outline.

5. Calculate the cell concentration as follows:

    number of cells/ml = average cell count per square x 10^4 x dilution factor

    Note: The dilution factor in this calculation should include the factor of two if cells were stained with trypan blue.

6. Calculate the total number of cells in the original suspension as follows:

total number of cells = number of cells/ml x total ml of cell suspension

7. If trypan blue exclusion was used to determine cell viability, calculate the cell viability as follows:

$$\text{cell viability} = \frac{\text{total number of viable cells}}{\text{total number of viable + nonviable cells}}$$

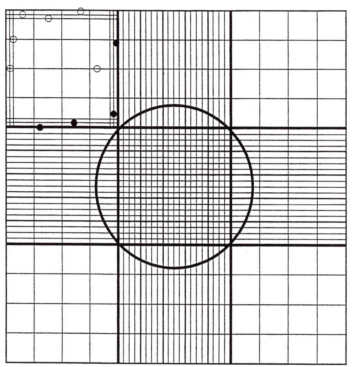

Figure 1 Standard hemocytometer chamber. The circle indicates the approximate area covered at 100x magnification (10x ocular and 10x objective). Count the cells touching the top and left central lines (*open circles*) but not those touching the bottom and right central lines (*closed circles*). Count the cells in the four large corner squares and the central large square for each chamber (one chamber is represented here). The corner square equals 1 mm.

STANDARD METHODS USED FOR ISOLATING DNA

This section describes several standard methods used during the purification of DNA. These protocols represent individual steps in the overall process of isolating high-quality DNA.

Extraction of DNA Samples with Organic Chemicals

Extractions with organic chemicals (phenol, chloroform, ether) are frequently used during the purification of DNA. Extraction of aqueous DNA solutions with phenol removes contaminating materials, particularly proteins (including degradative enzymes). Phenol must be equilibrated to a pH above 7.6; otherwise, the DNA tends to partition into the organic phase instead of the aqueous phase. DNA samples are generally extracted once or twice with phenol alone, phenol:chloroform (1:1), or phenol:chloroform:isoamyl alcohol (25:24:1). Typically, samples are then extracted once or twice with chloroform (or in some cases ether) to remove residual phenol, which may interfere with future manipulations of the DNA. Chloroform also denatures proteins. Isoamyl alcohol is often added to the chloroform to reduce foaming. The following precautions should be observed during the use of hazardous organic chemicals:

- Perform all extractions with organic chemicals in tubes constructed of material that is resistant to phenol and chloroform (e.g., glass and polypropylene). Polycarbonate tubes are not resistant to these organic reagents and should never be used.
- Perform all extractions with organic reagents in a chemical fume hood to avoid inhalation. Wear gloves, protective clothing, and safety glasses to avoid contact with the skin.
- Cap all tubes before mixing or centrifuging.
- Immediately rinse any areas of skin that come in contact with phenol with a large volume of H_2O and wash with soap and H_2O. Phenol causes severe burns to the skin and has an anesthetic effect, reducing the perceived severity of the injury. Seek immediate medical attention.
- Use extreme caution in handling flammable organic chemicals, particularly ether.
- Consult the local institutional safety officer regarding the proper disposal of organic chemical waste.

PREPARING EQUILIBRATED PHENOL

Phenol readily undergoes oxidation and must therefore be redistilled before use. Since redistilled phenol is acidic, a neutralization step with Tris base and then an equilibration step with TE (pH 8.0) or a similar buffer must be performed. Phenol that has been redistilled and preequilibrated is commercially available from many suppliers.

In the protocol below for equilibrating redistilled phenol, the antioxidant 8-hydroxyquinoline and *m*-cresol (optional) are added to extend the storage life of the phenol stock solution. The addition of 8-hydroxyquinoline also provides a yellow color, which aids in detecting the phenol phase during extractions.

1. In a chemical fume hood, thaw a 500-g bottle of **phenol** at room temperature or 37°C. Place the bottle of phenol in a clean beaker filled with clean distilled H_2O so that the phenol will leak into the H_2O if the phenol bottle breaks.

   phenol (see Appendix for Caution)

2. (*Optional*) Use a glass pipette or a glass graduated cylinder to add 70 ml of **m-cresol** to the thawed phenol. The total volume should now be approximately 0.8 liter. Stir gently with a magnetic stirring bar.

   m-cresol (see Appendix for Caution)

3. Add 0.8 g of **8-hydroxyquinoline** to make a final concentration of 0.1% (w/v).

   Note: This powder turns the phenol yellow.

   8-hydroxyquinoline (see Appendix for Caution)

4. Add approximately 50 ml of H_2O and stir to mix.

5. Stop stirring and allow the phases to separate.

6. Check the pH of the aqueous (upper) phase with a pH stick.

   Note: The aqueous phase should be very acidic.

7. Remove the aqueous phase by aspiration with a glass pipette.

8. Neutralize the phenol with Tris base as follows:

   a. Add approximately 0.8 liter of 1 M Tris base and stir to mix.

   b. Stop stirring and allow the phases to separate.

   c. Check the pH of the aqueous (upper) phase with a pH stick. If the pH is less than 5.0, remove the aqueous phase by aspiration and repeat steps a–c.

   d. Once the pH is above 5.0, completely remove the aqueous phase by aspiration.

9. Equilibrate the phenol with TE (pH 8.0) or a similar buffer by repeating step 8 at least three times using TE (pH 8.0) instead of Tris base.

10. Divide the phenol into aliquots and overlay the phenol with TE (pH 8.0). Store in dark bottles at 4°C for up to 1 month or in polypropylene tubes at −20°C for up to 6 months.

EXTRACTING WITH PHENOL, PHENOL:CHLOROFORM, OR PHENOL:CHLOROFORM:ISOAMYL ALCOHOL

Phenol alone can be used for extractions. A mixture of phenol and chloroform, with or without isoamyl alcohol can also be used in the protocol below. In all extractions, the phenol must first be equilibrated to a pH above 7.6 so that the DNA partitions into the aqueous phase.

1. Add 1 volume of **phenol** saturated with TE (pH 8.0) or a similar buffer (for preparation, see pp. 413–414) to the DNA sample. Mix gently by vortexing for 2–3 minutes or by inverting the capped tube for up to 5 minutes to form an emulsion.

   Notes: Either phenol:**chloroform** (1:1) or phenol:chloroform:isoamyl alcohol (25:24:1) can replace the phenol.

   Large volumes (2–20 ml) can be conveniently extracted in 15- or 50-ml polypropylene tubes. Small volumes (<0.5 ml) can be extracted in 1.5-ml microcentrifuge tubes. Volumes of 0.5–2 ml should be divided into aliquots in microcentrifuge tubes.

   For maximal recovery, the DNA should be dissolved in at least 100 µl of TE (pH 8.0).

   To recover RNA, use **DEPC**-treated H_2O to dissolve the RNA and to equilibrate the phenol.

   phenol, chloroform, DEPC (see Appendix for Caution)

2. For large volumes, centrifuge at 5000*g* at room temperature for 5 minutes to separate the organic and aqueous DNA-containing phases. For small volumes, centrifuge in a microcentrifuge at 12,000*g* for 5 minutes.

   Notes: Complete separation of the phases should be apparent. If not, repeat the centrifugation for a longer period of time and/or at a higher speed.

   Shorter periods of centrifugation may be sufficient for small volumes.

   The lower layer should be the organic phase and the upper layer should be the aqueous phase. The phases are often separated by a white or yellowish pellicle of precipitated protein and cellular debris.

3. Carefully remove the aqueous layer with a pipette and transfer it into a new tube for further extractions or precipitation of the DNA. Avoid touching the precipitated protein layer at the interface. Discard the organic phase.

   Notes: If the interface is noticeably cloudy (an indication of the presence of excess protein), a second extraction is recommended.

   Many investigators perform a series of extractions before precipitating the DNA with alcohol. Extraction with phenol:chloroform (1:1), phenol:chloroform:isoamyl alcohol (25:24:1), and then chloroform:isoamyl alcohol (24:1) is common for large amounts of DNA. Other investigators extract with phenol, phenol:chloroform (1:1), or phenol:chloroform: isoamyl alcohol (25:24:1) and then remove the residual phenol by extraction with chloroform (see p. 416).

   In some specialized protocols elsewhere in this manual, diethyl ether is recommended as an alternative to chloroform for removing phenol. Ether should be saturated with H_2O to prevent loss of volume from the aqueous phase. When extractions are performed with diethyl ether, the aqueous DNA-containing phase is the lower phase because of the lower density of ether. Since the DNA is in the lower phase, multiple extractions can be performed in the same tube. Ether in the upper phase causes an inverted meniscus at the interface, readily revealing the continued presence of ether. After removal of as much ether as possible with pipetting aids, residual traces of ether can be evaporated. The DNA can then be recovered by precipitation with alcohol.

EXTRACTING WITH CHLOROFORM TO REMOVE PHENOL

After extraction with phenol, chloroform is often used to remove residual phenol from the DNA. The DNA will remain in the aqueous (upper) phase. In general, one to two extractions with chloroform are performed before the DNA is recovered by precipitation with alcohol.

1. Add 1 volume of **chloroform** to the DNA sample and mix gently by vortexing for 1–2 minutes or by inverting the capped tube to form an emulsion.

   Notes: Chloroform:isoamyl alcohol (24:1) can replace the chloroform.

   Large volumes (2–20 ml) can be conveniently extracted in 15- or 50-ml polypropylene tubes. Small volumes (<0.5 ml) can be extracted in 1.5-ml microcentrifuge tubes. Volumes of 0.5–2 ml should be divided into aliquots in microcentrifuge tubes.

   chloroform (See Appendix for Caution)

2. For large volumes, centrifuge at 5000g at room temperature for 5 minutes to separate the organic and aqueous DNA-containing phases. For small volumes, centrifuge in a microcentrifuge at 12,000g for 5 minutes.

   Notes: The lower layer should be the organic phase and the upper layer should be the aqueous phase.

   Complete separation of the phases should be apparent. If not, repeat the centrifugation for a longer period of time and/or at a higher speed.

3. Carefully remove the aqueous layer with a pipette and transfer it into a new tube. Discard the organic phase.

Concentration of DNA Samples by Precipitation with Alcohols

CHOOSING THE ALCOHOL

DNA is readily recovered from aqueous solution by precipitation with alcohol in the presence of monovalent cations (e.g., sodium or ammonium). Both ethanol and isopropanol are widely used for this purpose, with the choice between them depending mainly on the volume of aqueous phase (2–2.5 volumes of ethanol is required, whereas only 1 volume of isopropanol is required). Precipitations with ethanol are usually effective at removing unwanted salt in DNA preparations.

CHOOSING THE SALT

If the existing monovalent cation concentration is low, either sodium acetate or ammonium acetate is added to the DNA before precipitation. Occasionally, NaCl is substituted. The choice between sodium or ammonium is determined by the subsequent use of the DNA. Although most of the salts should be removed by the end of the procedure, small amounts of a specific residual salt can cause problems. For example, DNA ligase can be inhibited by sodium ions, whereas CIP and bacteriophage T4 polynucleotide kinase can be inhibited by ammonium ions. Neither ion inhibits most other enzymes.

Washing precipitated DNA with 70% ethanol removes most of the salt. In situations where it is critical to remove all of the salt, the use of ammonium acetate as the source of salt in the precipitation should be considered. Residual ammonium acetate can be removed by drying the DNA pellet under vacuum. This procedure is often used with low-molecular-weight DNA but is not practical or desirable for large genomic DNA (see pp. 424–425) since the drying step would desiccate the genomic DNA, requiring a lengthy resuspension/dissolution period.

High concentrations of EDTA (10 mM) or phosphate (1 mM) in the DNA solution should be avoided since they may coprecipitate with the DNA. Dilution of the DNA solution or removal of these salts by use of spin columns must be performed before precipitation.

MAXIMIZING THE DNA YIELD

Temperature was once thought to be a key factor in efficiently precipitating DNA. Incubations at 4°C or in dry-ice/ethanol baths are not necessary but are still commonly used. The critical factor in the recovery of small amounts of DNA is the length of the centrifugation step. Longer periods of centrifugation will aid in the recovery of particularly small amounts of DNA.

If the DNA concentration is low, addition of *E. coli* tRNA can help in recovering the DNA. The tRNA should be extracted with phenol and boiled to remove contaminating DNases. tRNA specifically prepared for precipitating DNA is also commercially available.

COLLECTING THE DNA

For large-volume precipitations (>2 ml) in which the DNA concentration is high, DNA strands will form a visible precipitate, which collects into a compact mass of material that can easily be removed by "spooling." Spooling large genomic DNA separates the DNA from the bulk of the RNA, which has been copurified but remains in solution (this eliminates the need to add exogenous RNase, which may be contaminated with nucleases), and washing with 70% ethanol helps remove the majority of the salts. If the DNA is very dilute, the precipitated material may not be visible. In this case, the DNA must be recovered by centrifugation.

1. Add either 0.1 volume of 3 M sodium acetate (pH 5.2) or 0.3 volume of 10 M ammonium acetate to the DNA sample. Mix thoroughly by vortexing or by inverting the tube.

   Notes: The final concentration of sodium acetate should be approximately 300 mM.

   In most cases, the desired final concentration of ammonium acetate is 2–2.5 M. The stock solutions of ammonium acetate are usually at a pH of 7.0 or 7.4.

   For aqueous volumes of up to 0.5 ml, 1 μg of tRNA can be added before precipitation with alcohol. A carrier should only be added when its presence will not interfere with future uses of the DNA. For example, DNA ligation using bacteriophage T4 DNA ligase can be performed in the presence of tRNA carrier.

2. Add 2–2.5 volumes of absolute ethanol or 1 volume of isopropanol. Mix thoroughly by vortexing or by inverting the tube. Incubate for at least 15 minutes.

   Notes: Incubations can be performed at room temperature. Low-temperature incubations (on ice or at –20°C or colder) are not necessary but are still commonly used.

   The critical factor in the recovery of small amounts of DNA is the centrifugation step below, but use of prolonged precipitation periods may help maximize the DNA yield.

   If the concentration is high, a stringy white precipitate will appear almost immediately.

3. Collect the DNA precipitate by spooling if the DNA concentration is high and the size of the DNA is reasonably large; if the DNA is small or very dilute, collect the DNA by centrifuging.

   To collect visible DNA by spooling:
   a. Spool the stringy precipitated DNA on a pasteur pipette (i.e., slowly wind the DNA around the tip of a pasteur pipette).

   b. Wash the spooled DNA by repeatedly dipping it into a separate tube containing 70% ethanol. Do not dry the DNA.

   c. Place the DNA spooled on the pipette tip in a new tube containing a suitable volume of TE (pH 8.0) or another appropriate solution. Allow it to sit until the DNA is released from the pipette tip. Once the DNA has been released, mix gently by low-speed vortexing or by flicking the tube with a finger to help the DNA dissolve.

      Note: The DNA should dissolve quickly, but allowing it to sit at 4°C overnight is sometimes necessary.

   To collect DNA by centrifuging:
   a. For large volumes, centrifuge in a 15- or 30-ml Corex tube in a Sorvall SS34 or HB6 rotor (or equivalent) at 12,000*g* at 4°C for 30 minutes. For small volumes, centrifuge in a 1.5-ml microcentrifuge

tube in a microcentrifuge at 12,000*g* for 15 minutes. Decant the supernatant and invert the tube on a paper towel to drain.

Note: In general, centrifugation for 15 minutes is adequate to collect most DNA samples. Longer periods of centrifugation (e.g., 30 minutes) will aid in the collection of particularly small amounts of DNA.

b. To wash the DNA pellet, add 70% ethanol (5 ml for 15- or 30-ml tubes, 1 ml for microcentrifuge tubes) and gently rotate the tube, recentrifuge for 5 minutes, and decant the supernatant.

Note: This step should rinse the walls of the tube but keep the DNA pellet intact.

c. Dry the DNA pellet under vacuum for 5–10 minutes in a SpeedVac Concentrator or allow the DNA pellet to air dry.

Note: Small volumes of low-molecular-weight DNA are often dried under vacuum. Drying under vacuum is not practical or desirable for large genomic DNA since it would desiccate the genomic DNA and necessitate a lengthy resuspension/dissolution period (see p. 424).

d. Dissolve the DNA in TE (pH 8.0) or another appropriate solution.

4. Determine the DNA concentration (see pp. 421–422) and adjust to the appropriate concentration for storage or immediate use.

Quantitation of DNA

The accurate measurement of DNA concentration is essential for many applications. Several methods are in common usage for measuring DNA concentration, three of which are provided here. These methods are based largely on spectrophotometric measurement of UV absorbance or binding of fluorescent dyes.

UV ABSORBANCE

An advantage of the spectrophotometric method for DNA quantitation is that the amount of protein contamination in the sample can also be determined by measuring the OD_{280}. The disadvantage of this method is that it is sensitive to contaminating RNA, which can lead to an overestimation of the DNA concentration.

1. Set the spectrophotometer to a wavelength of 260 nm (in the UV spectrum). For a DNA sample dissolved in TE (pH 8.0), adjust the spectrophotometer to zero with TE (pH 8.0) in a quartz cuvette with a 1-cm path length.

2. Dilute the sample and measure the OD_{260}.

3. If the DNA solution is too dilute (i.e., the OD_{260} is <0.05), repeat the measurement with a more concentrated DNA sample.

   Note: Most spectrophotometers are accurate at an OD_{260} ranging from 0.05 to approximately 0.8.

4. Set the spectrophotometer to a wavelength of 280 nm and readjust to zero. Measure the OD_{280} of the sample.

   Note: Pure DNA will have a ratio of OD_{260}/OD_{280} of approximately 1.8. A ratio that is very different from 1.8 (i.e., <1.5 or >2) may be indicative of either residual protein or organic solvents in the DNA sample. In this case, extract the DNA sample with **phenol:chloroform** again and then precipitate with alcohol again.

   phenol, chloroform (see Appendix for Caution)

5. Calculate the DNA concentration as follows:

   $$\text{double-stranded DNA concentration in } \mu g/ml = \text{measured } OD_{260} \times \frac{50\ \mu g/ml}{1\ OD_{260}} \times \text{dilution factor}$$

   $$\text{single-stranded DNA concentration in } \mu g/ml = \text{measured } OD_{260} \times \frac{36\ \mu g/ml}{1\ OD_{260}} \times \text{dilution factor}$$

   Note: An OD_{260} of 1 corresponds to 50 µg/ml of double-stranded DNA or 36 µg/ml of single-stranded DNA.

BINDING OF ETHIDIUM BROMIDE

Ethidium bromide binds to double-stranded DNA by intercalation. It absorbs UV light at 260 nm and emits fluorescence at 590 nm. The amount of fluorescence is proportional to the amount of DNA.

DNA concentrations of dilute solutions or very small sample volumes that cannot be subjected to spectrophotometric quantitation can be estimated by binding of ethidium bromide. Samples are analyzed by agarose gel electrophoresis and compared with DNA samples of known concentration. This method can detect as little as 1–5 ng of DNA.

This method has the advantage that it is insensitive to contamination with RNA, which runs ahead of the DNA on the gel. To avoid misinterpretation of the DNA concentration due to binding of RNA to the dye, be sure to stain with ethidium bromide after running the gel instead of including the dye in the gel and electrophoresis running buffer.

1. Dilute a DNA standard with TE (pH 8.0) to make DNA concentrations of 2, 1, 0.5, 0.25, and 0.125 µg/ml.

    Notes: Be sure to prepare dilutions of the DNA standard covering a broad range of DNA concentrations so that they encompass the DNA samples of unknown concentration.

    For genomic DNA, uncut bacteriophage λ DNA is a good standard since it migrates in a conventional agarose gel at a size similar to that of genomic DNA.

2. Mix 10 µl of each diluted standard and the DNA sample of unknown concentration with 2 µl of a 6x gel-loading solution. Analyze on an agarose gel.

    Note: Choose gel conditions such that DNA samples migrate at limiting mobility; 0.7–1% agarose gels generally suffice. The gel can be run at high voltage (≥100 mA) for 30 minutes; the samples just need to enter the gel.

3. Place the gel in 1x electrophoresis buffer containing **ethidium bromide** at a final concentration of 0.2–0.5 µg/ml and stain for 1 hour to detect the bands.

    Note: Staining for 10–15 minutes is frequently sufficient to detect bands.

    ethidium bromide (see Appendix for Caution)

4. Place the gel in 1x electrophoresis buffer and destain for 1 hour.

5. Photograph the gel using a **UV** transilluminator.

    UV irradiation (see Appendix for Caution)

6. Estimate the concentration of the DNA sample by locating the diluted standard with the fluorescence intensity that most closely matches that of the sample.

7. Repeat the analysis with a broader range of DNA standard concentrations if the dilutions of the DNA standard do not encompass the sample of unknown concentration.

FLUORIMETRY

The fluorochrome Hoechst 33258 binds to DNA. DNA quantitation by fluorimetry takes advantage of the specific excitation of the DNA-bound fluorochrome with UV light at 365 nm and the subsequent emission at 458 nm. This protocol was developed for use with the Hoefer minifluorimeter model TKO 100 but other fluorimeters can be adapted for this purpose. DNA at concentrations greater than 10 ng/ml can be quantitated even in the presence of contaminating RNA and protein.

1. Turn on the fluorimeter 15 minutes before use and set the sensitivity dial to maximum.

2. Add 2 ml of 1x Hoechst 33258 solution to the cuvette supplied with the fluorimeter. Place the cuvette in the fluorimeter and adjust to zero.

1x Hoechst 33258 solution

Component and final concentration	Amount to add per 100 ml
1x TEN buffer	10 ml of 10x
H_2O	90 ml
0.1 μg/ml Hoechst 33258	10 μl of 1 mg/ml (in H_2O)

Prepare just before use. (The 1 mg/ml stock solution of Hoechst 33258 dye can be stored in a foil-wrapped container at 4°C for up to 6 months.)

10x TEN buffer

Component and final concentration	Amount to add per 100 ml
100 mM Tris-Cl	10 ml of 1 M (pH 7.5 at 25°C)
10 mM EDTA	2 ml of 0.5 M (pH 8.0)
1 M NaCl	20 ml of 5 M
H_2O	68 ml

Store at room temperature for up to 6 months.

3. Replace the contents of the cuvette with 2 ml of either salmon sperm DNA or bacteriophage λ DNA at a concentration of 0.5 mg/ml in 1x Hoechst 33258 solution. Shake the cuvette well.

   Note: The size of the standard DNA does not affect the amount of Hoechst dye that binds.

4. Place the cuvette in the fluorimeter and set the fluorimeter sensitivity to 500.

5. Replace the contents of the cuvette with 2 ml of 1x Hoechst 33258 solution.

6. Place the cuvette in the fluorimeter and set the fluorimeter to zero.

7. Replace the contents of the cuvette with 2 ml of the DNA sample diluted in 1x Hoechst 33258 solution. Shake the cuvette well.

8. Place the cuvette in the fluorimeter and take a reading within 2 seconds.

   Note: The reading must be taken immediately because the signal decays with time.

9. Calculate the DNA concentration by adjusting the reading (which corresponds to the DNA concentration in µg/ml) for the dilution in step 7.

Dialysis of DNA

Dialysis is a common procedure performed to remove unwanted salts in DNA samples. Standard dialysis in tubing can be used for large volumes, whereas spot dialysis (drop dialysis) can be used for very small volumes. Precipitations with ethanol can also be used to remove unwanted salt effectively (see pp. 417–419), but resuspension/dissolution of HMW genomic DNA (>50 kb) precipitated with ethanol can take a long time (hours to days).

STANDARD DIALYSIS OF DNA

Removal of salt from genomic DNA can best be achieved by standard dialysis. This gentle procedure minimizes shearing, thus maintaining the large size of HMW DNA.

1. Prepare sterile dialysis tubing as follows:
   a. Cut dry dialysis tubing into convenient lengths (10–20 cm).

   Note: Always wear gloves when handling dialysis tubing.

   b. Place the tubing in a 2-liter glass beaker containing 1 liter of 100 mM sodium bicarbonate/1 mM EDTA and boil on a hot plate for 15 minutes. Using a sterile blunt rod or a 10–25-ml glass pipette, occasionally submerge the tubing as it bubbles up in the boiling solution.

   Note: Do not overload the beaker.

   100 mM Sodium bicarbonate/1 mM EDTA

Component and final concentration	Amount to add per 1 liter
100 mM sodium bicarbonate	8.4 g
1 mM EDTA	2 ml of 0.5 M (pH 8.0)
H_2O	to make 1 liter

   Prepare just before use.

   c. Allow the solution to cool completely, and then thoroughly rinse the inside of each piece of tubing with sterile H_2O.
   d. Place the tubing in 1 liter of 1 mM EDTA (pH 8.0) and boil for 15 minutes.
   e. Store the dialysis tubing in 1 mM EDTA (pH 8.0) at 4°C for up to 1 year.

2. Just before use, wash the inside and the outside of the dialysis tubing with sterile H_2O.

3. Pour (or use a wide-bore pipette tip [at least ~2-mm internal diameter] to gently pipette) each genomic DNA sample into a separate piece of sterile dialysis tubing.

   Note: It is best to pour the DNA into a dialysis bag held above a clean, sterile glass beaker so that the DNA sample is not lost if it does not go into the bag.

4. Dialyze against 1 liter of TE (pH 8.0) at 4°C for 12 hours with three changes of buffer.

   Note: If the DNA sample contains residual proteinase K from the purification procedure, **PMSF** can be added to inhibit any trace amounts of proteinase K remaining from the initial lysis of cells. Dialyze against 1 liter of TE (pH 8.0) containing 100 μM PMSF (prepared just before use) at 4°C for 12 hours with two changes of buffer. Residual proteinase K would cause degradation of enzymes used in subsequent analysis of the DNA.

   PMSF (see Appendix for Caution)

SPOT DIALYSIS

Salts or other small molecules (e.g., from ligations of vector to insert DNA) can be removed by spot dialyzing (drop dialyzing) the samples.

1. Place 20–30 ml of TE (pH 8.0) in a 10-cm petri dish. Place the petri dish where it will be undisturbed by contact or vibration.

2. Using blunt forceps, gently place a filter (Millipore VSWP, 0.025-mm pore size, 47-mm diameter) on top of the TE with the shiny side facing up. Allow the filter to wet for approximately 1 minute.

   Note: If more than one sample will be dialyzed simultaneously, mark the filters with pencil or waterproof pen before applying the samples. Either a single sample of 20–400 μl or as many as four 50-μl samples can be dialyzed on a single filter.

3. Slowly pipette the sample onto the surface of the filter. Cover the dish and dialyze at room temperature for 20 minutes to 2 hours.

   Note: Brief dialysis (<1 hour) may increase the sample volume if the applied sample contains a high concentration of salt or sucrose. Prolonged dialysis (>4 hours) results in loss of sample volume.

4. Transfer the dialyzed sample into a 0.5-ml microcentrifuge tube and place on ice.

   Notes: Do not try to transfer the entire sample, since this is likely to submerge the filter and result in the loss of all of the remaining samples.
   The expected recovery is approximately 80–90% of the sample volume.

ASSESSING THE EXTENT OF RADIOLABELING IN DNA PROBES BY PRECIPITATION WITH TCA

Random priming (Feinberg and Vogelstein [1983] *Anal. Biochem. 132:* 6–13) is the most commonly used procedure to prepare radiolabeled DNA probes. Random hexameric oligonucleotides are used to prime the incorporation of radiolabeled dNTPs into the probe DNA. The high efficiency of incorporation of [α-^{32}P]dNTPs by the Klenow fragment of *E. coli* DNA polymerase I results in the synthesis of probes with specific activities of more than 10^9 cpm per microgram of DNA. Typically, 90% of the isotopically labeled dNTPs will be incorporated. The efficiency of incorporation of [^{32}P]dNTPs can be determined by the separation of the labeled DNA from the unincorporated dNTPs.

Unless extremely clean backgrounds are required, no special procedures to remove unincorporated dNTPs are needed. The probe sample can be added directly to hybridization mixtures. In cases where high sensitivity and low backgrounds are important, the unincorporated [^{32}P]dNTPs can be removed by the use of column chromatography or precipitations with spermine.

The protocol below for precipitation with TCA is a simple method for quantitating the efficiency of radiolabeling. To determine the percentage of [^{32}P]dNTPs incorporated into newly synthesized probe DNA, an aliquot of the reaction mixture is treated with TCA and the resulting precipitated DNA is captured on Whatman GF/C glass-fiber filters. The unincorporated dNTPs are washed away. The amount of radioactivity incorporated into the precipitated DNA can be determined by Cerenkov counting. The ratio of the cpm captured by precipitation with TCA to the total cpm in an equivalent aliquot can be used to calculate the efficiency of radiolabeling.

1. For each radiolabeling reaction, label two Whatman GF/C glass-fiber filters (2.4-cm diameter) with a blunt pencil so that they can be identified later. Pin the filters onto a Styrofoam block covered with aluminum foil.

2. Spot 1 μl of **radiolabeling reaction mixture** on each of the two filters. Allow the filters to dry completely at room temperature or under a heat lamp.

   radioactive substances (see Appendix for Caution)

3. Transfer one filter into a beaker containing 100 ml of ice-cold 10% TCA/100 mM sodium pyrophosphate. Wash on a platform shaker (or with swirling) for 3 minutes. Decant the washing solution. Repeat the wash two more times using fresh washing solution each time. Do not wash the second filter.

   Notes: The unincorporated dNTPs are eluted from the filter; the ^{32}P-labeled DNA is retained. If available, use a vacuum manifold to wash the filter.

10% TCA/100 mM sodium pyrophosphate

Component and final concentration	Amount to add per 1 liter
10% **TCA**	100 ml of 100% (w/v)
100 mM sodium pyrophosphate decahydrate	44.6 g
H_2O	to make 1 liter

Dissolve the sodium pyrophosphate in sufficient H_2O to make a final volume of 0.9 liter. Add the TCA and mix. Store at 4°C for up to 6 months.

TCA (see Appendix for Caution)

4. Place the filters in a beaker containing 90% ethanol and swirl a few times. Remove the filters and dry completely.

5. Place the washed filter and the unwashed filter in separate scintillation vials.

6. Measure the radioactivity in each vial for 1 minute using the ^3H channel on the scintillation counter.

   Notes: The efficiency of detection of Cerenkov radiation is a function of the instrument and the geometry of the vials. Completely dry filters should be detected at approximately 25% efficiency.

   The addition of a **toluene**-based scintillation fluid to the vials will give 100% efficiency of detection for ^{32}P. However, as long as both filters are being measured in the same manner, the calculation of the percentage incorporated will be identical, regardless of the counting method used.

   toluene (See Appendix for Caution)

7. Calculate the percentage of [^{32}P]dNTPs incorporated as follows:

$$\text{percentage incorporated} = \frac{\text{cpm on washed filter}}{\text{cpm on unwashed filter}} \times 100$$

8. Calculate the specific activity of the synthetic DNA probe as follows:

$$\text{cpm/µg of probe} = \frac{\text{cpm on washed filter} \times \text{µl of total labeling reaction}}{\text{µg of input DNA}}$$

Note: The specific activity of the probe is calculated by using the amount of input DNA. The high specific activity of the radiolabeled dNTPs results in an insignificant amount of total DNA synthesized relative to the input DNA.

DILUTION AND STORAGE OF OLIGONUCLEOTIDES

Newly synthesized oligonucleotides are best stored as lyophilized pellets in microcentrifuge tubes at −20°C. Simple procedures for suspending, quantitating, and storing oligonucleotides are provided here.

1. Suspend the oligonucleotide in H_2O to make a final concentration of 20 μM. The table below indicates the volume of H_2O to add to 1 OD unit of dried oligonucleotide of various lengths to obtain a 20 μM solution. (When measured in a quartz cuvette with a 1-cm path length, an OD_{260} of 1 corresponds to 33 μg/ml [33 ng/μl] of single-stranded oligonucleotide [unmodified].)

Length (nucleotides)	Concentration in ng/μl for a 20 μM solution	Volume of H_2O to add (μl)[a]
15	100	330
17	112	295
20	132	250
25	165	200
32	211	156
40	264	125

[a]The volumes required can be calculated as in the following example:
For a 20-mer, a 20 μM solution contains 132 μg/ml. Since an OD_{260} of 1 corresponds to 33 μg in 1 ml,

$$\frac{33 \text{ μg in 1 } OD_{260}}{132 \text{ μg/ml}} = 250 \text{ μl}$$

To prepare a 20 μM stock solution, suspend 1 OD unit of a 20-mer in 250 μl of H_2O.

2. To confirm the concentration of the resulting oligonucleotide stock solution, measure the OD_{260} of a 30-fold dilution in H_2O in a cuvette with a 1-cm path length and then perform the following calculation:

$$\text{concentration in μg/ml} = \text{measured } OD_{260} \times \frac{33 \text{ μg/ml}}{1 \text{ } OD_{260}} \times \text{dilution factor}$$

Note: For example, if the OD_{260} measurement is 0.133 for a 30-fold dilution of a 20-mer, the calculation is:

$$0.133 \text{ } OD_{260} \times \frac{33 \text{ μg/ml}}{1 \text{ } OD_{260}} \times 30 = 131.7 \text{ μg/ml}$$

This confirms that the concentration of the 20-mer oligonucleotide stock solution is 131.7 μg/ml or approximately 20 μM.

3. Store oligonucleotide stock solutions and dilutions prepared as working solutions at −20°C. These solutions are stable for at least 1–2 years.

STORAGE AND SHIPMENT OF BIOLOGICAL SAMPLES

This section provides commonly used methods for storing and shipping various types of biological samples: bacterial stocks, yeast stocks, mammalian tissue-culture cells, blood or tissue samples, and DNA. Compliance with local, state, and federal regulations for the shipment of biohazardous materials is the responsibility of the investigator. Consult the local institution for further guidelines.

Bacterial Stocks

Most bacterial stocks are frozen in 7% DMSO or 15% glycerol at -80°C for long-term storage. Viability of frozen cells depends on the specific strain and the health of the cells at the time of freezing. Cultures to be stored are typically started from a single colony and grown in a suitable medium with agitation overnight ($\sim$10–15 hours).

DMSO STOCKS

Transfer 1 ml of an overnight culture into a labeled 1.5-ml screwcap cryotube and add 80 µl of DMSO. (Use DMSO from a bottle specifically dedicated for bacterial stock preparation. Never pipette directly from the stock bottle; aseptically remove an aliquot from the bottle and use the aliquot of DMSO to prepare the cultures.) Cap the tube and mix gently. Store at -80°C. Long-term viability of stocks depends on the particular strain, but some bacterial stocks have been known to maintain good viability for up to 10 years after initial storage in DMSO.

GLYCEROL STOCKS

Transfer 0.5 ml of an overnight culture into a labeled 1.5-ml screwcap cryotube and add 0.5 ml of sterile 30% glycerol. Cap the tube and mix gently. Store at -80°C. Long-term viability of stocks depends on the particular strain.

Alternatively, grow bacteria in medium containing 8–10% glycerol in plastic multiwell plates and store at -80°C. This method is typically used for storing cosmid, bacteriophage P1, BAC, and cDNA libraries.

RETRIEVAL OF FROZEN BACTERIAL STOCKS

Never thaw frozen bacterial stocks in DMSO or glycerol. Use a sterile loop, sterile wooden stick, or sterile disposable pipette to scratch the surface of the stock. Streak appropriate agar plates (e.g., LB agar plates) for single colonies. Recap the frozen stock and return it to storage at -80°C. Incubate the plate overnight at 37°C. The colonies on a plate can be used for up to 1 week to inoculate cultures. Plates should be stored upside down at 4°C during this time.

SHIPMENT OF BACTERIAL STRAINS BY MAIL

In general, most bacterial strains can be shipped by several different methods that maintain good viability. Agar stab cultures have traditionally been used to store and ship bacterial strains. Parafilm-sealed petri dishes streaked for single colonies or sterile filter disks impregnated with bacterial culture can also be shipped. The latter are aseptically transferred to appropriate medium upon receipt and a fresh overnight culture is grown. Overnight cultures can be shipped in cryotubes at room temperature. Upon receipt, the culture is streaked on plates of appropriate selective medium and single colonies are isolated. Frozen DMSO and glycerol stocks can be shipped on dry ice.

In general, agar stab cultures of nonplasmid-containing strains can be stored at room temperature for many years. They are not appropriate for long-term storage of plasmid-bearing strains because of loss of the plasmid under nonselective conditions. However, plasmid-bearing strains can be conveniently shipped in agar stab cultures. Immediately upon receipt, the cells must be streaked on plates of selective medium and either DMSO or glycerol stocks should be made.

Agar stab cultures are prepared in 3-ml glass vials with rubber gaskets in the screwcaps (e.g., Wheaton) as follows: Place 2 ml of liquified LB top agar (0.7% agar; some investigators use 1–1.2% agar) in each vial. Autoclave the vials with the caps loosened. Allow to cool to room temperature, tighten the caps, and store at room temperature until needed. Use a sterile loop or sterile wooden stick to pick an isolated single colony and stab it through the center of the agar to the bottom of the vial. Tighten the cap.

Bacterial strains should only be shipped at room temperature in moderate weather conditions. Hot summer weather may prove to be lethal during shipment.

Yeast Stocks

Yeast stocks are typically frozen in 20% glycerol at –80°C for long-term storage. Viability of frozen yeast cells depends on the specific strain and the health of the cells at the time of freezing. Cultures to be stored are typically started from "patched out" clones and grown in a suitable medium with agitation overnight.

GLYCEROL STOCKS

Patch out clones on an appropriate plate. Incubate at 30°C for 2 days. Inoculate 6 ml of YPD medium with a "matchhead full" of cells from the plate (this is a large inoculum; the culture will be turbid before incubation). Incubate at 30°C with agitation overnight. Add 2 ml of sterile 80% glycerol and mix thoroughly. Transfer 0.5-ml aliquots into 1-ml freezer vials with O-ring seals. Thoroughly shake the freezer vials and freeze at –60°C or lower (typically –80°C). Yeast tend to die if frozen at temperatures above –55°C.

Yeast strains can be stored at –80°C indefinitely by using this method. Note that strains grown in YPD medium before freezing have better long-term viability than those grown in selective medium.

RETRIEVAL OF FROZEN YEAST STOCKS

Never thaw frozen yeast stocks. Use a sterile loop, sterile wooden stick, or sterile disposable pipette to scratch the surface of the stock. Streak appropriate agar plates (e.g., YPD or selective agar plates) for single colonies. Recap the frozen stock and return it to storage at −80°C. Incubate the plate at 30°C for 2 days. Yeast can be stored at 4°C for approximately 6 months on YPD agar plates or for approximately 2 months on selective plates (i.e., SC plates with added supplements). Plates should be stored upside down at 4°C during this time. For long-term storage, seal the plates or place them in bags to keep them from drying out. Supplementing YPD medium with adenine prevents the toxicity caused by the red pigment produced by *ade2* strains that are stored at 4°C.

SHIPMENT OF YEAST STRAINS BY MAIL

Patch out clones on an appropriate selective plate. Incubate at 30°C for 2 days. Inoculate 6 ml of YPD medium with a "matchhead full" of cells from the plate (this is a large inoculum; the culture will be turbid before incubation). Incubate the culture at 30°C with agitation overnight. Add 2 ml of sterile 80% glycerol and mix thoroughly. Transfer 0.5-ml aliquots into 1-ml freezer vials with O-ring seals. Ship the vials by regular first class mail. Upon receipt, each cell suspension should be streaked on a suitable agar plate and incubated at 30°C for 2–3 days. Alternatively, ship yeast on YPD plates or in tubes or vials (with loosened caps) containing solid YPD medium (YPD slants).

If *un*saturated cultures are prepared (e.g., by inoculating fresh YPD medium with cells and tightly capping the freezer vial) and sent in the mail, the tubes may explode during transit because of the buildup of pressure from actively fermenting yeast in a tightly sealed vessel.

Yeast strains should only be shipped at room temperature in moderate weather conditions. Hot summer weather may prove to be lethal during shipment.

Mammalian Tissue-culture Cells

Typically, mammalian tissue-culture cells are grown in one of a number of media (e.g., RPMI 1640) supplemented with 5–15% FBS/FCS and 1% L-glutamine. Cultures are grown at 37°C in a 5% CO_2 environment. The volume of medium in the flask can affect the growth of cells, since the surface-to-air ratio is important in maintaining the proper pH of the medium. Factors that can affect the growth characteristics of a cell line include a change in incubation temperature, a difference in the lot of FBS/FCS and/or medium, depletion of L-glutamine in the medium, contamination with Mycoplasma, and length of time in continuous culture. To control these factors, medium is warmed to 37°C before it is added to cells, new lots of FBS/FCS and medium are tested for at least 2 weeks with a control cell line before they are accepted for general use, and fresh glutamine is added to the medium as required.

Each cell line has specific growth requirements. The final cell concentration (typically 5×10^6 to 5×10^7 cells/ml) during frozen storage may affect viability. Follow specific handling and storage medium recommendations for each cell line.

DMSO STOCKS

Centrifuge freshly grown, healthy cells at 500g at 4°C for 10 minutes. Discard the medium and resuspend the cell pellet in FBS/FCS containing 8–10% DMSO at 4°C. Transfer 0.5–1-ml aliquots of cell suspension into cryotubes and freeze in a –80°C freezer. Samples can be stored at –80°C or transferred into a **liquid nitrogen** storage tank. Cells should be viable at –80°C for up to 1 year or at –185°C in liquid nitrogen for up to 10 years.

Since each aliquot of frozen mammalian cells should be thawed just before use, at least two separate batches of ten aliquots each should be prepared for each cell line. As a precaution in case of freezer failure, store the aliquots of each cell line in at least two different freezers or liquid nitrogen storage tanks.

liquid nitrogen (see Appendix for Caution)

RETRIEVAL OF FROZEN MAMMALIAN TISSUE-CULTURE CELLS

The DMSO must first be removed from the frozen cells and the cells revived. To do this, quickly thaw the frozen aliquot of tissue-culture cells by placing it in a water bath set at 37°C with the top of the vessel above the H_2O line. Clean the outside of the vessel with 70% ethanol. Use a sterile pipette to transfer the cells into a 15-ml centrifuge tube containing appropriate medium (typically, 10 ml of medium containing 10% FBS/FCS). Gently centrifuge the cell suspension at 200g at room temperature for 5 minutes. Discard the medium and resuspend the cell pellet in 10 ml of appropriate medium containing FBS/FCS (typically 10%). Transfer the cells into an appropriate-sized flask and incubate at 37°C in a 5% CO_2 environment. Each cell line will recover at a different rate.

SHIPMENT OF MAMMALIAN TISSUE-CULTURE CELLS

Mammalian tissue-culture cell lines can be shipped as frozen stocks or as growing cultures. Frozen stocks should be shipped on dry ice in a Styrofoam container for next-day delivery. To ship growing cultures, inoculate the medium with a small aliquot of cells in a tissue-culture flask so that the cells are approximately a quarter to half confluent on the next day. At the time of shipping, the cells should be in log growth phase but should not be too dense (dense monolayers tend to peel off during transit). Before shipping, fill the flask to the neck with culture medium, cap tightly, and cover the cap with Parafilm M to prevent leaking. Wrap the flask in paper towels or place in a sealable plastic bag and cushion with cotton balls (this prevents the flask from breaking and also absorbs any liquid in case of a leak). Cell lines that grow in suspension can be shipped in centrifuge tubes or flasks filled to the neck and sealed.

Seasonal factors (extreme hot or cold) must be considered in shipping mammalian tissue-culture cells. If extremely warm weather conditions are anticipated, ship live cultures in a Styrofoam container. If extremely cold weather is anticipated, do not ship live cultures. For international shipping, minimize delays at customs by properly stating the value and contents of the package (it is advisable to check with customs officials in advance for additional shipping information). Upon receipt, allow the cells to recover by incubating live cultures at 37°C overnight before unsealing. Additional information about the shipping of tissue-culture cell lines can be obtained from the Coriell Institute for Medical Research, 401 Haddon Avenue, Camden, New Jersey 08103 (E-mail ccr@arginine.umdnj.edu; phone 609-757-4847; Fax 609-757-9737).

Blood and Tissue Samples

Blood to be used for DNA isolation should be collected in EDTA (purple-top Vacutainers). Blood containing EDTA can be stored at 4°C for 2 months. If the white blood cells are to be immortalized, collect the blood in heparin (green-top Vacutainers) to prevent clotting and store at room temperature for a maximum of 4 days. Blood samples can be shipped on wet ice.

Mouse and **human tissue samples** to be used for DNA preparation should be frozen in **liquid nitrogen** and stored at −80°C for up to 6 months. Do not thaw samples slowly since this allows nucleases to degrade the DNA; immediately place them in a denaturing cell lysis solution at the time of DNA isolation. Frozen tissue samples can be shipped on dry ice.

human blood, blood products, and tissues; liquid nitrogen (see Appendix for Caution)

DNA

Impure DNA containing traces of chemicals used during isolation often does not store well. Contamination with heavy metals, free radicals as chemical breakdown products, and oxidation products of phenol degradation can cause breakage of phosphodiester bonds. UV irradiation causes the production of thymine dimers and cross-links, resulting in loss of biological activity. Ethidium bromide causes photooxidation with visible light and molecular oxygen. Nucleases found on human skin do not generally pose a major problem for DNA. (RNases are very stable, but most DNases are not; however, the use of gloves is always recommended.)

STORAGE OF DNA

As a general rule, the more highly purified the DNA, the longer it can be stored under any conditions. DNA in solution is typically stored in TE (pH 8.0). DNA for long-term storage should contain a high salt concentration (at least 1 M NaCl or other salt) and 10 mM EDTA (to chelate heavy metals). Always dissolve DNA pellets in low-ionic-strength solutions (e.g., TE) and then add more salt if desired. Dried DNA pellets can be stored at −20°C for up to 6 months and DNA precipitated with ethanol can be stored at −20°C indefinitely.

Storage at 4°C is the best condition for routine storage of highly purified DNA. For storage at −20°C, it is generally preferable to use a nonfrost-free freezer. (Single- and double-stranded breaks may occur when the DNA is subjected to frequent freeze/thaw cycles.) For long-term storage, −80°C is recommended.

SHIPMENT OF DNA

Highly purified DNA can be shipped as an aqueous solution at room temperature or 4°C (i.e., on ice) or frozen on dry ice. It can also be precipitated with ethanol and then shipped at room temperature either as a dried DNA pellet or as precipitated DNA in ethanol. When the purity of the DNA is in doubt, do not ship DNA at room temperature as an aqueous solution. Trace contamination with nucleases may result in significant degradation of the DNA during shipment.

APPENDIX 3

Safety Cautions

Appendix 3 briefly describes safety cautions that should be observed for specific chemicals used throughout this manual. This appendix should not be considered a comprehensive listing nor does it contain comprehensive safety information. Suppliers of hazardous chemicals are required by the Occupational Safety and Health Administration to provide Material Safety Data Sheets. Refer to these sheets for more complete information.

Acetonitrile is very volatile and extremely flammable. It is an irritant and a chemical asphyxiant that can exert its effects by inhalation, ingestion, or absorption through the skin. Treat cases of severe exposure as cyanide poisoning. Wear gloves and safety glasses and work in a chemical fume hood.

Concentrated acids should be handled with great care. Wear gloves and a face mask.

Unpolymerized **acrylamide** and **bisacrylamide** are potent neurotoxins and are absorbed through the skin (the effects are cumulative). Wear gloves and a face mask when handling powdered acrylamide and methylenebisacrylamide and, if possible, weigh in a chemical fume hood and do not breathe the dust. Wear gloves, safety glasses, and protective clothing when handling solutions containing these chemicals. Polyacrylamide is considered to be nontoxic, but it should be handled with care because it might contain small quantities of unpolymerized acrylamide.

Actinomycin D is a teratogen and a carcinogen. It is highly toxic and may be fatal if inhaled, swallowed, or absorbed through the skin. It may also cause irritation. Wear gloves, safety glasses, and a face mask and always work in a chemical fume hood.

Ammonium hydroxide (see concentrated bases).

Antipain may be harmful if inhaled, swallowed, or absorbed through the skin. Wear gloves and safety glasses.

Aprotinin may be harmful if inhaled, swallowed, or absorbed through the skin. It may also cause allergic reactions. Exposure may cause gastrointestinal effects, muscle pain, blood pressure changes, or bronchospasm. Wear gloves, safety glasses, and a face mask and do not breathe the dust.

Concentrated bases should be handled with great care. Wear gloves and a face mask.

Bisacrylamide (see acrylamide).

Human blood, blood products, and tissues may contain occult infectious materials such as hepatitis B virus and HIV that may result in laboratory-acquired infections. Investigators working with EBV-transformed lymphoblastoid cell lines are also at risk of EBV infection. Any human blood, blood products, or tissues should be considered a biohazard and should be handled accordingly. Wear disposable gloves, protective clothing, and goggles; use mechanical pipetting devices; work in a laminar-flow hood or biological safety cabinet; protect against the possibility of aerosol generation (e.g., during centrifugation or mixing by vortexing); and disinfect all waste materials before disposal. Autoclave contaminated plasticware before disposal; autoclave contaminated liquids or treat with bleach (10% [v/v] final concentration) for at least 30 minutes before disposal. Consult the local institutional safety officer for specific handling and disposal procedures.

BrdU is a mutagen. It may be harmful if inhaled, swallowed, or absorbed through the skin. It may cause irritation. Wear gloves and safety glasses and always work in a chemical fume hood.

n-**Butanol** and *sec*-**butanol** are irritating to the mucous membranes, upper respiratory tract, skin, and especially the eyes. Wear gloves, safety glasses, and a face mask and do not breathe the vapors. *n*-Butanol and *sec*-butanol are also highly flammable.

Chloroform is irritating to the skin, eyes, mucous membranes, and respiratory tract. It is a carcinogen and may damage the liver and kidneys. Wear gloves and safety glasses and always work in a chemical fume hood.

m-**Cresol** is extremely destructive to the mucous membranes of the respiratory tract, the eyes, and the skin. It may be fatal if inhaled, swallowed, or absorbed through the skin. Exposure can cause burns and may damage the kidneys and eyes. Wear gloves, protective clothing, and safety glasses and always work in a chemical fume hood.

DAPI is a possible carcinogen. It may be harmful if inhaled, swallowed, or absorbed through the skin. It may also cause irritation. Wear gloves, safety glasses, and a face mask and do not breathe the dust.

Diethylamine is extremely destructive to the skin, eyes, mucous membranes, and upper respiratory tract. It may be fatal if inhaled and is harmful if swallowed or absorbed through the skin. It may also cause allergic reactions. Wear gloves, protective clothing, safety glasses, and a face mask and always work in a chemical fume hood. Diethylamine is extremely flammable.

Diethyl ether is extremely volatile and extremely flammable. It is irritating to the eyes, mucous membranes, and skin. It is also a CNS depressant with anesthetic effects. Diethyl ether exerts its effects through inhalation, ingestion, or absorption through the skin. Wear gloves and safety glasses and always work in a chemical fume hood. Explosive peroxides can form in concentrated solutions during storage or on exposure to air or direct sunlight.

DEPC is a potent protein denaturant and a suspected carcinogen. Aim the bottle away from you when opening it; internal pressure can lead to splattering. Wear gloves and protective clothing and work in a chemical fume hood.

DMF is irritating to the eyes, skin, and mucous membranes. It can exert its toxic effects through inhalation, absorption through the skin, or ingestion. Chronic inhalation can cause liver and kidney damage. Wear gloves, safety glasses, and a face mask and work in a chemical fume hood.

Use of **EBV**-transformed lymphoblastoid cell lines poses a risk for EBV infection. Discarded cells or flasks should be considered a biohazard and should be handled accordingly. Consult the local institutional safety officer for specific handling and disposal procedures.

Ethidium bromide is a powerful mutagen and is moderately toxic. Wear gloves when working with solutions that contain this dye. Consult the local institutional safety officer for specific handling and disposal procedures.

Formaldehyde is toxic and is also a carcinogen. It is readily absorbed through the skin and is irritating or destructive to the skin, eyes, mucous membranes, and upper respiratory tract. Wear gloves and safety glasses and always work in a chemical fume hood.

Formamide is teratogenic. The vapor is irritating to the eyes, skin, mucous membranes, and upper respiratory tract. It may be harmful if inhaled, ingested, or absorbed through the skin. Wear gloves and safety glasses and always work in a chemical fume hood when using concentrated solutions of formamide. Keep working solutions covered as much as possible.

Glacial acetic acid is volatile. Concentrated acids must be handled with great care. Wear gloves, safety glasses, and a face mask and work in a chemical fume hood.

Glutaraldehyde is toxic. It is readily absorbed through the skin and is irritating or destructive to the skin, eyes, mucous membranes, and upper respiratory tract. Wear gloves and safety glasses and always work in a chemical fume hood.

Concentrated HCl is volatile. Concentrated acids should be handled with great care. Wear gloves, safety glasses, and a face mask and work in a chemical fume hood.

Hydroxylamine is extremely destructive to the skin, eyes, mucous membranes, and upper respiratory tract. It may be fatal if inhaled and is harmful if swallowed or absorbed through the skin. Wear gloves, protective clothing, and safety glasses and always work in a chemical fume hood.

8-Hydroxyquinoline is irritating to the eyes, skin, mucous membranes, and upper respiratory tract. It may be harmful if inhaled, ingested, or absorbed through the skin. Wear gloves, safety glasses, and a face mask and do not breathe the dust.

Leupeptin may be harmful if inhaled, swallowed, or absorbed through the skin. Wear gloves and safety glasses.

LIDS is harmful if inhaled. Wear gloves and a face mask when weighing LIDS.

Liquid nitrogen's temperature is −185°C. Handle frozen samples with extreme caution. Seepage of liquid nitrogen into frozen vials can cause the vial to explode when it is removed from the liquid nitrogen. Use vials with O-rings when possible. Wear thermal gloves and a face mask when working with liquid nitrogen.

β-Mercaptoethanol may be fatal if inhaled or absorbed through the skin and is harmful if swallowed. High concentrations are extremely destructive to the skin, eyes, mucous membranes, and upper respiratory tract. Wear gloves and safety glasses and work in a chemical fume hood.

Methanol is poisonous and can cause blindness. Adequate ventilation is necessary to limit exposure to vapors.

Methotrexate is a carcinogen and a teratogen. It may be harmful if inhaled, ingested, or absorbed through the skin. Exposure may cause gastrointestinal effects, bone marrow suppression, or liver or kidney damage. It may also cause irritation. Wear gloves and safety glasses and always work in a chemical fume hood.

Methylenebisacrylamide (see acrylamide).

Osmium tetroxide (osmic acid) is very toxic if inhaled, swallowed, or absorbed through the skin. Vapors can react with corneal tissues and cause blindness. Wear gloves and safety glasses and always work in a chemical fume hood.

Pepstatin A may be harmful if inhaled, swallowed, or absorbed through the skin. Wear gloves and safety glasses.

Phenol is highly corrosive and can cause severe burns. Wear gloves, protective clothing, and safety glasses and always work in a chemical fume hood. Rinse any areas of skin that come in contact with phenol with a large volume of H_2O and wash with soap and H_2O; do not use ethanol!

p-**Phenylenediamine** is harmful if swallowed, inhaled, or absorbed through the skin. Wear gloves and safety glasses.

Piperidine is corrosive to the eyes, skin, respiratory tract, and gastrointestinal tract. It reacts violently with acids and oxidizing agents. Keep piperidine away from heat or flames. Wear gloves and safety glasses and do not breathe the vapors.

PMSF is a highly toxic cholinesterase inhibitor. It is extremely destructive to the mucous membranes of the respiratory tract, the eyes, and the skin. It may be fatal if inhaled, swallowed, or absorbed through the skin. Wear gloves and safety glasses and always work in a chemical fume hood. In case of contact, immediately flush eyes or skin with copious amounts of H_2O and discard contaminated clothing.

KOH should be handled with great care. Wear gloves and a face mask.

Propidium iodide is harmful if inhaled, swallowed, or absorbed through the skin. It is irritating to the eyes, skin, mucous membranes, and upper respiratory tract. It is mutagenic and possibly carcinogenic. Wear gloves, safety glasses, and protective clothing, and always work in a chemical fume hood.

Pyridine is extremely destructive to the skin, eyes, mucous membranes, and upper respiratory tract. It is harmful if inhaled, swallowed, or absorbed through the skin. It is also a possible mutagen. Keep pyridine away from heat and flames. Wear gloves and safety glasses and always work in a chemical fume hood.

Wear gloves, protective clothing, and safety glasses when handling **radioactive substances**. Consult the local safety office for further guidance in the appropriate use and disposal of radioactive materials.

SDS is harmful if inhaled. Wear a face mask when weighing SDS.

Human semen should be treated with the same precautions as human blood, blood products, and tissues (see above).

Sigmacote may be harmful if inhaled, swallowed, or absorbed through the skin. The vapor is irritating to the eyes, skin, mucous membranes, and upper respiratory tract. Sigmacote is also flammable. Wear gloves and safety glasses and always work in a chemical fume hood.

Silane is harmful if inhaled. It is extremely flammable. Avoid contact with eyes and skin. Wear gloves and safety glasses and always work in a chemical fume hood.

Sodium azide is highly poisonous. It blocks the cytochrome electron transport system. Solutions containing sodium azide should be clearly marked. Wear gloves and handle sodium azide with great care.

Sodium deoxycholate is irritating to mucous membranes and the respiratory tract and is harmful if swallowed. Wear gloves, safety glasses, and a face mask when handling the powder and do not breathe the dust.

Solid **NaOH** is caustic and should be handled with great care. Wear gloves and a face mask. Concentrated bases should be handled in a similar manner.

Sodium perchlorate is irritating to the eyes, mucous membranes, and upper respiratory tract. It may be harmful if inhaled or swallowed. Wear gloves, protective clothing, and safety glasses and always work in a chemical fume hood. Sodium perchlorate is a strong oxidizing agent and may cause fires when in contact with other materials.

TCA should be handled with great care. Wear gloves and a face mask.

Concentrated TFA is volatile. Concentrated acids must be handled with great care. Wear gloves and a face mask and work in a chemical fume hood.

Toluene vapors are irritating to the eyes, skin, mucous membranes, and upper respiratory tract. Toluene can exert harmful effects by inhalation, absorption through the skin, and ingestion. Wear gloves, safety glasses, and a face mask and do not breathe the vapors. Toluene is extremely flammable.

Triethylamine is flammable. It is extremely corrosive to the mucous membranes, upper respiratory tract, eyes, and skin. It may be harmful if inhaled, ingested, or absorbed through the skin. Wear gloves and safety glasses and work in a chemical fume hood.

UV radiation is dangerous, particularly to the eyes. To minimize exposure, make sure that the UV light source is adequately shielded. Wear protective goggles or a full safety mask that efficiently blocks UV light. Wear protective gloves when holding materials under the UV light source. UV radiation is also mutagenic and carcinogenic.

Xylene is flammable and may cause a narcotic effect, lung irritation, chest pain, and edema. Wear gloves and safety glasses and work in a chemical fume hood.

APPENDIX 4

Useful Facts

COMPILED BY JOAN KOBORI

Appendix 4 contains information commonly used in the analysis of genomes. The following are included in this appendix:

- Conversion factors for calculations involving nucleic acids and proteins
- Genome comparisons
- Average sizes of DNA fragments generated by cleavage with restriction enzymes
- Single-letter abbreviations for amino acids
- Genetic code
- Codon facts
- Isotope information

CONVERSION FACTORS

Nucleic Acid Weight Conversions

1000 bases $= 6.5 \times 10^5$ daltons of double-stranded DNA (sodium salt)
1000 bases $= 3.3 \times 10^5$ daltons of single-stranded DNA (sodium salt)
1000 bases $= 3.4 \times 10^5$ daltons of single-stranded RNA (sodium salt)

Average molecular weight of a dNMP = 324.5 daltons
Average molecular weight of a base pair of DNA = 649 daltons

Nucleic Acid Molar Conversions

1 μg/ml solution of DNA = 3.08 μM phosphate
1 μg/ml solution of 1-kb DNA = 3.08 nM 5′ ends
1 μg of 1-kb DNA = 1.54 pmole (3.08 pmole of ends)

Nucleic Acid Spectrophotometric Conversions

In a quartz cuvette with a 1-cm path length, an OD_{260} of 1 corresponds to:

- a 33 μg/ml solution of single-stranded oligonucleotide (unmodified)
- a 36 μg/ml solution of single-stranded DNA
- a 50 μg/ml solution of double-stranded DNA
- a 40 μg/ml solution of single-stranded RNA

Protein Conversions

1 kb of DNA can encode a 333-amino-acid protein with a m.w. of 37,000.
2.7 kb of DNA can encode a 900-amino-acid protein with a m.w. of 100,000.

GENOME COMPARISONS

Genome	Number of haploid chromosomes	Approximate size of haploid genome		Approximate amount of DNA per cell (pg)	Ploidy
		kb	kD		
Homo sapiens (human being)	23	3,300,000	2.1×10^{12}	6	diploid
Mus musculus (mouse)	20	2,700,000	1.75×10^{12}	5.4	diploid
D. melanogaster (fruit fly)	4	165,000	1.1×10^{11}	0.33	diploid
C. elegans (nematode)	6	100,000	6.5×10^{10}	0.2	diploid
Arabidopsis thaliana (plant)	5	100,000	6.5×10^{10}	0.2	diploid
S. cerevisiae (yeast)	16	14,000	9.1×10^{9}	0.014	haploid
E. coli (bacteria)	1	4700	3.05×10^{9}	0.0047	haploid
Bacteriophage λ (bacteriophage)	1	48.5	3.1×10^{7}	n.r.	n.r.

(n.r.) Not relevant.

AVERAGE SIZES OF DNA FRAGMENTS GENERATED BY CLEAVAGE WITH RESTRICTION ENZYMES

The table below lists those restriction enzymes that are known or predicted to cleave infrequently in ten commonly studied genomes (*Arabidopsis thaliana, C. elegans, D. melanogaster, E. coli,* human being, mouse, *Rhodobacter sphaeroides, S. cerevisiae, Staphylcoccus aureus,* and *Xenopus laevis*). Factors affecting the ability of restriction enzymes to cleave a particular genome include (1) percentage G + C content, (2) specific dinucleotide, trinucleotide, and/or tetranucleotide frequencies, and (3) methylation. Using available information on percentage G + C content, dinucleotide frequencies, and a few kilobases of DNA sequence, predictions can be made about potential cleavage with restriction enzymes. (Reprinted, with permission, from New England Biolabs.)

Enzyme	Sequence	ATH	CEL	DRO	ECO	HUM	MUS	RSS	YSC	STA	XEL
Apa I	GGGCCC	25,000	40,000	6,000	15,000	2,000	3,000	8,000	20,000	70,000	5,000
Asc I	GGCGCGCC	400,000	400,000	60,000	20,000	80,000	100,000	4,000,000	500,000	600,000	200,000
Avr II	CCTAGG	15,000	20,000	20,000	150,000	8,000	7,000	10,000	20,000	20,000	15,000
*Bam*H I	GGATCC	6,000	9,000	4,000	5,000	5,000	4,000	5,000	9,000	15,000	5,000
Bgl I	GCCN$_5$GGC	20,000	25,000	4,000	3,000	3,000	4,000	10,000	15,000	30,000	6,000
Bgl II	GCGCGC	2,000	4,000	4,000	6,000	3,000	3,000	3,000	4,000	6,000	3,000
*Bss*H II	GCGCGC	50,000	30,000	6,000	2,000	10,000	15,000	80,000	30,000	40,000	20,000
Dra I	TTTAAA	2,000	1,000	1,000	2,000	2,000	3,000	2,000	1,000	1,000	2,000
Eag I	CGGCCG	10,000	20,000	3,000	4,000	10,000	15,000	60,000	20,000	50,000	15,000
*Eco*R I	GAATTC	4,000	2,000	4,000	5,000	5,000	5,000	3,000	3,000	4,000	4,000
Hind III	AAGCTT	1,000	3,000	4,000	5,000	4,000	3,000	6,000	3,000	2,000	3,000
Nae I	GCCGGC	6,000	15,000	3,000	2,000	4,000	6,000	40,000	15,000	20,000	6,000
Nar I	GGCGCC	10,000	15,000	3,000	2,000	4,000	6,000	50,000	15,000	15,000	7,000
Nhe I	GCTAGC	10,000	30,000	10,000	25,000	10,000	10,000	15,000	10,000	10,000	10,000
Not I	GCGGCCGC	200,000	600,000	30,000	200,000	100,000	200,000	2,000,000	450,000	1,000,000	200,000
Pac I	TTAATTAA	70,000	20,000	25,000	50,000	60,000	100,000	100,000	15,000	9,000	50,000
Pme I	GTTTAAAC	60,000	40,000	40,000	40,000	70,000	80,000	50,000	50,000	25,000	50,000
Rsr II	CGGWCCG	25,000	50,000	15,000	10,000	60,000	60,000	50,000	60,000	150,000	70,000
Sac I	GAGCTC	3,000	4,000	4,000	10,000	3,000	3,000	3,000	9,000	10,000	4,000
Sac II	CCGCGG	10,000	20,000	5,000	3,000	6000	8,000	70,000	20,000	40,000	15,000
Sal I	GTCGAC	6,000	8,000	5,000	5,000	20,000	20,000	25,000	10,000	15,000	15,000
Sfi I	GGCCN$_5$GGCC	400,000	1,000,000	60,000	150,000	30,000	40,000	150,000	350,000	2,000,000	100,000
*Sgr*A I	CXCCGGXG	30,000	100,000	20,000	8,000	70,000	80,000	100,000	90,000	150,000	90,000
Sma I	CCCGGG	10,000	30,000	10,000	6,000	4,000	5,000	30,000	50,000	50,000	5,000
Spe I	ACTAGT	8,000	8,000	9,000	60,000	10,000	15,000	6,000	6,000	6,000	8,000
Sph I	GCATGC	10,000	15,000	5,000	4,000	6,000	6,000	7,000	10,000	9,000	6,000
Srf I	GCCCGGGC	400,000	1,000,000	90,000	50,000	50,000	90,000	1,000,000	600,000	2,000,000	100,000
Sse I	CCTGCAGG	100,000	200,000	50,000	40,000	15,000	15,000	60,000	150,000	200,000	30,000
Ssp I	AATATT	3,000	1,000	1,000	2,000	2,000	3,000	2,000	1,000	1,000	2,000
Swa I	ATTTAAAT	50,000	9,000	15,000	40,000	30,000	60,000	25,000	15,000	6,000	30,000
Xba I	TCTAGA	5,000	4,000	9,000	70,000	5,000	8,000	7,000	4,000	5,000	6,000
Xho I	CTCGAG	4,000	5,000	4,000	15,000	7,000	7,000	6,000	15,000	25,000	10,000

Average size fragments predicted for ATH Arabidopsis thaliana, CEL Caenorhabditis elegans, DRO Drosophila melanogaster, ECO Escherichia coli, HUM Human, MUS Mouse, RSS Rhodobacter sphaeroides, YSC Saccharomyces cerevisiae, STA Staphylococcus aureus, and XEL Xenopus laevis.

SINGLE-LETTER ABBREVIATIONS FOR AMINO ACIDS

Amino acid	One-letter code
Alanine	A
Arginine	R
Aspartic acid	D
Asparagine	N
Cysteine	C
Glycine	G
Glutamic acid	E
Glutamine	Q
Histidine	H
Isoleucine	I
Leucine	L
Lysine	K
Methionine	M
Phenylalanine	F
Proline	P
Serine	S
Threonine	T
Tryptophan	W
Tyrosine	Y
Valine	V

GENETIC CODE

The first base in each mRNA codon is in the center of the circle, the second base is in the second circle from the center, and the third base is in the third circle from the center. The standard three-letter abbreviation for the amino acid encoded by each codon is on the outside. UAG, UAA, and UGA stop codons are designated by the word STOP.

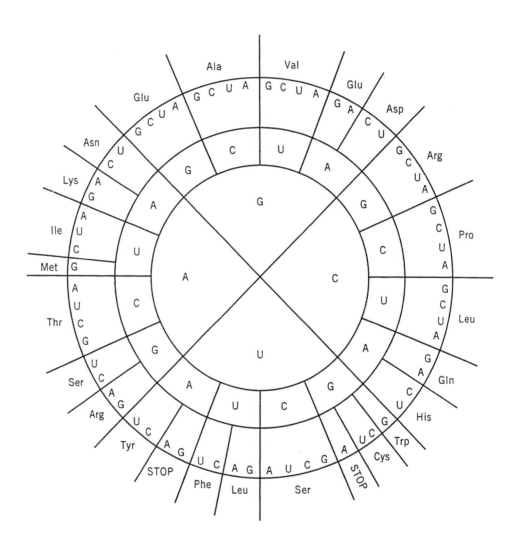

CODON FACTS

Codon Redundancy

Amino acid	Number of synonymous codons
W, M	1
N, D, C, Q, E, H, K, F, Y	2
I	3
A, G, P, T, V	4
R, L, S	6

Termination Codons

UAA	ochre
UAG	amber
UGA	opal

ISOTOPE INFORMATION

Isotope Half-lives

Isotope	Emitted particle	Half-life
^{14}C	beta	5730 years
^{3}H	beta	12.4 years
^{125}I	gamma	60 days
^{32}P	beta	14.3 days
^{35}S	beta	87.4 days

Radioactivity Conversions

1 Ci = 1000 mCi
1 mCi = 1000 μCi
1 μCi = 2.2 x 10^6 dpm

1 Bq = 1 disintegration/second
1 μCi = 3.7 x 10^4 Bq
1 Bq = 2.7 x 10^{-5} μCi

APPENDIX 5

Suppliers

With the exception of those suppliers listed in the text with their addresses, all suppliers mentioned in this manual can be found in the BioSupplyNet Source Book and on the Web site at:

```
http://www.biosupplynet.com
```

If a copy of BioSupplyNet Source Book was not included with this manual, a free copy can be ordered by using any of the following methods:

- Complete the Free Source Book Request Form found at the Web site at:

```
http://www.biosupplynet.com
```

- E-mail a request to info@biosupplynet.com
- Fax a request to 516-349-5598

Index